Oncology: Genomics, Precision Medicine and Therapeutic Targets

Hardeep Singh Tuli • Mükerrem Betül Yerer Aycan
Editors

Oncology: Genomics, Precision Medicine and Therapeutic Targets

Springer

Editors
Hardeep Singh Tuli
Department of Biotechnology
Maharishi Markandeshwar
(Deemed to be University)
Ambala, Haryana, India

Mükerrem Betül Yerer Aycan
Department of Pharmacology
Erciyes University
Kayseri, Kayseri, Türkiye

ISBN 978-981-99-1528-6 ISBN 978-981-99-1529-3 (eBook)
https://doi.org/10.1007/978-981-99-1529-3

This Springer imprint is published by the registered company Springer Nature Singapore Pte Ltd.
The registered company address is: 152 Beach Road, #21-01/04 Gateway East, Singapore 189721, Singapore

Preface

Personalized medicine, which simply refers to choosing the course of action that is most suitable for a given patient, entails integrating and using a number of cutting-edge technologies to the clinical treatment of patients. The scope is significantly broader than what the title suggests. Since numerous non-genomic elements are taken into account while creating tailored genomic medicine. The nucleic, cytogenetic, and bioinformatic techniques that are anticipated to have an impact on the management and therapy of cancer are covered in this book. We also provide a summary of the molecular insights into noncoding RNAs' possible clinical applications in cancer. The description of genetic variants and how they relate to drug interactions are also examined in terms of pharmacokinetics and pharmacodynamics, along with therapeutic drug monitoring. Another feature of the current book is the improvements in molecular identification methods like genomics, transcriptomics, and proteomics as well as the capacity to correlate with drug development toward multi-therapy approaches. Interest in personalized medicine is rising. Although molecular biology and biotechnology have made significant strides toward personalized medicine, there are still some myths that persist in both the academic and business worlds. A reliable source of information that covers both the foundations and applications of customized medicine is lacking. The current book is primarily concerned with the difficulties in the existing cancer therapy, as well as the function of translational research, current promising clinical trials, and the importance of big data in this field.

Ambala, Haryana, India — Hardeep Singh Tuli
Kayseri, Kayseri, Turkey — Mükerrem Betül Yerer Aycan

Acknowledgments

Dr. Hardeep Singh Tuli would like to thank Maharishi Markandeshwar (Deemed to be University), Mullana-Ambala, for providing requisite facilities and Dr. Mükerrem Betül Yerer Aycan, Erciyes University, Kayseri, for providing requisite facilities to conduct the work.

Contents

About the Editors

Hardeep Singh Tuli is an Associate Professor at the Department of Biotechnology, Maharishi Markandeshwar (Deemed to be University), Mullana, India. He previously served as a Lecturer at the Department of Applied Sciences, Institute of Science and Technology, Klawad, Yamunanagar (2009–2011), India. His research interests are focused on the isolation, characterization, and biochemical evaluation of natural metabolites as anticancer agents. He has served as a reviewer for several international journals and as a guest editor for the journal *Frontiers in Pharmacology*. He has more than 10 years of teaching and research experience in pharmacology, mammalian physiology, and natural products. He has published more than 100 papers in peer-reviewed international journals and authored a number of book chapters. He edited six international books in the area of cancer biology and chemoprevention. Dr. Tuli is editorial board member of various national and international scientific research journals. In addition, he has been placed in the top 2% of highly cited researchers in the world in the area of pharmacology (https://elsevier.digitalcommonsdata.com/data.../btchxktzyw/3).

Mükerrem Betül Yerer Aycan graduated from Hacettepe University Faculty of Pharmacy in 1999 and completed her master's and doctorate degrees at Erciyes University in 2006. She carried out the experimental parts of his doctoral thesis at the University of Mallorca Balearic Islands in Spain as a Tübitak 2214 scholar and completed her postdoctoral studies at the Basel University Neuropsychopharmacology Department as a Tübitak 2219 scholar in Switzerland. After being appointed as the founding Head of the Department of Pharmacology, Faculty of Pharmacy at Erciyes University in 2007, she served as the Deputy Dean in charge of education for 8 years between 2008 and 2016 and acted as the accreditation coordinator of the Faculty. Continuing her studies at the University of Colorado, Denver, USA, as the coordinator of the international Tubitak-1011 project in 2011, Aycan was awarded the title of Associate Professor in the field of pharmacology in 2012 and was appointed as a Professor in 2018. In the same year, she was appointed as the Dean of the Faculty of Pharmacy at Erciyes University and as the Director of the Erciyes University

Pharmaceutical Application and Research Center. Accrediting the Pharmaceutical Application and Research Center in eight different categories, Prof. Dr. M. Betül Aycan has more than 100 publications, more than 70 of which are SCI, on the use of natural compounds in neuropsychopharmacology and cancer research, many international book/book chapters, and national and international awards.

Abbreviations

5-FdUMP	Deoxythymidylic acid
5-FU	5-Fluorouracil
ADC	Antibody-drug conjugate
ADCC	Antibody-dependent cell-mediated cytotoxicity
ADCP	Antibody-dependent cellular phagocytosis
ADT	Androgen decreasing therapy
AGO	Argonaute
ANN	Artificial neural network
APC	Antigen-presenting cells
AR	Androgen receptor
BCa	Breast cancer
BCC	Basal cell cancer
BCMA	B-cell maturation antigen
BSSeq	Bisulfite sequencing
BTK	Bruton tyrosine kinase
BWA	Burrows-Wheeler Aligner
CAR	Chimeric antigen receptor
CDK	Chemistry development kit
CDK	Cyclin-dependent kinase
CEA	Carcinoembryonic antigen
ChIP	Chromatin immunoprecipitation
ChIP-seq	Chromatin immunoprecipitation sequencing
CRC	Colorectal cancer
CSF	Cerebrospinal fluid
CSF1R	Colony stimulating factor 1 receptor
CTC	Circulating tumor cells
CTLA-4	Cytotoxic T-lymphocyte antigen 4
DHEA	Dehydroepiandrosterone
DHFU	Dihydrofluorouracil
DHT	Dihydrotestosterone
dUMP	Deoxyuridine monophosphate

EGFR	Epidermal growth factor receptor
EMBL	European molecular biology laboratory
EMT	Epithelial-mesenchymal transition
EpCAM	Epithelial cell adhesion molecules
ErbB	Epidermal growth factor receptor
FAF1	Fas-associated factor 1
FCAS	Flow cytometry
FDA	Food and Drug Administration
FDPs	Fibrin degradation products
FGFR	Fibroblast growth factor receptor
FKBP-12	FK506 Binding Protein-12
FLT3	FMS-like tyrosine kinase 3
FPPE	Formalin-fixed paraffin-embedded
GAS5	Growth arrest-specific 5
GD2	Glycolipid disialoganglioside
GEF	Nucleotide exchange factors
GWAS	Genome-wide association studies
HBV	Hepatitis B virus
HCC	Hepatocellular carcinoma
HCV	Hepatitis C virus
HDAC	Histone deacetylase
HER2	Hormonal estrogen receptor 2
HGFR	Hepatocyte growth factor receptor
Hh	Hedgehog pathway
HIF-2α	Hypoxia-inducible factor 2 alpha
HLA	Human leucocyte antigen
HNC	Head and neck cancer
HONEY	Hyperbaric OxygeN therapy on brEast cancer patients with late radiation toxicity
HOTAIR	HOX transcript antisense intergenic RNA
HPV	Human papillomavirus
HSCT	Hematopoietic stem cell transplantation
hTERT mRNA	Telomerase reverse transcriptase mRNA
IDH2	Isocitrate dehydrogenase 2
IGF-1R	Insulin-like growth factor-1 receptor
IHC	Immunohistochemical
IL	Interleukin
IMMUNOGAST	IMMUNOtherapy-based regimens in recurrent GASTric adenocarcinoma
INMEX	Integrative meta-analysis of expression data
INR	International normalized ratio
IOMA	Integrative omics-metabolic analysis
Itk	Interleukin-2 receptor-inducible T-cell kinase
LB	Liquid biopsy

LCa	Lung cancer
Lck	Lymphocyte-specific protein tyrosine kinase
LOH	Loss of heterozygosity
mAb	Monoclonal antibodies
MALAT1	Metastasis-associated lung adenocarcinoma transcript 1
MEG3	Maternally expressed gene 3
miRISC	miRNA-induced silencing complex
miRNA	micro-RNA
MKL	Multiple kernel learning
ML	Machine learning
MMAE	Monomethyl auristatin E
MMR	Mismatch repair
MREs	miRNA response elements
mRNA	Messenger RNA
MS	Mass spectrometry
MSI	Microsatellite instability
MTHFR	Methylene tetrahydrofolate reductase
mTOR	Mammalian target of rapamycin
N^2M^2	NCT Neuro Master Match
NGS	Next-generation sequencing
NHL	Non-Hodgkin lymphoma
NMR	Nuclear magnetic resonance
NSCLC	Non-small cell lung cancer
OncomiRs	Oncogenic miRNAs
PARP	Poly (ADP-ribose) polymerase
PCa	Prostate cancer
PCR	Principal component regression
PDGFR	Platelet-derived growth factor receptor
PD-L1	Programmed cell death ligand 1
PGx	Pharmacogenomics
PHI	Prostate health index
PI3K	Phosphatidylinositol-3-kinase
PLS	Partial least squares
PSA	Prostate specific-antigen
PSMA	Prostate-specific membrane antigen
PTC	Papillary thyroid carcinoma
QSAR	Quantitative structure-property relationships
RAINBO	Refining Adjuvant treatment IN endometrial cancer Based On molecular features
RANKL	Receptor activator of nuclear factor kappa-B ligand
RF-RFE	Random forest-recursive feature elimination
RGIFE	Rule-guided iterative feature elimination
RNA-seq	RNA sequencing
RON	Recepteur Origine Nantais

RT	Retention time
RXR	Retinoid X receptor
SCC	Squamous cell cancer
SCLC	Small cell lung carcinoma
SCLO2B1	Solute carrier organic anion transporter family member 2B1
SJS	Steven-Johnson syndrome
SOM	Self-organizing maps
SRA	Sequence read archive
SVM	Support vector machine
SVM-RFE	Support vector machine recursive feature elimination
TCGA	The cancer genome atlas
TDM	Therapeutic drug monitoring
TNBC	Triple negative breast cancer
TRKA	Tropomyosin receptor tyrosine kinases
TSGs	Tumor suppressor genes
TS-miRs	Tumor suppressor miRNAs
VEGF	Vascular endothelial growth factor
WES	Whole-exome sequencing
WGBS	Whole-genome bisulfite sequencing
ZEB1	Zinc finger E-box binding homeobox 1

Oncology: Way Forward from Clinical and Molecular Diagnosis to Treatment

Saumya Srivastava and Anjana Pandey

1 Introduction

Cancer is a hereditary disease. A crucial event in the early stages of tumour growth is the manifestation of oncogenesis. Oncogenes are triggered by one of two mechanisms: either cellular proto-oncogenes—which are often normal—mutate into oncogenes or tumour viruses invade cells (Imran et al. 2017). Genetic examination is no longer restricted to obvious chromosomal defects in karyotypes; cells' DNA may now be examined down to the level of individual base pairs. The chances of personalised treatment for certain cancers is increased by our intricate understanding of the genetics of cancer. Cancer diagnosis has traditionally been divided into three categories: anatomic origin, microscopic morphology, and protein-based testing like immunohistochemistry (Ong et al. 2012; Sherman et al. 2017). Cell surface indicators for leukaemia and lymphoma, specific cytokine production, and other nonspecific markers, such as Ig clonality in lymphoid malignancies, are additional helpful tools for diagnosis and monitoring. Based on these traits, the tumor's stage and the degree of tumour dissemination, medical oncologists choose the best course of treatment. Laboratory medicine, which depends on the identification of specific biologic molecules, includes molecular diagnostics (Sokolenko and Imyanitov 2018). Onco haematologists were the first to identify the potential of molecular genetic tools since certain chromosomal translocations may greatly benefit in the diagnosis of different leukemias and lymphomas (Fey and Wainscoat 1988). The development of user-friendly molecular analysis techniques is primarily responsible for the advent of practical molecular oncology applications. Clinical DNA testing had a significant advance with the development of PCR (polymerase chain reaction).

S. Srivastava · A. Pandey (✉)
Department of Biotechnology, Motilal Nehru National Institute of Technology (MNNITA), Prayagraj, Uttar Pradesh, India
e-mail: anjanap@mnnit.ac.in

H. S. Tuli, M. B. Yerer Aycan (eds.), *Oncology: Genomics, Precision Medicine and Therapeutic Targets*, https://doi.org/10.1007/978-981-99-1529-3_1

PCR-based methods are very compatible with clinical practise and only require minimal amounts of biological material and relatively basic infrastructure and instruments.

In the middle of the twentieth century, a technique called immunohistochemistry (IHC), which allows the viewing of a particular antigen within the tissue, was developed (Coons and Kaplan 1950). IHC was modified for the clinical assessment of the degree of oestrogen receptor (ER) expression more than 30 years ago; this was a true historical advancement in personalised oncology since it altered medical perceptions of the most prevalent oncological illness, i.e. breast cancer (BC), by customising endocrine therapy to a laboratory test (Coombes et al. 1987). The topic of molecular diagnostics is now infrequently mentioned when it comes to some traditional protein-targeted procedures, such as IHC or the identification of tumor-specific serum indicators (PSA, CA-125, etc.). Both, DNA- or RNA-based assays and some recent sophisticated proteomic technologies fall under the latter term (Pal et al. 2022).

Molecular diagnostics have been integrated into standard patient care in two different ways (Table 1). First, clinical oncology now routinely identifies patients who have hereditary malignancies. Second, a variety of tests are available to aid in choosing the best course of action based on the molecular makeup of tumour tissues or other biological indicators of malignant illness (Malone et al. 2020). There are some more applications as well, although they are still under development.

2 Cancer Diagnosis and Treatment Based on Molecular Pathology

Getting the right diagnosis is one of the biggest obstacles in the clinical care of cancer patients. Numerous technologies have been created as a result, and they are now often used to subtype molecular malignancies. These include fluorescent in situ hybridization, in situ hybridization, and immunohistochemistry, as well as the examination of DNA and RNA taken from the lesion (FISH) (Imran et al. 2017). Following that, the cancer material is subtyped utilising several molecular biology techniques, such as Sanger sequencing, pyrosequencing, and allele-specific PCR. Next-generation sequencing, mass spectroscopy-based assays, and snapshot assays are used to genotype cancer. Next-generation sequencing has made it possible to define recurrent mutations that may be combated with new drugs while also revealing the full variety of malignancies. The development of next-generation sequencing is enabling the identification of recurrent mutations that can be treated with novel medicines as well as revealing the real variety of malignancies. Such genomic-level investigations will have an effect for years to come (Khan et al. 2015; Lopez-Chavez et al. 2015).

For the treatment of cancer, several methodologies and treatments have been used over time. Surgery, radiation treatment, chemotherapy, hormone therapy,

Table 1 Molecular diagnostics in oncology

Hereditary cancer syndromes	Assays	• Recurrent mutations (PCR)
		• Single gene analysis
		• Multigene panels
		• Whole exome sequencing
	Cancer patients	• Risk of second malignancy
		• Choice of treatment
	Healthy people	• Identification of subjects at risk
Predictive markers	Assays	• DNA
		• RNA
		• Proteins
		• Cells
		• Tissues slices
		• PDKs
	Molecular targets	• HER2
		• EGFR-MUT
		• BRAF-MUT
		• ALK
		• ROS
	Tumour phenotypes	• MSI-H
		• BRCAness
		• Mutation Burden
Circulating tumour fragments	Assays	• CTCs
		• ctDNA
		• RNA
		• Proteins
	Cancer patients	• Control of tumour eradication
		• Monitoring of tumour burden
		• Choice of therapy
	Healthy people	• Early diagnosis
Carcinomas of unknown primary	Assays	• Single markers
		• Integrative assays
	Tissue specific markers	• RNA
		• Proteins
	Tumour specific markers	• Point mutations
		• Gene rearrangements
		• Copy number variations

immunotherapy, adjuvant therapy, targeted growth signal inhibition, apoptosis-inducing medications, nanotechnology, RNA expression and profiling, and most recently CRISPR are some of the most frequently employed methods (Inchingolo et al. 2020; Iranzadasl et al. 2020; Sudhakar 2009). Gene replacements and oncogene knockouts are additional methods for killing cancer cells. Additionally, oncolytic viruses can be used with chemotherapeutic drugs to eradicate cancer cells (Ottolino-Perry et al. 2010; Russell et al. 2019).

2.1 Retroviral Therapy

Retroviruses (RVs) have been utilised in cancer therapy in addition to more traditional approaches. RVs have been utilised to transmit ring genes to mammalian cells previously. RVs produced from the Moloney Murine leukaemia virus are the most often used (MoMLV). In the past 20 years, gene therapy clinical trials, reliable siRNA administration, and the creation of transgenic animals have all been made possible by the artificial development of RVs. Recent studies on the successful clinical trials of gene therapy in individuals with severe immunodeficiency diseases discussed the enormous potential of RVs. However, there may be some inherent hazards associated with it (Barquinero et al. 2004; Vargas et al. 2016). By creating two groups of vectors, a comparative experiment was conducted. One had a helper retrovirus with it whereas the other was faulty and intrinsically incapable of replication. The replicative viruses transduced more than 85% of their targets under in vitro conditions, compared to less than 1% for the other viruses (Schambach and Morgan 2016).

2.2 Retroviral Vector Insertions

The study of retroviral vector insertions (insertional oncogenesis) has recently attracted more attention. Viral insertion sites have been explored for many years to find potential oncogenes and cancer signalling pathways. The availability of genetically engineered animals, insertion site cloning by high throughput PCR, and the completion of the mouse genome project have all expanded the breadth of this approach (Ellis et al. 2021; Irving et al. 2021; Nakamura 2005). However, hundreds of frequent locations have been found by different researchers. These integration sites in MoMLV-induced murine hematopoietic malignancies are typically linked to cancer genes (Kurata et al. 2021). The bulk of insertions often occur outside of coding areas. Therefore, only a small percentage of RISs (retroviral integration sites)—less than 10%—can be regarded as recognised tumour suppressor genes.

2.3 RNA Interference

RNA interference is one of the techniques used in cancer treatment. Small non-coding RNAs are used in RNA interference to prevent the translation of other mRNAs into proteins by binding with them. Gene function may be lost as a result of this. These RNAi can be employed in cancer treatment to disable the action of cancer genes that stop the disease from spreading (Cuciniello et al. 2021; Grillone et al. 2020). The homologous recombination approach was substituted by the RNAi method because it is quick, affordable, and very effective. However, it has flaws, such as partial knockdown and short-term suppression of gene activity. Additionally, it produces unexpected off-target effects. Recurrences of cancer were occasionally observed. These issues prompt researchers to look at novel strategies for modifying gene function.

2.4 *Editing of Genome*

The new and significantly superior method of treating cancer is genome editing. Engineered nucleases with specialised domains that can attach to the target spot and then cleave it were employed by scientists (Akram et al. 2022; Khadempar et al. 2019). These nucleases were able to cause double strand breaks (DSBs) in the target, which then caused the DNA repair systems to go into action. Both programmable nucleases like Zinc Finger Nucleases (ZFNs) and transcription-activator-like effector nucleases are employed (TALENs). With the use of these nucleases, many animal models of cancer were successfully treated by genome editing (Li et al. 2020).

- Use of Zinc finger proteins (ZNFs)

  The first nucleases utilised for gene editing are ZNFs. In eukaryotes, they are present as DNA binding domains. These are composed of 30 amino acid modules that have been organised to resemble a collection of Cys2-His2 DNA-binding zinc fingers. These modules are used by the FokI nuclease domain (Lee et al. 2019). The modules are made up of three to six zinc fingers that can recognise triplet nucleotides (Chen et al. 2020; Cui et al. 2021). To target any area of the genome, two zinc finger nucleases are required since the FokI nuclease only works as dimers. The sequence upstream of the portion of the genome that will be changed will be identified by one ZFN, and the downstream sequence by another (Bi et al. 2022). To cause a double-stranded break in a particular location, these arrays attach to neighbouring DNA sequences that are in the opposing strands. The breaks are subsequently fixed using a variety of techniques that may result in point mutations, indels, or translocations in the targeted area. The ZNFs are customized and can detect any type of DNA region as well as every potential nucleotide (Bi et al. 2022; Chen et al. 2020; Cui et al. 2021).
- Modification via Transcription activator-like effector nucleases (TALENs)

  Since TALENs require the same DNA binding motifs and nuclease to modify the genome, they are comparable to zinc finger nucleases. The TALENs domain recognises a single nucleotide rather than a triplet, which is the difference in how nucleotides are recognised. When contrast to ZNFs, the interactions between

TALEN domains and their target sites are greater. Comparatively speaking, designing TALENs is simpler than developing ZNFs (Bi et al. 2022; Lyu et al. 2020). The approach of using TALENs for cancer treatment is quite successful. Two specially designed TALENs that can recognise the DNA sequences on the opposing strands of the target gene are required. It cleaves the target gene's sequence by dimerizing the TALENs' FokI nuclease cleavage domain (Das and Paudel 2021; Harale et al. 2022; Li et al. 2020). The targeted gene experiences double-stranded DNA breaks. The end-joining DNA repair mechanism corrects the lesion caused by the DNA break. The reading frame alteration has an impact on the target gene. The existing mutations can also be eliminated using this technique. Since this gene editing method may target any gene in the genome, it can be effectively employed to cure cancer cell lines. TALENs can also be used to address complex cancer genes (Lan et al. 2022).

- Utilization of CRISPR/CAS9 domain

  All previous genome-engineering methods are currently being eclipsed by a potent genome-editing tool known as Clustered regularly interspaced palindromic sequences, or CRISPR. This innovative technique enables researchers to carry out targeted manipulation in any gene (DNA sequence) in the entire genome of any organism in vitro or, as of recently, even directly in endogenous genome, assisting in the understanding of the functional organisation of the genome at systems level and identifying incidental genetic variations (Bi et al. 2022).

2.5 *Epigenetics*

Epigenetics s defined as DNA sequence-independent covalent alterations to nucleic acids and histones, which may be passed down by cell replication and are crucial for determining cell destiny (Halabian et al. 2021). The accessibility of chromatin to transcription factors or other DNA binding proteins is controlled by chemical changes such as DNA methylation, histone acetylation, and histone methylation. Writers are the conventional name for the enzymes that deposit these chemical changes. The interaction of the molecular markers with chromatin remodelers and noncoding RNAs, which mediate downstream effects, is regulated by "reader" proteins that identify certain epigenetic markings. Since the chemical alterations may be undone with "erasers," the process of epigenetic changes is reversible (Feng et al. 2022). These epigenetic modifiers cooperate to guarantee that gene expression is regulated both spatially and temporally. Inhibitors of DNA methyltransferase (DNMT) and histone deacetylase (HDAC) are a few of the epigenetic medicines that have been produced to date that target the three types of epigenetic proteins—writers, readers, and erasers. These compounds are used to treat various hematologic malignancies.

1. Identification of epigenetics writers in cancer

   The process of adding active and repressive marks to DNA or histones is catalysed by epigenetic writers. Among the writer proteins, which also include histone acetyltransferases (HATs) and histone methyltransferases (HMTs), DNMTs, and disruptor of telomeric silencing 1-like (DOT1L) have proven to be the most alluring targets for drug development. In therapeutic studies, several inhibitors that target these proteins are being studied (Bhan et al. 2017; Kanwal et al. 2015).

   - Inhibition of DNA methyltransferase

     Numerous DNA methylation changes, including localised DNA hypermethylation at CpG-rich locations and genome-wide hypomethylation, are frequently seen in cancer cells (Smith et al. 2020). Oncogenes are activated and tumor-suppressor genes are silenced as a result of aberrant DNA methylation. A successful cancer treatment now targets DNMTs to return the epigenome of cancer cells to a "normal" condition. Numerous DNMT inhibitors have been identified throughout the years, and they may be divided into two groups: nucleoside analogues and non-nucleoside DNMT inhibitors. Nucleoside analogues can be introduced into freshly manufactured DNA and covalently linked to DNMTs by modifying the cytosine ring. As a result, the progeny cells' hemimethylated DNA cannot be methylated by DNMTs, which results in universal demethylation and proteasomal destruction of the DNMTs (Chattopadhyaya and Ghosal 2022; Minarovits et al. 2016).

     The two most researched nucleoside analogues, 5-aza-20-deoxycytidine (decitabine) and 5-aza-5-azitidine (azacitidine), promote DNA demethylation and reawaken silenced genes in human malignancies. Azacitidine is a ribonucleoside that phosphorylates to preferentially integrate into RNA. However, the ribonucleotide reductase route allows for a lesser degree of incorporation into DNA. Decitabine induces DNA demethylation at significantly lower doses than azacitidine because it is a deoxyribose analogue that integrates directly into DNA after phosphorylation. For the treatment of acute myeloid leukaemia (AML), chronic myelomonocytic leukaemia (CMML), and myelodysplastic syndrome (MDS), the US Food and Drug Administration (FDA) has approved the use of both medications (Bell et al. 2019; Van Der Helm et al. 2011).In two phase 3 trials comparing azacitidine or decitabine to standard treatment regimens, older AML patients showed better response rates and higher overall survival (Dombret et al. 2015).

     These medications suffer from poor chemical and metabolic stability, high toxicity, and efficient cancer cell targeting. The prodrugs of decitabine or azacitidine, which have been created as next-generation DNMT inhibitors and display improved pharmacokinetic profiles, have been produced in response to the benefits and disadvantages of azacitidine and decitabine. For instance, the dinucleotide decitabine-*p*-deoxyguanosine guadecitabine (SGI-110), which has a lower peak plasma concentration than decitabine and a longer plasma half-life, results in reduced toxicity and a longer exposure time (Daher-Reyes et al. 2019).

For the treatment of several malignancies, guadecitabine has begun clinical studies (Daher-Reyes et al. 2019; Fey and Wainscoat 1988). Primary results of a phase 3 trial of guadecitabine in a population of AML patients who had never received treatment and were not eligible for intensive chemotherapy suggest that guadecitabine monotherapy may improve survival compared to other treatment options in patients who had at least four or six cycles of therapy. While genomic instability and mutagenesis risk are two major adverse effects of nucleoside analogues, it is hypothesised that recently produced non-nucleoside analogues might mitigate these consequences. These non-nucleoside analogues specifically target the catalytic sites of certain DNMT enzymes, in contrast to nucleoside analogues that integrate into DNA (Roboz et al. 2019).

- Inhibition of Histone methyltransferases

  Numerous efforts have been undertaken to create epigenetic drugs that target HMTs, including as EZH2, DOT1L, G9A, and NSD2, and inhibitors of these enzymes are now being tested in laboratories to treat cancer. The H3K27 trimethylation repressive mark is deposited by EZH2, a catalytic subunit of the polycomb repressive complex 2 (PRC2) (H3K27me3). Gain-of-function mutations in EZH2 have been identified in follicular lymphoma (FL) and diffuse large B-cell lymphoma (DLBCL) (Oricchio et al. 2017; Zhou et al. 2015). Additionally, in several cancers, including prostate, breast, ovarian, and lung cancer, overexpression of EZH2 is linked to a poor prognosis and tumour stage.

2. Identification of epigenetic erasers in cancer

   Epigenetic erasers remove alterations to DNA or histones that writers have added to control gene expression. Eraser proteins, including histone demethylases (HDMs), HDACs, and ten-eleven translocation (TET) enzymes, have been linked to cancer (Cheng et al. 2019). Lysine specific demethylase 1 (LSD1, also known as KDM1A) and HDAC inhibitors have received a lot of attention, and several have reached the clinical development stage.

   Histone demethylases inhibition

   Histone lysine demethylases are intriguing therapeutic targets since abnormal expression of these enzymes has been observed in a variety of solid and hematologic cancers. The Jumonji C (JmjC) class and the LSD1 and LSD2 demethylases are the two categories that can be distinguished among the demethylases (Karakaidos et al. 2019). Too far, LSD1 has been the target of most inhibitors in clinical studies that target HDMs. LSD1 selectively demethylates H3K4me1/2 and H3K9me1/2 and is involved in cellular functions such as cell division and proliferation, the shift from epithelial to mesenchymal tissue, and malignant transformation. Numerous malignancies, including leukaemia, bladder, lung, and colorectal cancer (CRC), have been linked to overexpression of LSD1 (Cheng et al. 2019; Karakaidos et al. 2019; Valdespino and Valdespino 2015).

- Inhibition of Histone deacetylase (HDACs)

  Numerous solid and haematological cancers have been linked to HDAC dysregulation. In some malignancies, overexpression of HDACs is associated with a poor prognosis and progressive illness. There are several HDAC members, but class I (HDAC 1, 2, 3, and 8) and class II (HDAC 4, 5, 6, 7, and 10) have drawn the most attention for drug development (Atadja 2011). The majority of FDA-approved medications for treating particular haematological malignancies target HDACs, such as vorinostat (SAHA), romidepsin (FK228), belinostat (PXD101), and panobinostat (LBH589) (Bondarev et al. 2021). Most HDAC inhibitors from the first generation are nonselective. The first HDAC inhibitor licenced by the FDA to treat patients with cutaneous T-cell lymphoma (CTCL) was vorinostat, a hydroxamic acid-based medication. At low nanomolar doses, vorinostat is non-specific and suppresses all zinc-dependent HDACs (McClure et al. 2017). In addition to suppressing metastasis and promoting apoptosis, vorinostat has also been shown to make cancer cells more susceptible to radiation and chemotherapy. It is now being studied in the clinic to treat a variety of solid tumours. Belinostat and panobinostat, two further nonselective HDAC inhibitors, have just received approval for the treatment of multiple myeloma and peripheral T-cell lymphomas, respectively (Shah 2019).

3. Identification of epigenetic readers in cancer

   Certain post-translational changes on DNA or histones are recognised by domains found in epigenetic readers. These alterations include methylated CpGs, acetylated histone residues, methyl-CpG-binding domains (MBDs), and methylated lysines. The extra terminal domain (BET) family, which has been the subject of reader protein drug development, plays important functions in carcinogenesis. There are four members of the BET family, which include the proteins 2, 3, 4, and t. (BRD2, BRD3, BRD4, and BRDt). A rare but incredibly aggressive epithelial cancer called NUT midline carcinoma (NMC) is caused by the fusion of BRD4 with the Nut protein, which is brought on by BRD4's chromosomal translocation (French 2018; Yan et al. 2011). It should be noted that the BET inhibitor JQ1 separates the BRD4 fusion oncoprotein from chromatin, causing squamous differentiation, decreased proliferation, and tumour regression in NMC xenograft models (Eagen and French 2021). These preclinical investigations have established a compelling case for the administration of BET inhibitors to NMC patients. Orally bioavailable BET inhibitor OTX015 (MK-8628), which inhibits tumour growth in an NMC model with BRD4-NUT fusion, inhibits the proliferation of several cancer cell lines and induces tumour regression (Coudé et al. 2015).

2.5.1 Epigenetics in Solid Tumors

Breast Cancer

The tumours that make up BC are diverse, meaning that each group has a different prognosis and therapeutic response. Although the mortality rate from BC has decreased due to modern therapies, it is still one of the leading causes of cancer-related deaths in females globally (Azamjah et al. 2019; Yedjou et al. 2019). Intense research is being done to comprehend the molecular and cellular underpinnings of tumour heterogeneity that are important for BC diagnosis, prognosis, and treatment. BC is often classified based on the size of the tumour, the histological subtype and grade, the lymph node status, and the expression of the oestrogen receptor (ER), progesterone receptor (PR), and human epidermal growth factor receptor 2 (HER2) (Guido and Gomez-Fernandez 2020). An increasing collection of molecular data acquired from cutting-edge genomic, transcriptomic, and epigenomic methods has produced prognostic and predictive models for precision medicine and information that improves BC subtyping accuracy.

In 2012, TCGA published the results of an integrated genome-wide study of 466 primary BCs, which showed that genetic and epigenetic changes converge phenotypically into four major disease subgroups with molecular heterogeneity: luminal-A, luminal-B, basal-like, and HER2-enriched (HER2E) (Nebbioso et al. 2018). Although the overall mutation rate was higher in the basal-like and HER2E subtypes, TCGA discovered unique altered genes that were implicated in BC. These mutations were shown to be significantly more varied and common among luminal-A/B subgroups. Additionally, TCGA found five DNA methylation clusters. Group 3 had the greatest levels of DNA methylation, the fewest PIK3CA, MAP3K1, and MAP2K4 mutations, and the lowest expression of Wnt pathway genes. This group mostly consisted of luminal-B subtypes. The majority of the basal-like tumours in Group 5 had the highest incidence of TP53 mutations and the lowest levels of DNA methylation.

Rhabdoid Tumor

Atypical teratoid/rhabdoid tumour (AT/RT), also known as malignant rhabdoid tumour (MRT), is an uncommon paediatric soft tissue sarcoma that can develop anywhere on the body but most frequently does so in the brain (Halcrow et al. 2016; Momota et al. 2015). With the exception of a few rare instances caused by SMARCA4 deletion, the loss of SMARCB1 in conjunction with abnormalities of the chromosomal band 22q11.2 in limited neural precursors is the primary cause of AT/RT. Members of the SWI/SNF complex for chromatin remodelling include SMARCB1 and SMARCA4. SMARCB1 deletion mice models demonstrated its crucial function in cancer prevention (Feng et al. 2022). Tazemetostat was used to demonstrate that SMARCB1 deletion causes I cell cycle progression through p16/INK4A down-regulation and up-regulation of cyclin D1 and E2F, (ii) abnormal activation of the Hedgehog-Gli pathway and -catenin/TCF targets, (iii) suppression of brain development, and (iv) higher expression of EZH2 leading to bring apoptosis and differentiation in SMARCB1-deleted MRT cells (Feng et al. 2022).

2.6 *Clinician-Level Molecular Diagnostics*

As we learn more about the complex molecular mechanisms that cause a normal cell to become aberrant along with the dysregulation of alternate complimentary pathways, clinical molecular diagnostics and biomarker discoveries are continually improving in cancer. The age of personalised and predictive medicine has defined how clinical practise is evolving today and foretells how it will be practised in the future due to the rapid advancement of genetic and genomic technologies that are revolutionising our approach to prognosis, screening, and therapy targeting (Ong et al. 2012).

The foundation of personalised medicine is the clinical molecular assessment of biomarkers, which can be prognostic, predictive, pharmacodynamic, or diagnostic (Kariyawasam et al. 2019; Morganti et al. 2019). Independent of therapy, prognostic biomarkers have a relationship with clinical outcomes like overall survival or recurrence-free survival. Clinical choices are made using predictive biomarkers, which evaluate the expected value of a certain medication for a particular patient. The development of monoclonal antibodies and small-molecule tyrosine kinase inhibitors that target EGFR in colorectal malignancies and non-small-cell lung carcinoma (NSCLC) was made possible by utilising this understanding of biomarkers (Meador et al. 2021). The predicted use of poly-ADP (ribose) polymerase inhibitors in BRCA1/2-deficient cancers is another example of the usefulness of predictive biomarkers (Meador et al. 2021).

The vast amount of money spent on cancer medications in healthcare as well as the projected savings from patient stratification and selection based on the findings of these biomarker diagnostic tests on predictive biomarkers with established clinical value underscore the significance and necessity of these biomarkers. Predictive diagnostics and patient stratification have the added benefit of preventing the toxic side effects of the therapeutic regimen, hence minimising morbidity, as was the case with trastuzumab and cardiotoxicity in the treatment of breast cancer (Ong et al. 2012).

2.6.1 Personalized Medicine Treatment

The age of personalised medicine in clinical cancer has recently begun with the clinical molecular testing of predictive pharmacogenetic biomarkers of great therapeutic usefulness.

- Vemurafenib and BRAF V600E mutation in melanoma

  The recent discovery of targeted therapy for particular subgroups of patients with the BRAF V600E mutation and metastatic melanoma is a result of significant breakthroughs into the molecular aetiology of melanoma (Menzer and Hassel 2022; Strashilov and Yardanov 2021). This activating mutation (V600E mutation) in exon 15 replaces valine at amino acid 600 with glutamic acid in 80–90% of instances. A more aggressive clinical course is thought to be related with several clinical changes in advanced melanomas that carry a BRAF mutation. It has been demonstrated that vemurafenib, a medication that specifically inhibits active BRAF, dramatically extends life in melanoma patients whose tumours have the V600E mutation in the BRAF gene (Giugliano et al. 2021). The majority of patients with V600E-mutant melanoma, including those with considerable disease-related symptoms and tumour burden, experience fast tumour regression after starting vemurafenib. Patients given vemurafenib had a considerably higher overall survival rate than those given dacarbazine (estimated 6-month survival rates: 84 vs. 64%; hazard ratio for death: 0.37; 95% CI: 0.26–0.55). Initial vemurafenib therapy resulted in substantially prolonged progression-free survival (median: 5.3 vs. 1.6 months; hazard ratio: 0.26; 95% CI: 0.20–0.33).
- Erlotinib and EGFR in NSCLC

  The EGF receptor (EGFR), a transmembrane protein with cytoplasmic kinase activity transmits significant growth factor signals from the extracellular environment to the cell. There are several chemotherapies utilised for advanced metastatic NSCLC. Tyrosine kinase inhibitors may be the most effective form of treatment for advanced NSCLC patients with tumours that are EGFR-mutant, according to studies (Meador et al. 2021). Erlotinib, a currently available EGFR tyrosine kinase inhibitor, has received FDA approval for use in the United States as monotherapy for NSCLC in the second-line setting and, in the repairs, setting, regardless of EGFR status, as shown by the SATURN maintenance study in the relapsed set—up (McLoughlin and Gentzler 2020). As a result, mutation testing has evolved into the standard care for identifying these individuals.
- *KRAS* and cetuximab or panitumumab in colorectal cancer

  The KRAS mutation in colorectal cancer screens against people who will not benefit from anti-EGFR receptor therapy, such as cetuximab or panitumumab, in contrast to the aforementioned biomarkers that determine whether patients will benefit from the use of molecularly targeted therapies (Ong et al. 2012). Although the KRAS oncogene is often mutated in many other cancers, such as pancreatic, lung, and ovarian cancer (20%), it is overexpressed in colorectal cancer. It serves as a crucial signal transducer for a number of cellular receptors and is an essential component of the MAPK, JAK-STAT, and PI3K cell-signaling pathways (Timar and Kashofer 2020). The majority of mutation detection techniques, such as nucleic acid sequencing, allele-specific PCR techniques, single-strand conformational polymorphism analysis, melt-curve analysis, and probe hybridization, use PCR to amplify the gene's exons 2 and 3, which helps to distinguish between wild-type and mutant sequences in crucial codons 12 and 13.

- ERBB2 (HER2/neu) and trastuzumab in breast cancer

  Worldwide, 14% of cancer-related fatalities in women are due to breast cancer. Five to ten percent of breast cancers are caused by mutations in BRCA1 and BRCA2 (Alkhayyat et al. 2022). Tamoxifen or raloxifene suppression treatment outcomes can be predicted based on the hormone receptor status of the tumour, whether it is oestrogen receptor (ER) or progesterone receptor (Allegretti et al. 2022). Since many years ago, tamoxifen has been employed as the first-line treatment because it competes with oestrogen for binding to the ER. The effectiveness of trastuzumab and lapatinib will be determined by the tumor's hEGF receptor 2 (HER2/neu) status if it is hormone receptor-negative. Tamoxifen has been proven to be less efficient than aromatase inhibitors at stopping tumour growth, regardless of the presence of HER2/neu (Kumar et al. 2022). Aromatase inhibitors combined with lapatinib have been demonstrated to significantly improve prognosis of disease prognosis in breast cancers that express both HER2/neu and hormone receptor.
- BCR–ABL and tyrosine kinase inhibitors in CML

  Myeloid cell growth in the bone marrow is a hallmark of CML, a clonal bone marrow stem cell illness. The BCR-ABL fusion gene is caused by the chromosomal aberration t (9;22) (q34;q11.2) in CML patients. Numerous signal transduction pathways are activated by BCR-increased ABL's tyrosine kinase activity (Bracco et al. 2021). As a result, CML cells develop the leukemic phenotype, which includes uncontrolled growth, diminished adherence to the bone marrow stroma, and a faulty apoptotic response to mutagenic stimuli. Routine karyotyping can detect the translocation in 90–95% of cases. The other occurrences include complicated or cytogenetically cryptic chromosomal rearrangements, and only FISH or RT-PCR may reveal the translocation. The response to therapy at the hematologic, cytogenetic, and molecular levels is the most significant prognostic indication. Imatinib now has an overall 5-year survival rate of between 80 and 95% and a full cytogenetic response rate of between 70 and 90%. Despite imatinib's effectiveness, it is not curative, and in the majority of patients, BCR-ABL transcripts are still detectable by quantitative RT-PCR (Bracco et al. 2021; Ong et al. 2012).

2.6.2 Liquid Biopsy: A Multimodal Diagnostic Technique

The discipline of medical oncology has been tremendously revitalised by the idea of precision medicine over the past several decades; the advent of patient-tailored medicines has greatly improved all quantifiable results. A ground-breaking method called liquid biopsy is providing heretofore unanticipated viewpoints. It entails the identification and isolation of circulating tumour cells, circulating tumour DNA, and exosomes in cancer patients as a source of genomic and proteomic data. Different types of circulating biomarkers are shown in Fig. 1. The focus of the newest branch of oncology research in recent years has been on blood-borne cancer-related substances. Six Cancer cells that have gone into apoptosis or necrosis produce exosomes

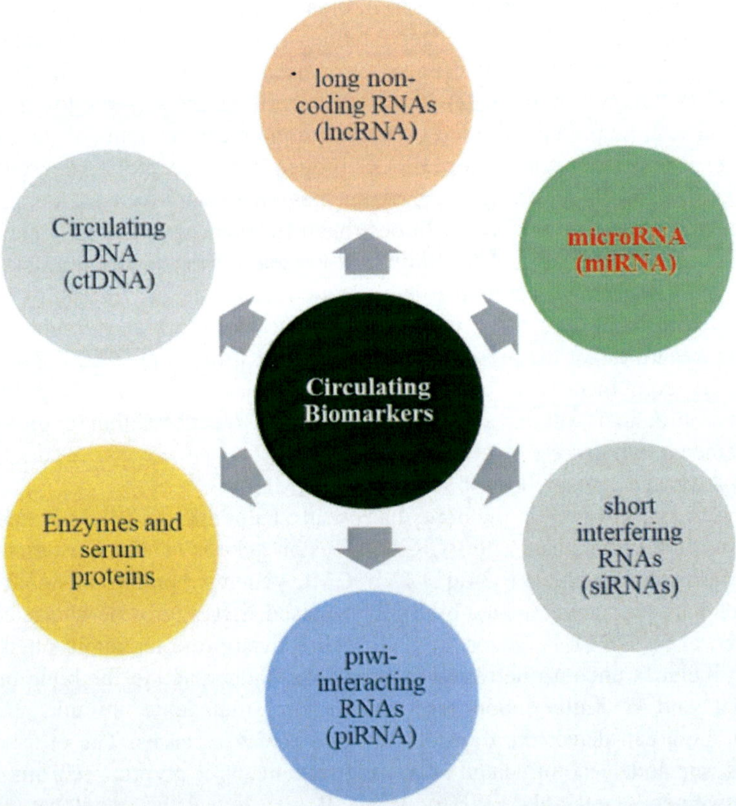

Fig. 1 Different types of circulating biomarkers in cancer

(EXOs), which are membrane-encapsulated subcellular structures that contain proteins and nucleic acids and are known as circulating tumour DNA (ctDNA). Five to 8 Vital cells that have been ingested into the circulation by circulating tumour cells can be exfoliated by primary tumour and metastatic locations as well (CTCs). The procedure known as "liquid biopsy" involves the separation of these tumor-derived components from peripheral blood and their genetic or proteomic analysis (Palmirotta et al. 2018).

Currently, liquid biopsy has a high specificity, making it possible to acquire reliable data using a blood sample in an easy, noninvasive manner. Although liquid biopsy is not currently a standard procedure in clinical practise, its potential applications are rapidly expanding. These include minimal residual disease quantification, monitoring surgical outcomes for radicality, evaluating feedback or reaction to systemic treatments, and diagnostic genomic profiling (Holdenrieder et al. 2022; Keup et al. 2022).

The establishment of distant metastases is caused by CTCs, which are shed from either main or secondary tumour sites and move into the bloodstream. CTCs are incredibly uncommon, occurring at rates as low as 1 CTC per 106–107 leukocytes and much less often in early stages of illness. CTCs are initially classified as nonleukocytic, nucleated cells of epithelial origin, however their morphological characteristics might change depending on the kind and stage of malignancy. Due to their better chance of survival and immune system and oxidative stress protection, CTCs may potentially group together with parental tumour cells or with fibroblasts, leukocytes, endothelial cells, or platelets, producing aggregates that are more likely to cause distant metastases than solitary CTCs (Palmirotta et al. 2018).

On the other hand, CTCs offer a perfect method for molecular cancer detection and therapy alternatives, and their study is widely used in cancer research. Although cytokeratins, such as CK8, CK18, CK1926 and particular tumour markers (TTF-1, PSA, HER-2, etc.) are also helpful for identification and separation, CTCs express epithelial cell adhesion molecule (EPCAM), which is recognised as a biomarker representing a risk factor for tumour recurrence. On the other hand, EPCAM is typically lost during the epithelial to mesenchymal transition (EMT), which is the process that promotes CTC extravasation and migration to secondary sites as well as resistance to anoikis and apoptosis, which has recently been linked to a poor prognosis in breast, colorectal, and nonsmall cell lung cancers (NSCLCs)(Petri and Klinge 2020; Xie et al. 2022).

The only CTC diagnostic method for metastatic breast, prostate, and colorectal cancer that the US Food and Drug Administration (FDA) has authorised to far is the CellSearch test (Menarini Silicon Biosystems, Firenze, Italy), which uses antibody-based immunomagnetic technology and image cytometry. However, the more current DEPArray system allows for both the detection and recovery of single CTCs by surface or cytoplasmic markers, as well as size and dielectroph. A different approach is based on the migratory abilities of CTCs, which enable the binding of these cells to synthetic substrates cotreated with certain complimentary molecules, or protein secretion.

- Circulating tumor DNA

  Mandel and Metais published the first experimental proof of cell-free DNA (cfDNA) in blood in 1948. There have been several hypothesised pathways for ctDNA release in the circulation up to this point. The fact that the majority of ctDNA fragments are 180–200 base pairs (bp) in length suggests that under normal circumstances, macrophages and other scavenger cells quickly phagocytose cfDNA from necrotic and apoptotic cells. When DNA is broken down into nucleosomal units, nuclear material of this bp size is produced (Cescon et al. 2020; Cheng et al. 2021). When DNA is broken down into nucleosomal units, nuclear material of this bp size is produced. This is typical of apoptosis. With a half life of between 16 and 2.5 h, plasma cfDNA concentration in healthy individuals ranges from less than 10 ng/ml to more than 100 ng/ml. High amounts of cfDNA, however, are frequently linked to other disorders including inflammation, trauma, or strenuous exercise rather than malignant illness. The amount of ctDNA in the total cfDNA of cancer patients ranges from less than 0.1% to over

10%, depending on the tumour load, disease stage, cellular turnover, and therapeutic response.

Several PCR-based methods, like BEAMing (beads, emulsion, amplification, and magnetic) and droplet digital PCR (ddPCR), which exhibit extremely high sensitivity (ranging from 1% to 0.001%), are appropriate for the detection of somatic point mutations (Romero et al. 2021; Singh 2020). For example, in BEAMing technologies, a single molecule amplification by PCR is carried out using magnetic beads in water-in-oil emulsions, and then a flow cytometry is employed to quantify the genetic variations. In droplet digital PCR, a subtype of emulsion-based PCR, the sample is broken up into tens of thousands of droplets, each of which represents a partition of a single molecule and is subjected to a PCR analysis using chosen primers against recognised ctDNA regions (Palmirotta et al. 2018).

Fluorescent probes that are bound to the amplified area measure the molecular changes. Utilizing specialised pipelines with various bioinformatics tools, NGS data are processed to find several kinds of genomic abnormalities, such as substitutions, insertions and deletions, CNVs, or gene fusions (e.g. GATK, Annovar, VarScan2). The reads from each sample are then mapped to the reference sequence hg19 (human genome version 19) using alligners software after the FASTQ files obtained from each sequencing system have undergone quality control (QC). The identified genomic changes are then reported, filtered, annotated, and related to clinical relevance in accordance with particular databases (such as dbSNP, COSMIC). The PARE (personalised analysis of rearranged ends) approach was recently developed to accurately detect somatic structural chromosomal changes.

- Exosomes

  EXOs are nanovesicles that range in size from 40 to 100 nm and are produced by cells. They may be found in most bodily fluids, including plasma, urine, saliva, and ascites. EXOs are the byproducts of the recycling endosomal process and arise from inward budding of the plasma membrane, in contrast to other extracellular vesicles (EVs) which directly bud off from the cell membrane, such as microvesicles (50–1000 nm in diameter) or apoptotic bodies (Kalluri and LeBleu 2020; Pegtel and Gould 2019).

  It has been shown that EXOs play a role in a number of physiological and pathological processes, including the growth and metastatic spread of cancer. Noteworthy, cancer releases EXOs cells, namely tumor-derived EXOs, stimulate angiogenesis and the creation of an immunosuppressive environment, promote EMT, and alter the proliferation, migration, and invasion of cancer cells. Additionally, EXOs are developing as a new chemoresistance strategy, principally relying on drug release via vesicle budding, neutralisation of antibody-based medications, and EXOmediated transfer of micro RNAs (miRNA). The genetic profile of tumor-derived EXOs may be examined as diagnostic indicators after their collection and examination by electron microscopy or flow cytometry techniques (Ni et al. 2020; Palmirotta et al. 2018). The detection of serum EXOs enriched with high amounts of migration inhibitory factor (MIF) can identify individuals with pancreatic ductal adenocarcinoma, potentially reflecting an adverse prognostic factor, while CAV1+ EXO have been postulated as prospective diagnostic markers of melanoma. Similar to this, prostate-specific membrane proteins such as PSMA, prostate-specific transglutaminase, and prostate stem cell antigen, whose serum and urine levels correspond with tumour burden, can identify EXOs from prostate cancer cells, more specifically prostasomes.

2.6.3 Artificial Intelligence in Clinical Oncology

Data being gathered to improve cancer care in clinical oncology is expanding quickly. With recent developments in artificial intelligence (AI), it is now possible to integrate and synthesise this expanding corpus of multidimensional data, identify trends, and forecast outcomes to enhance shared decision-making between patients and clinicians (Fig. 2). Despite the enormous potential, there are still substantial

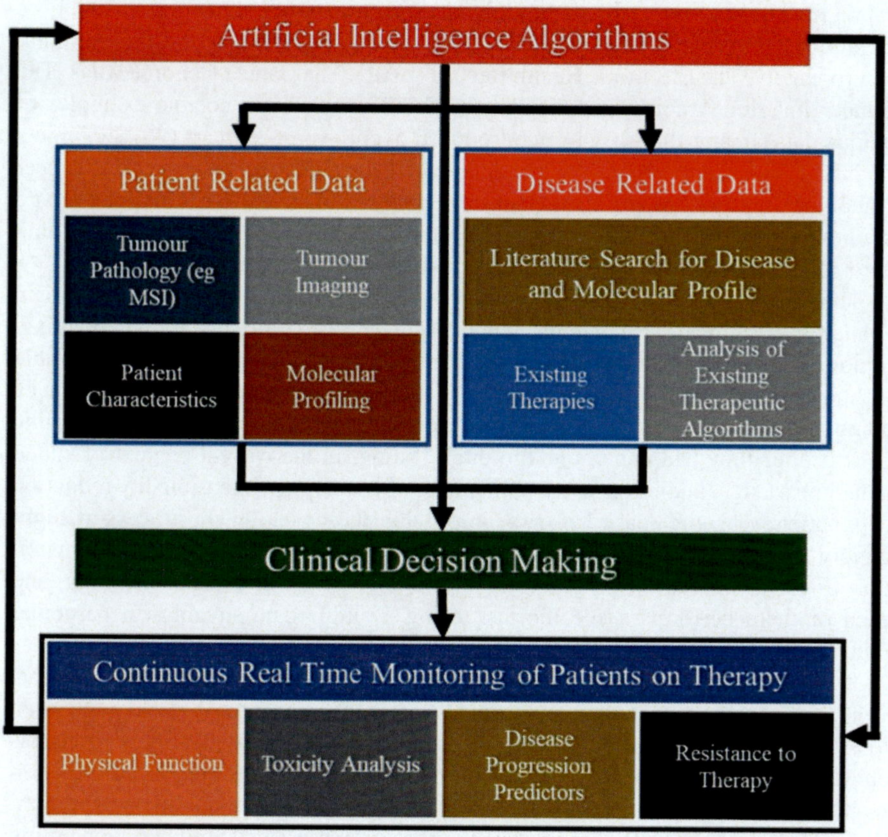

Fig. 2 Artificial Intelligence approach in clinical oncology

obstacles. In this regard, number of applications have been analysed and a pipeline of clinical cancer care touchpoints for narrow-task AI applications were suggested.

- A mathematical optimization for cancer treatment

  The patient care pathway through cancer risk prediction, screening, diagnosis, and therapy must be seen through a mathematical lens in order to fully comprehend the promise of AI applications for clinical oncology. The patient route is an optimization issue from the viewpoint of AI, where diverse data streams converge as inputs into a mathematical framework (i.e. machine learning techniques) (Kann et al. 2021).

The first version of AJCC TNM staging, which used the three generally accessible inputs of tumour size, nodal involvement, and the presence of metastases (TNM), established the standard of care for risk classification and decision-making in oncology in 1977. Predictive nomograms might be created using straightforward linear models in later decades with the introduction of additional discrete data points, and they have found application in some circumstances.

More recently, the construction of increasingly complicated and targeted stratification models has been facilitated by advances in current data extraction and analysis techniques, new data streams, and a better knowledge of inter- and intra-tumoral heterogeneity. The Electronic Health Record (EHR), The Cancer Genome Atlas, The Cancer Imaging Archive, and the Project GENIE project are notable examples of unique data streams that have been introduced over the past 20 years (AACR Project GENIE Consortium 2017). The prostate cancer Decipher score and breast cancer OncotypeDx score are two important examples of advanced risk stratification and prediction models that use discrete genomic data and shallow machine-learning algorithms to create clinically useful predictive models (Kann et al. 2021).

The following useful cancer data sources are broadly listed in order of historical availability: clinical presentation, tumour stage, histology, qualitative imaging, tumour genomics, quantitative imaging, liquid biopsies, patient genomics, wearable technology, and digital behaviour. These specific AI applications might then be combined over the course of a patient's treatment to offer a quantifiable, therapeutic benefit once they had undergone thorough testing. In the current context of siloed data, knowledge, and infrastructure, this type of AI-driven dimensionality reduction of a patient's feature space helps for improving the development process of high-quality AI applications. Currently, there are around 20 Food and Drug Administration (FDA)-approved AI programmes specifically aimed at clinical oncology, and each of them performs a specific task using a single data stream at a particular touchpoint for cancer care. (Benjamens et al. 2020; Hamamoto et al. 2020).

A model is frequently assessed in the general order of internal validation, external validation, prospective testing, and local testing in the relevant real-world population to establish clinical validity (Park et al. 2020). To ensure reproducibility, transparency, and methodologic rigour, it is important to adhere to recently created recommendations like FAIR data, CONSORT/SPIRTAI, and the (in preparation) TRIPOD-AI checklists (Liu et al. 2019). These principles are significant in defining

AI model development routes and provide a foundation for judging the methodological rigour of AI studies.

3 Conclusion

It has always been, and to a large part still is, a massive job to unravel the genetic and epigenetic modifications involved in cancer development. More knowledge is being gained on the role of the epigenome in cell type identification, development, and disease. Personalized medicine in cancer has advanced significantly with the development and application of clinical molecular diagnostics, with predictive biomarkers guiding both therapy and surveillance of disease progression or remission. The development of clinical diagnostic laboratory tests that are paired with advances in biomarker technology will help medicine continue to evolve, where the standard of care will continue to be defined by the personalised and customised therapy that is appropriate for each individual patient. Additionally, by engaging at specific touchpoints throughout the cancer treatment system with narrow-task apps, and developments in computational algorithms have positioned AI to improve clinical oncology. Although there are some potential AI applications for clinical cancer under development, there are still significant obstacles to overcome before clinical translation can occur. The clinical effectiveness may be increased by selecting the appropriate patients, adjusting the trial design, and changing the dose regimens.

References

AACR Project GENIE Consortium (2017) AACR Project GENIE: powering precision medicine through an International Consortium. Cancer Discov 7(8):818–831. https://doi.org/10.1158/2159-8290.CD-17-0151. Epub 2017 Jun 1. PMID: 28572459; PMCID: PMC5611790

Akram F, Sahreen S, Aamir F, Ul Haq I, Malik K, Imtiaz M, Naseem W, Nasir N, Waheed HM (2022) An insight into modern targeted genome-editing technologies with a special focus on CRISPR/Cas9 and its applications. Mol Biotechnol 65:227. https://doi.org/10.1007/S12033-022-00501-4

Alkhayyat S, Khan M, Ahmad T, Haroon, Tariq H, Baig M (2022) A bibliometric analysis of the top 100 most cited papers and research trends in breast cancer related BRCA1 and BRCA2 genes. Medicine 101:E30576. https://doi.org/10.1097/MD.0000000000030576

Allegretti M, Cesta MC, Zippoli M, Beccari A, Talarico C, Mantelli F, Bucci EM, Scorzolini L, Nicastri E (2022) Repurposing the estrogen receptor modulator raloxifene to treat SARS-CoV-2 infection. Cell Death Differ 29:156. https://doi.org/10.1038/s41418-021-00844-6

Atadja PW (2011) HDAC inhibitors and cancer therapy. Prog Drug Res 67:175–195. https://doi.org/10.1007/978-3-7643-8989-5_9

Azamjah N, Soltan-Zadeh Y, Zayeri F (2019) Global trend of breast cancer mortality rate: a 25-year study. Asian Pac J Cancer Prev 20:2015–2020. https://doi.org/10.31557/APJCP.2019.20.7.2015

Barquinero J, Eixarch H, Perez-Melgosa M (2004) Retroviral vectors: new applications for an old tool. Gene Ther 11:S3. https://doi.org/10.1038/sj.gt.3302363

Bell JA, Galaznik A, Pompilus F, Strzok S, Bejar R, Scipione F, Fram RJ, Faller DV, Cano S, Marquis P (2019) A pragmatic patient-reported outcome strategy for rare disease clinical trials: application of the EORTC item library to myelodysplastic syndromes, chronic myelomonocytic leukemia, and acute myeloid leukemia. J Patient Rep Outcomes 3:35. https://doi.org/10.1186/s41687-019-0123-4

Benjamens S, Dhunnoo P, Meskó B (2020) The state of artificial intelligence-based FDA-approved medical devices and algorithms: an online database. NPJ Digit Med 3(1):1–8

Bhan A, Deb P, Mandal SS (2017) Epigenetic code: histone modification, gene regulation, and chromatin dynamics. In: Gene regulation, epigenetics and hormone signaling. https://doi.org/10.1002/9783527697274.ch2

Bi R, Li Y, Xu M, Zheng Q, Zhang D-F, Li X, Ma G, Xiang B, Zhu X, Zhao H, Huang X, Zheng P, Yao Y-G (2022) Direct evidence of CRISPR-Cas9-mediated mitochondrial genome editing. Innovation 3:100329. https://doi.org/10.1016/j.xinn.2022.100329

Bondarev AD, Attwood MM, Jonsson J, Chubarev VN, Tarasov V, Schiöth HB (2021) Recent developments of HDAC inhibitors: emerging indications and novel molecules. Br J Clin Pharmacol 87:12. https://doi.org/10.1111/bcp.14889

Bracco E, Shahzad Ali M, Magnati S, Saglio G (2021) The paradigm of targeting an oncogenic tyrosine kinase: lesson from BCR-ABL. Adv Precision Med Oncol. IntechOpen. https://doi.org/10.5772/intechopen.97528

Cescon DW, Bratman SV, Chan SM, Siu LL (2020) Circulating tumor DNA and liquid biopsy in oncology. Nat Cancer 18:114. https://doi.org/10.1038/s43018-020-0043-5

Chattopadhyaya S, Ghosal S (2022) DNA methylation: a saga of genome maintenance in hematological perspective. Hum Cell 35:448–461. https://doi.org/10.1007/S13577-022-00674-9

Chen CD, Rudy MA, Zeldich E, Abraham CR (2020) A method to specifically activate the Klotho promoter by using zinc finger proteins constructed from modular building blocks and from naturally engineered Egr1 transcription factor backbone. FASEB J 34:7234–7246. https://doi.org/10.1096/fj.202000171R

Cheng Y, He C, Wang M, Ma X, Mo F, Yang S, Han J, Wei X (2019) Targeting epigenetic regulators for cancer therapy: mechanisms and advances in clinical trials. Signal Transduct Target Ther 4:62. https://doi.org/10.1038/s41392-019-0095-0

Cheng ML, Pectasides E, Hanna GJ, Parsons HA, Choudhury AD, Oxnard GR (2021) Circulating tumor DNA in advanced solid tumors: clinical relevance and future directions. CA Cancer J Clin 71(2):176–190

Coombes RC, Berger U, Mcclelland RA, Trott PA, Powles TJ, Wilson P, Gazet JC, Ford HT (1987) Prediction of endocrine response in breast cancer by immunocytochemical detection of oestrogen receptor in fine-needle aspirates. Lancet 330:701–703. https://doi.org/10.1016/S0140-6736(87)91071-3

Coons AH, Kaplan MH (1950) Localization of antigen in tissue cells; improvements in a method for the detection of antigen by means of fluorescent antibody. J Exp Med 91:1–13. https://doi.org/10.1084/jem.91.1.1

Coudé MM, Braun T, Berrou J, Dupont M, Bertrand S, Masse A, Raffoux E, Itzykson R, Delord M, Riveiro ME, Herait P, Baruchel A, Dombret H, Gardin C (2015) BET inhibitor OTX015 targets BRD2 and BRD4 and decreases c-MYC in acute leukemia cells. Oncotarget 6:17698–17712. https://doi.org/10.18632/oncotarget.4131

Cuciniello R, Filosa S, Crispi S (2021) Novel approaches in cancer treatment: preclinical and clinical development of small non-coding RNA therapeutics. J Exp Clin Cancer Res 40:383. https://doi.org/10.1186/S13046-021-02193-1

Cui Z, Liu H, Zhang H, Huang Z, Tian R, Li L, Fan W, Chen Y, Chen L, Zhang S, Das BC, Severinov K, Hitzeroth II, Debata PR, Jin Z, Liu J, Huang Z, Xie W, Xie H, Lang B, Ma J, Weng H, Tian X, Hu Z (2021) The comparison of ZFNs, TALENs, and SpCas9 by GUIDE-seq in HPV-targeted gene therapy. Mol Ther Nucleic Acids 26:1466–1478. https://doi.org/10.1016/j.omtn.2021.08.008

Daher-Reyes GS, Merchan BM, Yee KWL (2019) Guadecitabine (SGI-110): an investigational drug for the treatment of myelodysplastic syndrome and acute myeloid leukemia. Expert Opin Investig Drugs 28:835–849. https://doi.org/10.1080/13543784.2019.1667331

Das BD, Paudel N (2021) A review on reliability and validity of CRISPR/Cas9 technology for gene editing. Jordan J Biol Sci 14(3)

Dombret H, Seymour JF, Butrym A, Wierzbowska A, Selleslag D, Jang JH, Kumar R, Cavenagh J, Schuh AC, Candoni A, Récher C, Sandhu I, Del Castillo TB, Al-Ali HK, Martinelli G, Falantes J, Noppeney R, Stone RM, Minden MD, McIntyre H, Songer S, Lucy LM, Beach CL, Döhner H (2015) International phase 3 study of azacitidine vs conventional care regimens in older patients with newly diagnosed AML with >30% blasts. Blood 126:291–299. https://doi.org/10.1182/blood-2015-01-621664

Eagen KP, French CA (2021) Supercharging BRD4 with NUT in carcinoma. Oncogene 40:1396. https://doi.org/10.1038/s41388-020-01625-0

Ellis GI, Sheppard NC, Riley JL (2021) Genetic engineering of T cells for immunotherapy. Nat Rev Genet 22:427. https://doi.org/10.1038/s41576-021-00329-9

Feng S, Carvalho D, Daniel D, Feng S, De Carvalho DD (2022) Clinical advances in targeting epigenetics for cancer therapy. FEBS J 289:1214–1239. https://doi.org/10.1111/febs.15750

Fey MF, Wainscoat JS (1988) Molecular diagnosis of haematological neoplasms. Blood Rev. 2:78–87. https://doi.org/10.1016/0268-960X(88)90028-8

French CA (2018) NUT carcinoma: clinicopathologic features, pathogenesis, and treatment. Pathol Int 68:583–595. https://doi.org/10.1111/PIN.12727

Giugliano F, Crimini E, Tarantino P, Zagami P, Uliano J, Corti C, Trapani D, Curigliano G, Ascierto PA (2021) First line treatment of BRAF mutated advanced melanoma: does one size fit all? Cancer Treat Rev 99:102253. https://doi.org/10.1016/j.ctrv.2021.102253

Grillone K, Riillo C, Riillo C, Scionti F, Rocca R, Rocca R, Tradigo G, Guzzi PH, Alcaro S, Di Martino MT, Tagliaferri P, Tagliaferri P, Tassone P (2020) Non-coding RNAs in cancer: platforms and strategies for investigating the genomic "dark matter". J Exp Clin Cancer Res 39:117. https://doi.org/10.1186/S13046-020-01622-X

Guido LP, Gomez-Fernandez C (2020) Advances in the molecular taxonomy of breast cancer. Arch Med Res 51:777. https://doi.org/10.1016/j.arcmed.2020.08.001

Halabian R, Arshad V, Ahmadi A, Saeedi P, Azimzadeh Jamalkandi S, Alivand MR (2021) Laboratory methods to decipher epigenetic signatures: a comparative review. Cell Mol Biol Lett 26:46. https://doi.org/10.1186/S11658-021-00290-9

Halcrow PW, Dancer M, Panteah M, Walden C, Ohm JE (2016) Molecular changes associated with tumor initiation and progression of soft tissue sarcomas: targeting the genome and epigenome. Prog Mol Biol Transl Sci 144:323–380. https://doi.org/10.1016/bs.pmbts.2016.10.001

Hamamoto R, Suvarna K, Yamada M, Kobayashi K, Shinkai N, Miyake M, Takahashi M, Jinnai S, Shimoyama R, Sakai A, Takasawa K (2020) Application of artificial intelligence technology in oncology: towards the establishment of precision medicine. Cancers 12(12):3532

Harale G, Pardeshi S, Majumdar P, Ganger S (2022) Transcription activator-like effector nucleases (TALENs): genome editing tool to explore enhanced activity of antidiabetic plants. In: Antidiabetic potential of plants in the era of omics. Apple Academic Press, Point Pleasant, pp 403–428. https://doi.org/10.1201/9781003282860-19

Holdenrieder S, Bronkhorst A, Keup C, Kimmig R, Kasimir-Bauer S (2022) Combinatorial power of cfDNA, CTCs and EVs in oncology. Diagnostics 12:870. https://doi.org/10.3390/diagnostics12040870

Imran A, Qamar HY, Ali Q, Naeem H, Riaz M, Amin S, Kanwal N, Ali F, Sabar MF, Nasir IA (2017) Role of molecular biology in cancer treatment: a review article. Iran J Public Health 46:1475

Inchingolo F, Santacroce L, Ballini A, Topi S, Dipalma G, Haxhirexha K, Bottalico L, Charitos IA (2020) Oral cancer: a historical review. Int J Environ Res Public Health 17:3168. https://doi.org/10.3390/ijerph17093168

Iranzadasl M, Mansouri P, Kamalinejad M, Gorji N (2020) Review of skin cancers terminology, etiology and treatment from ancient persian medicine view point. J Med Plant 19:27–36. https://doi.org/10.29252/jmp.1.73.27

Irving M, Lanitis E, Migliorini D, Ivics Z, Guedan S (2021) Choosing the right tool for genetic engineering: clinical lessons from chimeric antigen receptor-T cells. Hum Gene Ther 32:1044–1058. https://doi.org/10.1089/hum.2021.173

Kalluri R, LeBleu VS (2020) The biology, function, and biomedical applications of exosomes. Science 367:eaau6977. https://doi.org/10.1126/SCIENCE.AAU6977

Kann BH, Hosny A, Aerts HJWL (2021) Artificial intelligence for clinical oncology. Cancer Cell 39:916. https://doi.org/10.1016/j.ccell.2021.04.002

Kanwal R, Gupta K, Gupta S (2015) Cancer epigenetics: an introduction. Methods Mol Biol 1238: 3–25. https://doi.org/10.1007/978-1-4939-1804-1_1

Karakaidos P, Verigos J, Magklara A (2019) Lsd1/kdm1a, a gate-keeper of cancer stemness and a promising therapeutic target. Cancers (Basel) 11:1821. https://doi.org/10.3390/cancers11121821

Kariyawasam DST, D'silva A, Lin C, Ryan MM, Farrar MA (2019) Biomarkers and the development of a personalized medicine approach in spinal muscular atrophy. Front Neurol 10:898. https://doi.org/10.3389/FNEUR.2019.00898/FULL

Keup C, Kimmig R, Kasimir-Bauer S (2022) Multimodality in liquid biopsy: does a combination uncover insights undetectable in individual blood analytes? J Lab Med 46:255–264. https://doi.org/10.1515/LABMED-2022-0009/HTML

Khadempar S, Familghadakchi S, Motlagh RA, Farahani N, Dashtiahangar M, Rezaei H, Gheibi Hayat SM (2019) CRISPR–Cas9 in genome editing: Its function and medical applications. J Cell Physiol 234:5751–5761. https://doi.org/10.1002/JCP.27476

Khan MT, Afzal S, Rehman AU, Zeb T (2015) Interleukin 10 (IL-10) promoter-1082 A > G polymorphism and risk of cancer: metaanalysis. Adv Life Sci 2:67–73

Kumar S, Gupta S, Maurya AP, Singh R, Nigam S (2022) Hormonal and targeted treatments in breast cancer. Breast Cancer 443–463. https://doi.org/10.1007/978-981-16-4546-4_21

Kurata M, Onishi I, Takahara T, Yamazaki Y, Ishibashi S, Goitsuka R, Kitamura D, Takita J, Hayashi Y, Largaesapda DA, Kitagawa M, Nakamura T (2021) C/EBPβ induces B-cell acute lymphoblastic leukemia and cooperates with BLNK mutations. Cancer Sci 112:4920–4930. https://doi.org/10.1111/cas.15164

Lan T, Que H, Luo M, Zhao X, Wei X (2022) Genome editing via non-viral delivery platforms: current progress in personalized cancer therapy. Mol Cancer 21:71. https://doi.org/10.1186/S12943-022-01550-8

Lee JH, Wang JH, Chen J, Li F, Edwards TL, Hewitt AW, Liu GS (2019) Gene therapy for visual loss: opportunities and concerns. Prog Retin Eye Res 68:31. https://doi.org/10.1016/j.preteyeres.2018.08.003

Li H, Yang Y, Hong W, Huang M, Wu M, Zhao X (2020) Applications of genome editing technology in the targeted therapy of human diseases: mechanisms, advances and prospects. Signal Transduct Target Ther 5:1. https://doi.org/10.1038/s41392-019-0089-y

Liu X, Faes L, Calvert MJ, Denniston AK (2019) Extension of the CONSORT and SPIRIT statements. Lancet 394(10205):1225

Lopez-Chavez A, Thomas A, Rajan A, Raffeld M, Morrow B, Kelly R, Carter CA, Guha U, Killian K, Lau CC, Abdullaev Z, Xi L, Pack S, Meltzer PS, Corless CL, Sandler A, Beadling C, Warrick A, Liewehr DJ, Steinberg SM, Berman A, Doyle A, Szabo E, Wang Y, Giaccone G (2015) Molecular profiling and targeted therapy for advanced thoracic malignancies a biomarker-derived, multiarm, multihistology phase II basket trial. J Clin Oncol 33:1000–1007. https://doi.org/10.1200/JCO.2014.58.2007

Lyu P, Wang L, Lu B (2020) Virus-like particle mediated crispr/cas9 delivery for efficient and safe genome editing. Life 10:366. https://doi.org/10.3390/life10120366

Malone ER, Oliva M, Sabatini PJB, Stockley TL, Siu LL (2020) Molecular profiling for precision cancer therapies. Genome Med 14:8. https://doi.org/10.1186/s13073-019-0703-1

McClure JJ, Inks ES, Zhang C, Peterson YK, Li J, Chundru K, Lee B, Buchanan A, Miao S, Chou CJ (2017) Comparison of the deacylase and deacetylase activity of zinc-dependent HDACs. ACS Chem Biol 12:1644–1655. https://doi.org/10.1021/ACSCHEMBIO.7B00321

McLoughlin EM, Gentzler RD (2020) Epidermal growth factor receptor mutations. Thorac Surg Clin 30:127. https://doi.org/10.1016/j.thorsurg.2020.01.008

Meador CB, Sequist LV, Piotrowsk Z (2021) Targeting egfr exon 20 insertions in non–small cell lung cancer: recent advances and clinical updates. Cancer Discov 11:2145. https://doi.org/10.1158/2159-8290.CD-21-0226

Menzer C, Hassel JC (2022) Targeted therapy for melanomas without BRAF V600 mutations. Curr Treat Options Oncol 23:831–842. https://doi.org/10.1007/s11864-022-00946-4

Minarovits J, Banati F, Szenthe K, Niller HH (2016) Epigenetic regulation. Adv Exp Med Biol 879: 1–25. https://doi.org/10.1007/978-3-319-24738-0_1

Momota H, Ogino J, Takahashi A, Hasegawa T, Wakabayashi T (2015) Rhabdoid glioblastoma: an aggressive variaty of astrocytic tumor. Nagoya J Med Sci 77:321–328

Morganti S, Tarantino P, Ferraro E, D'Amico P, Duso BA, Curigliano G (2019) Next generation sequencing (NGS): a revolutionary technology in pharmacogenomics and personalized medicine in cancer. Adv Exp Med Biol 1168:9–30. https://doi.org/10.1007/978-3-030-24100-1_2

Nakamura T (2005) Retroviral insertional mutagenesis identifies oncogene cooperation. Cancer Sci 96:7–12. https://doi.org/10.1111/j.1349-7006.2005.00011.x

Nebbioso A, Tambaro FP, Dell'Aversana C, Altucci L (2018) Cancer epigenetics: moving forward. PLoS Genet 14:e1007362. https://doi.org/10.1371/journal.pgen.1007362

Ni Z, Zhou S, Li S, Kuang L, Chen H, Luo X, Ouyang J, He M, Du X, Chen L (2020) Exosomes: roles and therapeutic potential in osteoarthritis. Bone Res 8:25. https://doi.org/10.1038/s41413-020-0100-9

Ong FS, Das K, Wang J, Vakil H, Kuo JZ, Blackwell WLB, Lim SW, Goodarzi MO, Bernstein KE, Rotter JI, Grody WW (2012) Personalized medicine and pharmacogenetic biomarkers: progress in molecular oncology testing. Expert Rev Mol Diagn 12:593. https://doi.org/10.1586/erm.12.59

Oricchio E, Katanayeva N, Donaldson MC, Sungalee S, Joyce PP, Béguelin W, Battistello E, Sanghvi VR, Jiang M, Jiang Y, Teater M, Parmigiani A, Budanov AV, Chan FC, Shah SP, Kridel R, Melnick AM, Ciriello G, Wendel HG (2017) Genetic and epigenetic inactivation of SESTRIN1 controls mTORC1 and response to EZH2 inhibition in follicular lymphoma. Sci Transl Med 9:eaak9969. https://doi.org/10.1126/SCITRANSLMED.AAK9969

Ottolino-Perry K, Diallo JS, Lichty BD, Bell JC, Andrea McCart J (2010) Intelligent design: combination therapy with oncolytic viruses. Mol Ther 18:251. https://doi.org/10.1038/mt.2009.283

Pal M, Muinao T, Boruah HPD, Mahindroo N (2022) Current advances in prognostic and diagnostic biomarkers for solid cancers: detection techniques and future challenges. Biomed Pharmacother 146:112488. https://doi.org/10.1016/j.biopha.2021.112488

Palmirotta R, Lovero D, Cafforio P, Felici C, Mannavola F, Pellè E, Quaresmini D, Tucci M, Silvestris F (2018) Liquid biopsy of cancer: a multimodal diagnostic tool in clinical oncology. Ther Adv Med Oncol 10:1758835918794630. https://doi.org/10.1177/1758835918794630

Park Y, Jackson GP, Foreman MA, Gruen D, Hu J, Das AK (2020) Evaluating artificial intelligence in medicine: phases of clinical research. JAMIA Open 3(3):326–331

Pegtel DM, Gould SJ (2019) Exosomes. Annu Rev Biochem 88:487–514. https://doi.org/10.1146/ANNUREV-BIOCHEM-013118-111902

Petri BJ, Klinge CM (2020) Regulation of breast cancer metastasis signaling by miRNAs. Cancer Metastasis Rev 39:837–886. https://doi.org/10.1007/S10555-020-09905-7

Roboz GJ, Döhner H, Gobbi M, Kropf PL, Mayer J, Krauter J, Robak T, Kantarjian HM, Novak J, Jedrzejczak W, Thomas X, Ojeda-Uribe M, Miyazaki Y, Min YH, Yeh S-P, Brandwein JM, Gercheva L, Demeter J, Griffiths EA, Yee KWL, Issa J-P, Hao Y, Azab M, Fenaux P (2019) Results from a global randomized phase 3 study of Guadecitabine (G) vs Treatment Choice (TC) in 815 patients with Treatment Naïve (TN) AML unfit for Intensive Chemotherapy (IC) ASTRAL-1 study: analysis by number of cycles. Blood 134:2591–2591. https://doi.org/10.1182/blood-2019-127253

Romero A, Jantus-Lewintre E, Garcia-Pelaez B, Royuela A, Insa A, Cruz P, Collazo A, Perez Altozano J, Vidal OJ, Diz P, Cobo M, Hern Andez B, Vazquez Estevez S, Ben Itez G, Guirado M, Majem M, Bernab R, Ortega AL, Blasco A, Bosch-Barrera J, Jurado JM, Garcia Gonzalez J, Viteri S, Giron CG, Massuti B, Martin Lopez A, Rodriguez-Festa A, Calabuig-Farinas S, Molina-Vila MA, Provencio M, Romero CA, Jantus-Lewintre E, Molina-Vila MA, Biopsia L, Iquida L (2021) Comprehensive cross-platform comparison of methods for non-invasive EGFR mutation testing: results of the RING observational trial. Mol Oncol 15: 43–56. https://doi.org/10.1002/1878-0261.12832

Russell L, Peng KW, Russell SJ, Diaz RM (2019) Oncolytic viruses: priming time for cancer immunotherapy. BioDrugs 33:485–501. https://doi.org/10.1007/S40259-019-00367-0

Schambach A, Morgan M (2016) Retroviral vectors for cancer gene therapy. Recent Results Cancer Res 209:17–35. https://doi.org/10.1007/978-3-319-42934-2_2

Shah RR (2019) Safety and tolerability of histone deacetylase (HDAC) inhibitors in oncology. Drug Saf 42:235–245. https://doi.org/10.1007/S40264-018-0773-9

Sherman ME, Troester MA, Hoadley KA, Anderson WF (2017) Morphological and molecular classification of human cancer. In: Schottenfeld and Fraumeni cancer epidemiology and prevention, 4th edn. Oxford University Press, Oxford, pp 19–42. https://doi.org/10.1093/oso/9780190238667.003.0003

Singh RR (2020) Next-generation sequencing in high-sensitive detection of mutations in tumors: challenges, advances, and applications. J Mol Diagnostics 22:994. https://doi.org/10.1016/j.jmoldx.2020.04.213

Smith J, Sen S, Weeks RJ, Eccles MR, Chatterjee A (2020) Promoter DNA hypermethylation and paradoxical gene activation. Trends Cancer 6:392. https://doi.org/10.1016/j.trecan.2020.02.007

Sokolenko AP, Imyanitov EN (2018) Molecular diagnostics in clinical oncology. Front Mol Biosci 5:76. https://doi.org/10.3389/FMOLB.2018.00076/FULL

Strashilov S, Yardanov A (2021) Aetiology and pathogenesis of cutaneous melanoma: current concepts and advances. Int J Mol Sci 22:6395. https://doi.org/10.3390/ijms22126395

Sudhakar A (2009) History of cancer, ancient and modern treatment methods. J Cancer Sci Ther 1: i–iv. https://doi.org/10.4172/1948-5956.100000e2

Timar J, Kashofer K (2020) Molecular epidemiology and diagnostics of KRAS mutations in human cancer. Cancer Metastasis Rev 39:1029–1038. https://doi.org/10.1007/S10555-020-09915-5

Valdespino V, Valdespino PM (2015) Potential of epigenetic therapies in the management of solid tumors. Cancer Manag Res 7:241–251. https://doi.org/10.2147/CMAR.S70358

Van Der Helm LH, Alhan C, Wijermans PW, Van M, Kooy M, Schaafsma R, Biemond BJ, Beeker A, Hoogendoorn M, Van Rees BP, De Weerdt O, Wegman J, Libourel WJ, Luykx-De Bakker SA, Minnema MC, Brouwer RE, Croon-De Boer F, Eefting M, Jie K-SG, Van De Loosdrecht AA, Koedam J, Veeger NJGM, Vellenga E, Huls G, Huls G (2011) FDA drug approval summary: Azacitidine (5-azacytidine, Vidaza™) for injectable suspension. Oncologist 155:599–606. https://doi.org/10.1111/j.1365-2141.2011.08893.x

Vargas JE, Chicaybam L, Stein RT, Tanuri A, Delgado-Cañedo A, Bonamino MH (2016) Retroviral vectors and transposons for stable gene therapy: advances, current challenges and perspectives. J Transl Med 14:288. https://doi.org/10.1186/S12967-016-1047-X

Xie X, Li Y, Lian S, Lu Y, Jia L (2022) Cancer metastasis chemoprevention prevents circulating tumour cells from germination. Signal Transduct Target Ther 7:341. https://doi.org/10.1038/s41392-022-01174-w

Yan J, Diaz J, Jiao J, Wang R, You J (2011) Perturbation of BRD4 protein function by BRD4-NUT protein abrogates cellular differentiation in NUT midline carcinoma. J Biol Chem 286:27663–27675. https://doi.org/10.1074/jbc.M111.246975

Yedjou CG, Sims JN, Miele L, Noubissi F, Lowe L, Fonseca DD, Alo RA, Payton M, Tchounwou PB (2019) Health and racial disparity in breast cancer. Adv Exp Med Biol 1152:31–49. https://doi.org/10.1007/978-3-030-20301-6_3

Zhou Z, Gao J, Popovic R, Wolniak K, Parimi V, Winter JN, Licht JD, Chen YH (2015) Strong expression of EZH2 and accumulation of trimethylated H3K27 in diffuse large B-cell lymphoma independent of cell of origin and EZH2 codon 641 mutation. Leuk Lymphoma 56:2895–2901. https://doi.org/10.3109/10428194.2015.1006220

Oncogenic Genomic Changes in Cancer

Abdulbaki Yildirim, Hilal Akalin, and Munis Dundar

1 Introduction

Protooncogenes are genes responsible for the synthesis of many proteins that function in the signal transduction mechanism of cells. Proteins that function in the regulation of normal cell growth are encoded by protooncogenes and oncogenes (Batra et al. 2021). Mutations of protooncogenes (*cyclin D*, *CDK4*, etc.) encoding proteins in cell signalling pathways result in excessive production of growth factors, uncontrolled stimulation of the pathways between the cell membrane and nucleus, increased synthesis of transcription factors, and inability to prevent cell division (Batra et al. 2021).

When a protooncogene is mutated or misexpressed and contributes to the development of cancer, it is called an oncogene (Pierotti et al. 2003). Some of the oncogenes protect the cell from apoptosis, which is programmed cell death, while others reduce the growth factor requirement and ensure the continuous proliferation of the cell. Oncogenes are usually identified by three letters (like *Myc*, *Ras*) (Pierotti et al. 2003).

When they are first defined, according to the origin of oncogenes, v- suffixes are used for those of viral origin, and c- suffixes are used for those of cell or chromosomal origin (Burrell et al. 2017).

A. Yildirim · H. Akalin · M. Dundar (✉)
Department of Medical Genetics, School of Medicine, Erciyes University, Kayseri, Turkey
e-mail: dundar@erciyes.edu.tr

H. S. Tuli, M. B. Yerer Aycan (eds.), *Oncology: Genomics, Precision Medicine and Therapeutic Targets*, https://doi.org/10.1007/978-981-99-1529-3_2

2 Activation of Protooncogenes Mechanisms

Point mutation, insertion, translocation or gene amplification activates protooncogenes (Botezatu et al. 2016).

2.1 *Insertion*

With the introduction of a new pair between the nucleotide pairs in the cell genome, a change occurs in the number and order of nucleotides in the genome. Such a mutation results in a change in the protein synthesized by the proto-oncogene.

2.2 *Chromosomal Translocation*

When a chromosome segment carrying a protooncogene is displaced and integrated in front of or behind another gene with strong promoter activity, that gene enters the control of a strong promoter-effective gene and becomes overexpressed.

The first example of oncogene activation by chromosome translocation is Burkitt's lymphoma, which is the malignancy of antibody-producing B lymphocytes (Schmitz et al. 2014). By translocation, the *c-MYC* gene can be under the influence of the promoter and over-stimulated, resulting in excessive expression of the c-MYC protein, which can cause a tumourigenic phenotype with an oncogenic effect (Schmitz et al. 2014).

The second example due to translocation is the Philadelphia chromosome. In these patients, the *c-ABL* protooncogene located on chromosome 9 is separated from here and localized to the *BCR* gene region, which localizes on chromosome 22 and has very strong promoter sequences (Sampaio et al. 2021). The resulting *BCR/ABL* hybrid gene is overexpressed and the synthesis of the oncogenic fusion protein with tyrosine kinase activity increases (Sampaio et al. 2021).

2.2.1 Cyclins and Cyclin-Dependent Kinases

These molecules are molecules that direct the cell cycle by phosphorylating critical target proteins with the complex they form. Activating mutations of cyclins and cyclin-dependent kinases (CDK) involved in the cell cycle mimic oncogene, whereas CDK inhibitors mimic tumour suppressor gene function (Peyressatre et al. 2015).

2.2.2 Cyclin D1 and Cyclin E Protooncogenes

Cyclins form complexes with CDK molecules and regulate their activities. The CDK/Cyclin complexes formed are regulatory molecules that are important at every stage of the cell cycle.

Many cyclin genes are known to be associated with cancer development. For example, the gene encoding Cyclin D1 has been amplified in cancers of the breast, bladder, lung, and oesophagus (Alao 2007). High levels of cyclin D1 protein are thought to contribute to uncontrolled entry into the S phase (Alao 2007).

In certain parathyroid tumours and B cell lymphomas, the Cyclin D1 gene is also found in chromosomal abnormalities such as translocations. It is thought that these changes in the cyclin D1 gene may cause the emergence of abnormal gene expressions (Alao 2007).

The cyclin E gene is overexpressed in some leukaemias, and breast and colon cancers. Overexpression or periodic loss of these key cell cycle regulators ensures that cells are kept ready to continue the cell cycle, and cells are no longer able to leave the cycle, enter the G0 dormant stage, or undergo cell differentiation (Fagundes and Teixeira 2021).

2.2.3 Telomeres and Telomerase Oncogene

The regions that make up the end of the chromosomes are called telomeres. Telomeres consist of thousands of copies of the 6-nucleotide TTAGGG sequences in humans (Jafri et al. 2016).

Since DNA replication is bidirectional in humans and runs only in the 5′-3′ direction, a certain amount of telomere shortening is inevitable during each division. On the other hand, in normal cell cultures, after a certain number of passages, division stops and cells enter the replicative senescence stage. It has been determined that this phenomenon, also called the ‘Hayflick limit’, is parallel to the shortening of telomeres. Telomerase enzyme prevents this shortening.

This enzyme consists of a catalytic protein (hTERT) and a short RNA molecule (hTER). hTERT is an RNA-dependent reverse transcriptase that synthesizes DNA from hTER, thereby preventing telomeric shortening (Jafri et al. 2016).

Telomerase is active only in cells that need to be constantly proliferating, such as lymphocytes, basal keratinocytes, and intestinal crypt cells.

In tumours that normally develop from cells without telomerase activity, it is observed that telomeric sequences do not shorten despite uncontrolled proliferation, and telomerase activity is detected in these tumours. Therefore, telomerase shows a kind of oncogene effect in these tumours.

As a clinical reflection of this information, immunotherapy, gene therapy and oligonucleotide-based treatments to suppress telomerase activity are at the trial stage. It has also been stated that detecting telomerase activity in tissue samples and demonstrating the presence of hTERT by immunohistochemical staining can be used in cancer diagnosis and prognosis prediction.

2.3 *Gene Amplification*

Gene amplification is a mechanism that can cause an increase in oncogenic protein by replicating a specific chromosome region many times (DNA amplification). Gene amplification, which is an important event programmed for the development of eukaryotic organisms, breaks down and becomes an unscheduled disorder.

Oncogenes such as *c-MYC*, *c-ABL*, c-*ERBB*, and *c-RAS*, which are detected in various tumours, are amplified and duplicated in the cell. Gene amplification plays a role in gaining excess malignancy.

2.3.1 Epidermal Growth Factor Receptor (ErbB)

The *ErbB* gene family encodes transmembrane receptor proteins that belong to the tyrosine kinase receptors superfamily (Jafri et al. 2016). They play an important role in the proliferation and differentiation of normal cells. The epidermal growth factor receptor family plays an important role in cell cycle regulation in conjunction with the Ras–MAPK pathway (Jafri et al. 2016).

There are four members of the ErbB receptor gene family: *ErbB1*, *ErbB2* (*Her2*: 'human epidermal growth factor receptor'), *ErbB3* and *ErbB4* (Tebbutt et al. 2013). EGF receptor (ErbdB1) is overexpressed in 80% of lung squamous cell carcinoma, 50% or more in glioblastoma, and 80–100% in head and neck epithelial tumours (Tebbutt et al. 2013). HER2/NEU (ERBB2) receptors are increased by 25–30% in breast cancer, lung, ovarian and salivary gland adenocarcinoma (Tebbutt et al. 2013).

2.3.2 *ErbB2* Gene (*Her2/neu*)

In the ErbB2 oncoprotein activated by gene amplification, the tyrosine kinase is constantly active and continuously transmits cleavage signals to the nucleus. It has been determined that the *ErbB2* oncogene is associated with breast, ovarian and gastric cancer (Herter-Sprie et al. 2013). The use of Herceptin with ErbB2 monoclonal antibody is beneficial in breast cancer cases with *ErbB2* overexpression and progression despite standard chemotherapy.

2.3.3 *Myc* Gene

The Myc protein is a transcription factor located in the nucleus that controls the expression of genes associated with cell growth. Abnormal myc expression is associated with abnormal growth. It has been determined that the *c-Myc* gene is amplified in vitro in small-cell lung carcinomas and that these tumour cells proliferate faster than the cells without amplification (Dang 2012).

It has been determined that the n-myc gene is amplified in neuroblastomas (Dang 2012). In addition, n-myc and l-myc genes are amplified in small-cell lung cancer (Dang 2012).

2.4 *Point Mutation*

Mutations develop as a result of a base entering or leaving the structure of proto-oncogenes. Changes occur in the products of these mutated genes, resulting in the synthesis of oncoproteins with different activities and functions. For example, when there is a base difference in the DNA base sequence of the c-ras oncogene (change in the 12th position of p21), GTPase activity in the cell decreases and adenylate cyclase activity increases by chronically stimulating it.

This causes an increase in metabolic activity in cells. It is stated that most tumours in humans arise from point mutations in the RAS proto-oncogene. For example, it has been reported that approximately 90% of pancreatic cancers and 50% of colon and thyroid cancers are caused by mutations in the RAS proto-oncogene (Pylayeva-Gupta et al. 2011).

2.4.1 *Ras* Gene

The *Ras* protooncogene family is localized to the inner surface of the cell membrane. It has GTPase activity. By binding to either GDP or GTP, they form a cyclical structure between the inactive or active state. The Ras proteins (Kras, Hras, and Nras) normally allow signals that stimulate cell division in response to external growth factors to pass from the cell membrane to the nucleus (Miller and Miller 2012).

It has three homologous forms, Hras, Kras, and Nras. These genes are mutated in 40% of human tumours (Miller and Miller 2012). When a cell encounters a growth factor, the growth factor receptors on the cell membrane bind to the growth factor, and the cytoplasm-facing portion of the growth factor receptor undergoes autophosphorylation. This causes proteins known as nucleotide exchange factors to become embedded in the plasma membrane. Nucleotide exchange factors (GEF) cause Ras to leave and release GDP and bind to GTP. Thus Ras is activated. The GTP-bound active form of Ras sends the signal necessary for the phosphorylation of several proteins in the cytoplasm, and transcription factors in the nucleus are stimulated. When it sends Ras signals to the nucleus once again, GTP is hydrolyzed to GDP and goes into an inactive form.

Specific *Ras* gene mutations are frequently observed in many different cancer cells such as colon, lung, breast and bladder cancers and are detected at codons 12, 13 and 61 in almost all cases (Pylayeva-Gupta et al. 2011). These mutations prevent the hydrolysis of GTP to GDP, thus keeping the Ras protein in an active form, and encouraging the cell to divide continuously.

2.4.2 *Braf* Gene

Another oncogene that is activated after point mutations encode a tyrosine kinase. After activation; Activation of MEK (MAPKK) and ERK (MAPK) kinases, respectively, is observed. After this chain of activation, the expression of transcription factors increases. Amino acid changes are occurring in the kinase domain of the Braf gene in 70% of malignant melanomas and many of the different tumour types (Davies et al. 2002).

Codon 600 mutations are detected in most of the cases (Davies et al. 2002). With this; sometimes it can also be activated after a gene fusion.

2.5 Transduction Activation

When a cell is infected with a virus, part of the cellular DNA sequences can be incorporated into the viral genome by recombination. The cellular array that becomes part of the viral genome can replicate with the virus and thus spread to other cells. Such activation can lead to the stimulation of protooncogenes and tumourigenesis. Cellular protooncogenes taken up by the virus undergo various types of mutations (point mutation, deletion, insertion, substitution, etc.) and become oncogenes under the influence of a strong promoter (Fig. 1).

3 Activation of Oncogenes

3.1 Physical and Chemical Factors

During inflammation, cancerous cells enable various leukocytes such as neutrophils, macrophages, and lymphocytes to move to the tumour site with the cytokines and chemokines they secrete. Tumour-Associated Macrophages arriving at the cancerous site are an important component of inflammation. According to their function, the tumour can have two different processes; survival or degeneration. If Tumour-Associated Macrophages are activated by interleukin (IL)-2, IL-12, and interferons,

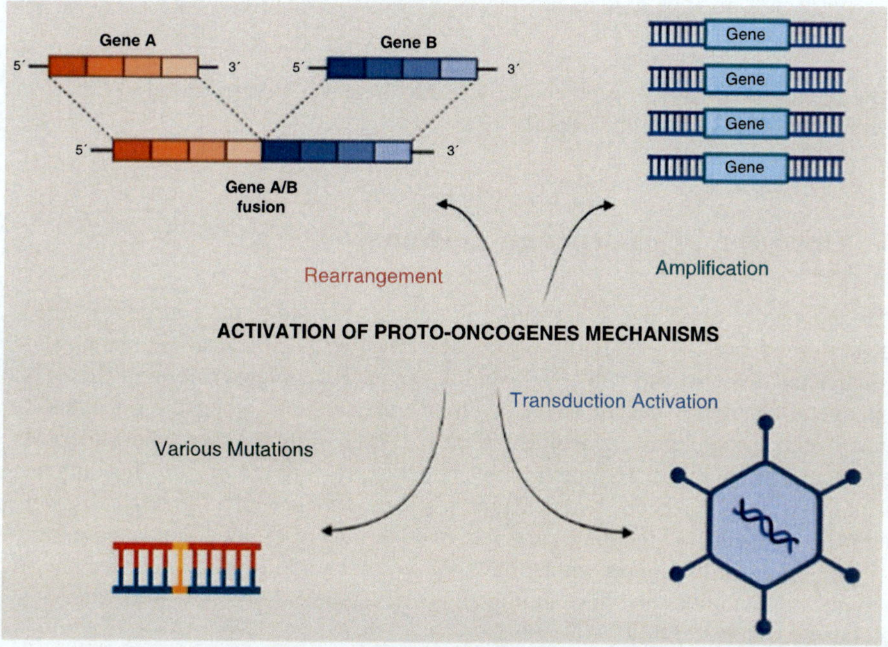

Fig. 1 Activation of protooncogenes occurs due to various mutations such as insertions, deletions, amplifications, rearrangements such as gene fusion, and genomic effects caused by viruses

they can participate in killing neoplastic cells (Lin et al. 2019). On the other hand, they can potentiate cancer development by producing growth factors associated with angiogenesis.

Polycyclic aromatic hydrocarbons, drugs such as diethylstilbestrol, and inorganic compounds such as arsenic, asbestos and aflatoxins can be given as examples. It affects oncogenes and tumour suppressor genes that serve as critical molecular targets in aflatoxin carcinogenesis. Numerous studies have shown that aflatoxin produces mutations in all three types of c-ras oncogenes (Hras, Kras and Nras) (Sinha et al. 1988) (McMahon et al. 1990).

3.2 *Biological Factors*

According to the International Agency for Research on Cancer (IARC), 4 of the most important infectious pathogens classified among group one carcinogens are; *Helicobacter pylori* including high-risk human papillomavirus (HPV), hepatitis B virus (HBV), hepatitis C virus (HCV). It has been stated that these are responsible for 90% of all infection-related cancers in the world (International Agency for Research On Cancer 2020).

3.3 Genetic Factors

Among genetic factors, it is evaluated whether there is a predisposition to cancer in the family history of the individual.

4 Oncogene Dependence on Tumours

The emergence of oncogene activation in tumours has led to the concept of 'oncogene dependence'. Accordingly, oncogene activation is critical in maintaining the 'malignant' state of the cell. For example, transgenic mice expressing the *MYC* oncogene develop T-cell or acute myeloid leukemia, and when the oncogene is inactivated, proliferation in leukaemic cells ceases, differentiates, and undergoes apoptosis. In another transgenic mouse model expressing the *Bcr-abl* fusion gene, leukaemia developed, but when expression was switched off, tumour cells were rapidly apoptized and the mice survived, even in mice with advanced disease.

These and similar observations have shown that certain types of cancer have a unique vulnerabilities and have prompted the development of 'targeted' therapies to address this vulnerability.

5 Tumour Suppressor Genes (TSGs)

Tumour suppressor genes work by preventing the cell from multiplying uncontrollably. It achieves this by inhibiting the cell cycle, stimulating cell differentiation or directing the cell to apoptosis (Gamudi and Blundell 2010). In addition to preventing uncontrolled cell proliferation, tumour suppressor genes also control cell proliferation by inhibiting the overactivation of oncogenic signalling pathways (Gamudi and Blundell 2010).

Loss of function (mutation, chromosomal deletions, etc.) that may occur in genes can cause cells to proliferate uncontrollably, gain resistance to apoptosis and trigger neoplastic differentiation.

Tumour suppressor proteins (oncostatic) synthesized by tumour suppressor genes also require cell surface receptors, signal transducers and nuclear transcription regulatory factors.

The first tumour suppressor gene identified in humans is the *Rb* gene. RB proteins negatively regulate cell proliferation, which is responsible for the progression of the cell cycle from the G1 phase to the S phase. The Rb protein is controlled by phosphorylation (Giacinti and Giordano 2006). When the Rb protein is phosphorylated, this negative regulatory effect of Rb disappears. As a result, tumourigenesis begins with the triggering of uncontrolled cell proliferation.

Inactivation of the RB protein occurs by both mutations and chromosome deletions. It is known that in most retinoblastomas, there is a deletion in the chromosome locus (13q14) where the *Rb* gene is located. Such changes directly cause the loss of the gene.

The tumour suppressor gene identified as the second after *Rb* is *p53*. *p53* is involved in many biological processes such as cell cycle arrest, apoptosis, differentiation and senescence.

They exert their TSGs effects, which suppress cell proliferation, in two different ways:

5.1 Gatekeeper Genes

They show their effect of keeping the cell cycle under control by directly suppressing cell proliferation. Examples of these genes are the *APC, p53,* and *Rb* genes.

5.2 Caretaker Genes

They are genes that have an indirect effect on cell proliferation. They are responsible for DNA repair and play an active role in the stability of the genome by preventing mutation formation, such as *BRCA1, BRCA2, MLH1 and MSH2*.

6 Tumour Suppressor Gene Inactivation

Two changes (double hit) required for the inactivation of TSGs can be gained in succession in life; one of them can be passed on through generations, causing cancer predisposition and familial cancer syndromes to be observed in these individuals.

Today, different mechanisms that cause the inactivation of TSGs, which play a role in cancer development, have been revealed.

6.1 Point Mutation in Tumour Suppressor Genes

The most common inactivation mechanism in tumour suppressor genes (TSGs) is mutations in these genes. Genes such as *p53, BRCA1, MSH2, PTEN, NF1* and NF2 in TSGs are inactivated after mutations (Gamudi and Blundell 2010). With this; Genes such as *BRCA2, MLH1, APC* and *RB1* can be inactivated after both mutation and methylation changes (Gamudi and Blundell 2010). For TSGs to be inactivated, every two copies (allele) of a gene in the cell must be out of function. Cancer development can be observed by mutating both alleles during life. However, this is not a very common situation.

Usually, the first mutation is passed on to the next generation familial. People carrying the mutation carry these changes in all their cells. In other words, an allele of the affected gene is inactive in all cells. After a second mutation that occurs in the other allele of the same gene in somatic (body) cells during life, the intact allele is also inactivated and the suppressor effect on cell proliferation disappears, and cancer development specific to the tissue region of the cell is observed.

In some cases, two different inactivation mechanisms can be seen together. Initially, the inactivation of one allele by mutation and then inactivation of the second allele by hypermethylation may cause the complete abolition of the tumour suppressor effect.

6.2 Loss of Heterozygosity (LOH)

Loss of heterozygosity (LOH) is one of the inactivation mechanisms of TSGs, which plays an important role in cancer development. It often occurs after an isolated deletion (interstitial deletion) of the chromosomal locus where TSG is located. It can sometimes occur during mitotic recombination or due to chromosomal loss due to mitotic nondisjunction.

Although the best-known example of LOH is the *RB1* gene region that causes retinoblastoma, LOH is also detected in other TSGs such as *P53* (Zhang and Sjöblom 2021).

Especially in colorectal cancers, LOH is frequently observed due to the loss of the short arm (17p) of the 17th chromosome and the long arm (18q) of the 18th chromosome (Zhang and Sjöblom 2021).

6.3 Methylation

Another important mechanism that causes inactivation is the methylation of the promoter region. This mechanism is observed in the promoter gene region containing CpG islands.

In normal cells, most of the CpG islets are unmethylated. In tumour cells, CpG islands in the promoter region of some genes that should normally be unmethylated are observed to be methylated. Methylation of CpG islets inhibits gene expression, causing inactivation of the gene of interest (Deaton and Bird 2011).

The susceptibility of gene regions to inactivation by methylation varies. Some genes, such as *MSH2*, are only inactivated by mutation. While some regions, such as *MLH1*, frequently lose their functions with point mutations, the methylation mechanism may also be effective as an alternative. *RASSF1A* and *HIC1* genes are inactive only with methylation changes.

6.4 Microsatellite Instability (MSI)

Microsatellite Instability (MSI) is not one of the inactivation mechanisms of tumour suppressor genes but occurs as a result of mutations in DNA repair genes such as *MLH1*, and *MSH2*, which are in the caretaker genes and ensure the stability of the genome by preventing the formation of mutations (Bonadona et al. 2011).

During DNA replication, situations that cause replication errors such as base mismatch, insertion and deletion can be observed. This results in the development of microsatellite instability (MSI).

Genes called DNA repair genes (MMR-mismatch repair) (*MLH1*, *MSH2*) enable the synthesis process to be restarted by detecting replication errors by the cell. Mutations in DNA repair genes, which are divided into two main groups, MutS and MutL, are characterized by disruption of the repair mechanism (Qiu et al. 2015). After all; It manifests itself with colorectal adenocarcinoma (Lynch Type I) or additionally with endometrial, ovarian, gastric, ureter, and renal pelvis cancers (Lynch Type II) (Table 1).

7 Nonprotein-Coding Protooncogenes and Tumour Suppressor Genes: Micro-RNAs (miRNAs)

In the human genome, some DNA sequences are transcribed into RNA, but not translated into protein.

It has been understood that some of these non-coding RNAs play a role in gene expression regulation. It has been determined that these molecules, which are called

Table 1 Common genetic changes seen in some cancers

Cancer type	Affected gene	Genomic changes	References
Colorectal cancer	*BRAF*	V600E	Agarwal and Saif (2014), Malats et al. (1997)
	KRAS	G12D, G12V, G12C	
Cervical cancer	*EGFR*	S703F	Zaanan et al. (2015), Tornesello et al. (2014)
	KRAS	Missense mutations in codon G12	
Pancreatic cancer	*KRAS*	Missense mutations in codon G12	Malats et al. (1997)
Thyroid cancer	*HRAS-NRAS*	Mutations in Codon 61	Vasko et al. (2003)
Melanoma	*BRAF*	V600E	Maldonado et al. (2003)
CML and acute leukemia	*c-ABL*	t(9:22) (q34:q11)	Pierotti et al. (2003)
Breast cancer	*MYC*, *EGFR2*	Approximately 20% amplification	Pierotti et al. (2003)

microRNA (miRNA) because they consist of short sequences of 22 nucleotides, are overexpressed or their expression decreased in various tumours.

Different cancer types appear to have their unique miRNA expression profiles. In general, miRNA expression was found to be decreased in tumour tissue and the expression profile was found to correlate with tumour type and differentiation.

In various malignancies such as chronic lymphocytic leukaemia, breast, colorectal, ovarian, liver, lung, pancreatic, and prostate cancer, certain miRNAs have been found to correlate with various clinical characteristics of the disease, such as stage, invasiveness, metastatic potential, histological subtype, and survival (Peng and Croce 2016).

In other words, it has been stated that certain miRNA expression profiles function as a kind of 'tumour signature' and that they can be used as biomarkers that can be used both in diagnosis and in prognosis prediction.

8 Conclusion

Both the onset and progression of cancer can be associated with many factors such as environmental, genetic and infectious agents. Inactivation of tumour suppressor genes and conversion of protooncogenes to oncogenes paves the way for cancer. An increase in the number of cells resulting from genomic changes alone does not cause cancer development. However, if this situation is accompanied by factors triggering DNA damage and growth factor activation, the risk of the neoplastic process increases. Chronic inflammation following tissue damage causes cells to potently transform and proliferate, contributing to the induction of cancer. In addition to what is known about cancer, many vague and unanswered questions still await extensive research.

References

Agarwal A, Saif MW (2014) KRAS in pancreatic cancer. J Pancreas 15:303–305. https://doi.org/10.6092/1590-8577%2F2660

Alao JP (2007) The regulation of cyclin D1 degradation: roles in cancer development and the potential for therapeutic invention. Mol Cancer 6:24. https://doi.org/10.1186/1476-4598-6-24

Batra N, Ghag I, Babu K, Divanji T (2021) Reviewing oncogenes and proto-oncogenes. Int J Sci Res Sci Technol 8(3):458–479. https://doi.org/10.32628/ijsrst2183100

Bonadona V, Bonaïti B, Olschwang S, Grandjouan S, Huiart L, Longy M, Guimbaud R et al (2011) Cancer risks associated with germline mutations in MLH1, MSH2, and MSH6 genes in lynch syndrome. *JAMA* 305:2304–2310. https://doi.org/10.1001/jama.2011.743

Botezatu A, Iancu IV, Popa O, Plesa A, Manda D, Huica I, Vladoiu S, Anton G, Badiu C (2016) Mechanisms of oncogene activation. In: New aspects in molecular and cellular mechanisms of human carcinogenesis. IntechOpen, London. https://doi.org/10.5772/61249

Burrell CJ, Howard CR, Murphy FA (2017) Mechanisms of viral oncogenesis. In: Fenner and White's medical virology. Academic Press, London, pp 121–134. https://doi.org/10.1016/b978-0-12-375156-0.00009-6

Dang CV (2012) MYC on the path to cancer. Cell 149:22–35. https://doi.org/10.1016/j.cell.2012.03.003

Davies H, Bignell GR, Cox C, Stephens P, Edkins S, Clegg S, Teague J et al (2002) Mutations of the BRAF gene in human cancer. Nature 417:949–954. https://doi.org/10.1038/nature00766

Deaton AM, Bird A (2011) CpG islands and the regulation of transcription, vol 25. Genes Dev, pp 1010–1022. https://doi.org/10.1101/gad.2037511

Fagundes R, Teixeira LK (2021) Cyclin E/CDK2: DNA replication, replication stress and genomic instability. Front Cell Dev Biol 9:774845. https://doi.org/10.3389/fcell.2021.774845

Gamudi D, Blundell R (2010) Tumour suppressor genes. Res J Med Sci 4:280–284. https://doi.org/10.3923/rjmsci.2010.280.284

Giacinti C, Giordano A (2006) RB and cell cycle progression. Oncogene 25:5220. https://doi.org/10.1038/sj.onc.1209615

Herter-Sprie GS, Greulich H, Wong KK (2013) Activating mutations in ERBB2 and their impact on diagnostics and treatment. Front Oncol 3:86. https://doi.org/10.3389/fonc.2013.00086

International Agency for Research On Cancer (2020) https://gco.iarc.fr/causes/infections/help. Accessed 31 Mar 2023

Jafri MA, Ansari SA, Alqahtani MH, Shay JW (2016) Roles of telomeres and telomerase in cancer, and advances in telomerase-targeted therapies. Genome Med 8:69. https://doi.org/10.1186/s13073-016-0324-x

Lin Y, Jianxin X, Lan H (2019) Tumour-associated macrophages in tumour metastasis: biological roles and clinical therapeutic applications. J Hematol Oncol 12:76. https://doi.org/10.1186/s13045-019-0760-3

Malats N, Porta M, Corominas JM, Piñol JL, Rifà J, Real FX (1997) Ki-ras mutations in exocrine pancreatic cancer: association with clinico-pathological characteristics and with tobacco and alcohol consumption. Int J Cancer 70:661–667. https://doi.org/10.1002/(SICI)1097-0215(19970317)70:6<661::AID-IJC6>3.0.CO;2-T

Maldonado JL, Fridlyand J, Patel H, Jain AN, Busam K, Kageshita T, Ono T, Albertson DG, Pinkel D, Bastian BC (2003) Determinants of BRAF mutations in primary melanomas. *J Natl Cancer Inst* 95:1878–1880. https://doi.org/10.1093/jnci/djg123

McMahon G, Davis EF, Huber LJ, Kim Y, Wogan GN (1990) Characterization of c-Ki-ras and N-ras oncogenes in aflatoxin B1-induced rat liver tumours. Proc Natl Acad Sci U S A 87:1104–1108. https://doi.org/10.1073/pnas.87.3.1104

Miller MS, Miller LD (2012) RAS mutations and oncogenesis: not all RAS mutations are created equally. Front Genet 2:100. https://doi.org/10.3389/fgene.2011.00100

Peng Y, Croce CM (2016) The role of microRNAs in human cancer. Signal Transduct Target Ther 1:15004. https://doi.org/10.1038/sigtrans.2015.4

Peyressatre M, Prével C, Pellerano M, Morris MC (2015) Targeting cyclin-dependent kinases in human cancers: from small molecules to peptide inhibitors. Cancers 7:179. https://doi.org/10.3390/cancers7010179

Pierotti MA, Sozzi G, Croce CM (2003) Holland-Frei cancer medicine: mechanisms of oncogene activation. BC Decker, Hamilton

Pylayeva-Gupta Y, Grabocka E, Bar-Sagi D (2011) RAS oncogenes: weaving a tumourigenic web. Nat Rev Cancer 11:761. https://doi.org/10.1038/nrc3106

Qiu R, Sakato M, Sacho EJ, Wilkins H, Zhang X, Modrich P, Hingorani MM, Erie DA, Weninger KR (2015) MutL traps MutS at a DNA mismatch. Proc Natl Acad Sci U S A 112:10914–10919. https://doi.org/10.1073/pnas.1505655112

Sampaio MM, Santos MLC, Marques HS, de Souza Gonçalves VL, Araújo GRL, Lopes LW, Apolonio JS et al (2021) Chronic myeloid leukemia-from the Philadelphia chromosome to specific target drugs: a literature review. World J Clin Oncol 12:69–94. https://doi.org/10.5306/wjco.v12.i2.69

Schmitz R, Ceribelli M, Pittaluga S, Wright G, Staudt LM (2014) Oncogenic mechanisms in Burkitt lymphoma. Cold Spring Harb Perspect Med 4:a014282. https://doi.org/10.1101/cshperspect.a014282

Sinha S, Webber C, Marshall CJ, Knowles MA, Proctor A, Barrass NC, Neal GE (1988) Activation of ras oncogene in aflatoxin-induced rat liver carcinogenesis. *Proc Natl Acad Sci U S A* 85: 3673–3677. https://doi.org/10.1073/pnas.85.11.3673

Tebbutt N, Pedersen MW, Johns TG (2013) Targeting the ERBB family in cancer: couples therapy. Nat Rev Cancer 13:663. https://doi.org/10.1038/nrc3559

Tornesello ML, Annunziata C, Buonaguro L, Losito S, Greggi S, Buonaguro FM (2014) TP53 and PIK3CA gene mutations in adenocarcinoma, squamous cell carcinoma and high-grade intraepithelial neoplasia of the cervix. J Transl Med 12:255. https://doi.org/10.1186/s12967-014-0255-5

Vasko V, Ferrand M, Di Cristofaro J, Carayon P, Henry JF, De Micco C (2003) Specific pattern of RAS oncogene mutations in follicular thyroid tumours. J Clin Endocrinol Metabol 88:2745–2752. https://doi.org/10.1210/jc.2002-021186

Zaanan A, Okamoto K, Kawakami H, Khazaie K, Huang S, Sinicrope FA (2015) The mutant KRAS gene up-regulates BCL-XL protein via STAT3 to confer apoptosis resistance that is reversed by BIM protein induction and BCL-XL antagonism. J Biol Chem 290:23838–23849. https://doi.org/10.1074/jbc.M115.657833

Zhang X, Sjöblom T (2021) Targeting loss of heterozygosity: a novel paradigm for cancer therapy. Pharmaceuticals 14:57. https://doi.org/10.3390/ph14010057

MicroRNAs and Long Non-coding RNAs as Key Targets

Onur Obut, Pınar Akbaba, Mehmet Ali Balcı, Yakup Bakır, and Vahap Eldem

1 Introduction

Cancers are a group of diverse and complex diseases characterized by uncontrollable cell growth, proliferation and metastasis to various other organs. Cancers have been traditionally thought of as a group of diseases caused by an accumulation of genetic mutations, but today they are considered a highly heterogeneous disease exhibiting diverse clinical manifestations driven by molecular aberrations at the genomic, epigenomic, transcriptomic and proteomic levels (Turnbull et al. 2018; Cory and Adams 2019). It is important to characterize and profile cancer heterogeneity at the transcriptomic level because changes in expression, activity and regulation of coding and non-coding genes govern the cellular phenotype. In particular, large-scale RNA-Seq projects such as Cancer Genome Atlas (TCGA) Pan-Cancer RNA-seq datasets on human normal/tumor tissues have led to the discovery of a multitude of coding genes and non-coding RNA (ncRNA) species. Although most efforts to understand the drivers of cancer have focused on the spatiotemporal expression and functional analysis of protein-coding genes (Uhlen et al. 2017; Cieslik and Chinnaiyan 2018), transcriptomics approaches employing high-throughput sequencing technologies have revealed that eukaryotic cells produce a plethora of non-coding transcripts (Carninci et al. 2005; Gil and Ulitsky 2020; Anastasiadou et al. 2018). Accumulating evidence has shown that ncRNAs account for over 60% of transcriptional output in human cells and regulate various biological processes

O. Obut · P. Akbaba · M. A. Balcı · V. Eldem (✉)
Faculty of Sciences, Department of Biology, Istanbul University, Istanbul, Turkey
e-mail: onur.obut@ogr.iu.edu.tr; akbabapinar@ogr.iu.edu.tr; balci.mehmetali@ogr.iu.edu.tr; vahap.eldem@istanbul.edu.tr

Y. Bakır
Department of Plant Bioactive Metabolites, ACTV Biotechnology, Istanbul, Turkey
e-mail: yakupb@actvlab.com

H. S. Tuli, M. B. Yerer Aycan (eds.), *Oncology: Genomics, Precision Medicine and Therapeutic Targets*, https://doi.org/10.1007/978-981-99-1529-3_3

and pathways in developmental and pathological contexts (Iyer et al. 2015; Lin and He 2017). Various classes of ncRNAs have been reported to date, and ncRNAs are traditionally classified into four categories although consensus is lacking: short ncRNAs (miRNAs, piRNAs, tiRNAs), mid-size ncRNAs (snoRNAs, PASRs, TSSa-RNAs, PROMPTs), a heterogeneous group of long non-coding RNAs (lncRNAs, lincRNAs, T-UCRs) and circular RNAs (ciRNAs) (Anastasiadou et al. 2018; Esteller 2011; Vo et al. 2019). The functional relevance of ncRNAs is particularly evident for miRNAs, lncRNAs and ciRNAs.

MicroRNAs (miRNAs) have emerged as an important class of small single-stranded non-coding regulatory RNAs that orchestrate gene expression at the post-transcriptional level. As post-transcriptional regulator, miRNAs hybridize to the complementary sequences in the 3′ or 5′ untranslated regions of target mRNAs to induce mRNA degradation or translation repression (Hill and Tran 2021). As for lncRNAs, they appear to be a functionally diverse group of ncRNAs that are highly abundant in the transcriptome. LncRNAs have traditionally been defined as ncRNA molecules over 200 bases in length that regulate target gene expression by various cellular mechanisms, acting as guides, scaffolds, decoys and signals. The canonical function of the majority of lncRNAs remains unclear, but experimental studies show that lncRNAs are associated with other molecules (DNA, RNA species and proteins) that coordinate and regulate numerous physiological processes, such as molecular signalling pathways, and their dysfunction may impact tumorigenesis and diverse human diseases (Iyer et al. 2015; Hosono et al. 2017). In other words, dysregulation in tissue and time-specific expression of these non-coding genes (miRNAs and lncRNAs) have been linked to multiple human cancers. Because of this, miRNAs and lncRNAs have been recognized as potential actionable and druggable target genes for cancer intervention.

In this chapter, we aim to provide an up-to-date perspective on prognostic and targetable therapeutic options for miRNA and lncRNA genes in cancer and cancer therapeutics. First, we review the current literature to map out how miRNAs and lncRNAs can act as post-transcriptional regulators of coding gene expression in major cancer types to highlight potential future points of cancer intervention, with a special focus on tumorigenesis. We also review advances in our understanding of the potential diagnostic, prognostic and therapeutic relevance of miRNA-lncRNA interactions in cancer biology and their implications for the molecular and cellular processes driving tumorigenesis. Finally, trends in the literature and directions for future cancer research are discussed, including prospects for potential biomarkers and targetable oncogenic drivers.

2 Exploring the Promising Potential of miRNAs for Diagnosis and Therapeutics of Cancer

Biogenesis of miRNAs occurs in both the nucleus and cytoplasm and involves a variety of enzymes and processes. A gene is first translated into a primary transcript (pri-miRNA) to begin the process of miRNA synthesis. Subsequently, the DROSHA/DGCR8 microprocessor complex cleaves transcribed pri-miRNAs into pre-miRNA molecules approximately 85 nucleotides long. The pre-miRNA is later transported from the nucleus to the cytoplasm by the Ran/GTP/Exportin 5 complex. Once a pre-miRNA is transported, it is cleaved into an 18- to 25-nucleotide miRNA duplex by a DICER enzyme then separated into a passenger strand and guide miRNA strand. The guide miRNA then merges with a multi-protein complex composed mainly of Argonaute (AGO) proteins to form the miRNA-induced silencing complex (miRISC), which can inhibit gene expression by attaching to target mRNAs (O'Brien et al. 2018). Despite the vast majority of miRNAs being found within cells, most of them migrate from cells to biofluids (Wang et al. 2010). These miRNAs, known as circulating miRNAs, are passively released as a result of apoptosis or necrosis, or actively transported via microvesicles or exosomes into various bodily fluids such as blood, breast milk, saliva, seminal fluid, tears and urine (Wang et al. 2010; Weber et al. 2010).

The target specificity of miRNAs depends on the compatibility of the guide miRNA inside the miRISC and the complementary sequences located on the target mRNA, also known as miRNA response elements (MREs) (O'Brien et al. 2018). MiRNAs can change how genes are expressed by inducing mRNA degradation or translation inhibition, mainly by binding to MREs at the 3′ UTR region of their target mRNAs (O'Brien et al. 2018; Cao et al. 2018). The grade of complementarity between the MREs and the guide miRNAs determines whether translation inhibition or mRNA degradation will occur. Imperfect complementarity inhibits translation, while perfect complementarity results in mRNA degradation (Cao et al. 2018). Aside from inhibiting translation, miRNAs can also upregulate gene expression in specific conditions. Occasionally, the coding region, 5′ UTR or promoter region of the mRNA may contain binding sites (Yao et al. 2019). In binding to the 5′ UTR or coding region, some miRNAs can activate mRNA translation (Brummer and Hausser 2014; Orom et al. 2008). However, the association between miRNA binding regions and type of regulation needs further investigation. MiRNAs can alter the translation of up to 60% of the human genome's protein-coding genes (Friedman et al. 2009). Because some of these genes are cancer-related, miRNAs are intimately associated with oncogenesis by upregulating or downregulating the production of specific cancer-related or critical proteins (Tutar et al. 2018; Yuan et al. 2019; Zhang et al. 2007). Cancer cells can sustain proliferative signaling, evading growth suppressors, resisting apoptosis, enabling replicative immortality, initiating metastasis and inducing angiogenesis (Hanahan and Weinberg 2011). Some of these miRNAs act either as tumor suppressors (TS-miRs) through the repression of, or as oncogenes (OncomiRs) through the induction of the abilities mentioned above and by playing an active role in various forms of cancer (Calin et al. 2004; Otmani and Lewalle 2021; Cui et al. 2019a; Asakura et al. 2020).

Even though significant advancements have been made in our understanding of the biology and pathogenesis of cancer, as well as in the creation of new targeted therapies, advanced early detection and screening tests remain underdeveloped as current cancer diagnosis methods have poor sensitivity and low specificity (Hamfjord et al. 2012). The lack of accurate early detection methods leads to diagnoses at advanced stages, which then results in decreased survival rates (Hawkes 2019). For this reason, the search for particular molecular alterations that can identify early-stage cancer has become a significant area of research. In recent years, miRNAs have emerged as potential novel molecular biomarkers for diagnosis due to their general dysregulation in cancer patients and distinct patterns of aberrant expression in early-stage cancer patients compared to healthy individuals (Weber et al. 2010; He et al. 2019a; Yi et al. 2010). Changes in miRNA expression patterns may also be unique to particular types of cancer (Niveditha et al. 2020). Moreover, miRNAs are highly stable in biopsy samples such as archived formalin-fixed paraffin-embedded samples (FPPE) and body fluids because of their interaction with argonaute proteins or high-density lipoproteins, which offer resistance to extreme pH and temperature conditions, as well as RNase activity (Mitchell et al. 2008; Meng et al. 2013; Cortez et al. 2011).

In addition to different miRNA expression patterns between cancer and non-cancer patients, variations in miRNA patterns across various cancer subtypes have also been reported (Wang et al. 2018a). For instance, one study showed that miRNA expression patterns differ among various thyroid cancer subtypes, including papillary, follicular and anaplastic thyroid cancers (Santiago et al. 2020). Furthermore, some studies also indicate changes in levels of miRNAs between different stages of cancer. For example, decreased levels of hsa-miR-510, hsa-miR-509-5p and hsa-miR-508-3p, alongside increasing hsa-miR-483-5p was seen in stage-III ovarian serous cancer compared to stage-I (Yu et al. 2013), as well as reduced levels of hsa-miR-21 in stage III-IV diffuse large B-cell lymphoma compared to stage I–II (Chen et al. 2014a) in Chinese populations. These discoveries showcase the potential of miRNAs as biomarkers for determining cancer subtype and discriminating early from advanced stages of cancer, which may contribute to choosing a more effective therapeutic strategy (Wang et al. 2018a).

Metastatic cancers with unknown primary origin constitute a significant problem in clinical applications, and in most cases it is challenging to identify the primary source using current histopathologic techniques (Sokilde et al. 2014). MiRNA profiling is promising for classifying origin cancer tissue due to the high tissue specificity of intracellular miRNA expression patterns. The expression levels of specific miRNAs vary between tissue samples taken from various tumor tissues and metastases and these variations can be used to accurately classify the primary origin of the metastasis (Sokilde et al. 2014; Rosenfeld et al. 2008). Additionally, the use of biofluids for diagnosis can facilitate clinical applications as they are more easily accessible than invasive tissue biopsies (Wang et al. 2018a; Marrugo-Ramirez et al. 2018). A significant number of studies indicate the potential for cancer detection by profiling circulating miRNAs present in various bodily fluids (Sohel 2020; Cui et al. 2019a). For instance, the levels of miRNAs associated with bladder cancer, such as hsa-let-7b-5p, hsa-miR-149-5p, hsa-miR-146a-5p and hsa-miR-423-5p, are reported to be highly increased in the urine of bladder cancer patients (Lin and Tsai 2021). Consequently, liquid biopsies can be used to identify circulating miRNAs to detect certain cancers, simplifying the diagnostic process using a non-invasive procedure.

Although the reliability of miRNA use still needs validation by further large-scale research before being widely applied in clinical practices, it is essential to organize our knowledge of miRNAs' role in oncogenesis to develop novel diagnostic and therapeutic tools. In the following section, we examine some of the miRNAs suggested as potential biomarkers for diagnostics and therapeutics for lung, breast, colorectal, gastric, and prostate cancer.

Lung Cancer: hsa-miR-182, hsa-miR-183, hsa-miR-210 and hsa-miR-126 are novel potential biomarkers suggested for the early detection of non-small cell lung cancer (NSCLC) (Zhu et al. 2016). The researchers used RT-qPCR to profile circulating miRNA in serum isolated from 216 blood samples collected from 112 NSCLC patients, 23 pneumonia patients, 21 gastric cancer patients, 20 current smokers without lung cancer and 40 healthy controls. The analysis revealed that hsa-miR-182, miR183 and hsa-miR-210 were considerably upregulated, while hsa-miR-126 was significantly downregulated in NSCLC patients compared to other groups. Moreover, combined usage of these four miRNAs with carcinoembryonic antigen (CEA), a protein biomarker for prognosis (Grunnet and Sorensen 2012), showed statistically high diagnostic value for early cancer identification (Zhu et al. 2016). Another study analyzed serum samples to estimate the diagnostic value of eight NSCLC-associated miRNAs: hsa-miR-146b, hsa-miR-205, hsa-miR-29c, hsa-miR-31, hsa-miR-30b, hsa-miR-337, hsa-miR-411 and hsa-miR-708 (Yang et al. 2019). RT-qPCR expression evaluation showed that hsa-miR-146b, hsa-miR-205, hsa-miR-29c, hsa-miR-30b and hsa-miR-337 were significantly overexpressed in NSCLC patients. Further analysis showed that a 4-miRNA panel consisting of hsa-miR-146b, hsa-miR-205, hsa-miR-29c and hsa-miR-30b is promising for early diagnosis of NSCLC patients. Another RT-qPCR-based study suggested two more promising biomarkers, hsa-miR-33a-5p and hsa-miR-128-3p, for the early detection of lung cancer (Pan et al. 2018). Researchers showed that in contrast to nearby non-cancer tissues, hsa-miR-33-5p and hsa-miR-128-3p are noticeably downregulated in malignant tissues. Additionally, their expression levels in whole blood were relatively stable across diverse circumstances, indicating they can be identified by analyzing non-invasively obtained blood samples. Further statistical analysis revealed that the combination of these miRNAs could separate early-stage cancer patients from healthy controls with high accuracy, specificity and sensitivity, indicating a potential use for these two miRNAs combined as a new non-invasive biomarker for the early diagnosis of lung cancer. Two miRNA panels were proposed for differentiating between NSCLC and SCLC subtypes after analyzing 1132 blood samples in another study (Lu et al. 2018). One of the two proposed panels with three microRNAs (hsa-miR-17, hsa-miR-190b and hsa-miR-

375) demonstrated high diagnostic accuracy in distinguishing SCLC from NSCLC, while the other with six microRNAs (hsa-miR-17, hsa-miR-190b, hsa-miR-19a, hsa-miR-19b, hsa-miR-26b and hsa-miR-375) was useful in separating lung cancer patients from healthy controls.

Breast Cancer: One study isolated 514 plasma samples from breast cancer patients and non-cancer control groups and analyzed these samples for differentiated miRNA patterns (Li et al. 2019a). As a result, a five-miRNA panel consisting of let-7b-5p, hsa-miR-122-5p, hsa-miR-146b-5p, hsa-miR-210-3p and hsa-miR-215-5p was identified as a biomarker for breast cancer detection. In a later study, hsa-miR-1246, hsa-miR-206, hsa-miR-24 and hsa-miR-373 expression levels were significantly upregulated in early-stage breast cancer patients (Jang et al. 2021). The combined diagnostic value of these miRNAs was shown to have statistically high specificity, accuracy and sensitivity, suggesting that they are promising biomarkers for early breast cancer detection. Several miRNAs have also been suggested for the subtyping of breast cancer. Another study analyzed the expression of five miRNAs, hsa-miR-92a-3p, hsa-miR-342-3p, hsa-miR-16, hsa-miR-21 and hsa-miR-199a-5p, which are already linked to breast cancer. Further examination and validation showed that hsa-miR-199a-5p was significantly underexpressed in triple-negative breast cancer (TNBC) compared to non-TNBC, which suggested that these miRNAs serve to distinguish TNBC from other subtypes (Shin et al. 2015). As previously mentioned, miRNA patterns may also vary between different stages of cancer. Researchers examined miRNAs' diagnostic value for predicting breast cancer stage with an in silico experiment (Yerukala Sathipati and Ho 2018). They obtained the miRNA expression levels of 386 patients, 193 with advanced-stage breast cancer and 193 with early-stage breast cancer, from an external database, then classified the early and advanced stages of breast cancer using the support vector machine (SVM) machine learning model. As a result, they discovered a panel of 34 miRNAs that could distinguish early stage from advanced stage cancers with relatively high accuracy, sensitivity and specificity. To further validate the results, they examined the ten highest-ranked miRNAs (hsa-miR-200c, hsa-miR-503, hsa-miR-1307, hsa-miR-361, hsa-miR-212, hsa-miR-592, hsa-miR-1185-1, hsa-miR-146b, hsa-miR-1468 and hsa-miR-769) and recognized that four of them (hsa-miR-503, hsa-miR-1307, hsa-miR-212 and hsa-miR-592) were strongly associated with breast cancer survival rates.

Colorectal cancer: A study evaluating 369 blood plasma samples investigated the diagnostic value of hsa-miR-27a and hsa-miR-130a for detecting colorectal cancer, using RT-qPCR to validate expression levels retrieved from databases. Results showed that a combination of hsa-miR-27a and hsa-miR-130a can facilitate the early detection of colorectal cancer (Liu et al. 2018a). In another study, researchers identified an additional biomarker, a 7-miRNA signature, for colorectal cancer with an experiment on 271 participants consisting of 139 colorectal cancer patients and 132 healthy controls (Zhang et al. 2019a). They detected and compared miRNA levels in the blood and tissue samples from cancer and non-cancer participants, finding that hsa-miR-103a-3p, hsa-miR-127-3p, hsa-miR-17-5p, and hsa-miR-18a-5p were significantly upregulated in colorectal cancer tissues, while hsa-miR-17-5p,

Table 1 List of potential target human miRNAs whose expression significantly changed in major cancer types (▲ = Up-regulated, ▼ = Down-degulated)

Cancer type	MiRNA candidate	Sample type	Diagnostic potential	References
Lung cancer	hsa-miR-182(▲) hsa-miR-183(▲) hsa-miR-210(▲) hsa-miR-126(▼)	Serum	Early detection	Zhu et al. (2016)
	hsa-miR-146b(▲), hsa-miR-205(▲), hsa-miR-29c(▲), hsa-miR-30b(▲)	Serum	Early detection	Yang et al. (2019)
	hsa-miR-33a-5p(▼), hsa-miR-128-3p(▼)	Tissue Whole blood	Early detection	Pan et al. (2018)
	hsa-miR-17(▲), hsa-miR-190b(▲), hsa-miR-375(▲)	Plasma	Subtyping	Lu et al. (2018)
	hsa-miR-17(▼), hsa-miR-190b(▼), hsa-miR-19a(▼), hsa-miR-19b(▼), hsa-miR-26b(▼), hsa-miR-375(▲)	Plasma	Early detection	Lu et al. (2018)
	hsa-miR-30b-5p(▲), hsa-miR-99a-5p(▲)	Plasma	Early detection	Adam-Artigues et al. (2021a)
	hsa-miR-141(▲)	Plasma	Early detection	Arab et al. (2017)
	hsa-miR-196b(▲), hsa-miR-205(▲), hsa-miR-375(▲)	Tissue	Subtyping	Hamamoto et al. (2013)
	hsa-miR-106a-5p(▲), hsa-miR-20a-5p(▲), hsa-miR-93-5p(▲)	Serum	Cancer detection	Zhang et al. (2017a)
	hsa-miR-210-3p(▲)	Serum	Cancer detection	Switlik et al. (2019)
	hsa-miR-492(▲), hsa-miR-590-3p(▲), hsa-miR-631(▼)	Serum	Early detection	Duan et al. (2021)
Breast cancer	hsa-let-7b-5p(▲), hsa-miR-122-5p(▲), hsa-miR-146b-5p(▲), hsa-miR-210-3p(▲), hsa-miR-215-5p(▲)	Plasma	Early detection	Li et al. (2019a)
	hsa-miR-1246(▲), hsa-miR-206(▲), hsa-miR-24(▲), hsa-miR-373(▲)	Plasma	Early detection	Jang et al. (2021)
	hsa-miR-199a-5p(▼)	Tumor tissue Plasma	Subtyping	Shin et al. (2015)
	34-hsa-miRNA panel	–	Staging	Yerukala Sathipati and Ho (2018)
	hsa-miR-1246(▲), hsa-miR-1307-3p(▲), hsa-miR-4634(▲), hsa-miR-6861-5p(▼), hsa-miR-6875-5p(▼)	Serum	Early detection	Shimomura et al. (2016)
	hsa-miR-30b-5p(▼)	Tumor tissue Plasma	Early detection	Adam-Artigues et al. (2021b)
	hsa-miR-27b-3p(▲), hsa-miR-22-5p(▲), hsa-miR-145-5p(▲)	Plasma	Early detection	Jusoh et al. (2021)
	hsa-miR-424(▲), hsa-miR-423(▼), hsa-miR-660(▼), hsa-let7-i (▼)	Urine	Early detection	Hirschfeld et al. (2020)
	hsa-miR-127-3p(▼), hsa-miR-133a-3p(▼), hsa-miR-155-5p(▲), hsa-miR-199b-5p(▼), hsa-miR-342-5p(▼)	Tissue	Early detection	Bitaraf et al. (2020)
	hsa-miR-130a(▲), hsa-miR-145(▲), hsa-miR-425-5p(▲), hsa-miR-139-5p(▲)	Serum	Early detection	Itani et al. (2021)
Colorectal cancer	hsa-miR-27a(▲), hsa-miR-130a(▲)	Whole blood	Early detection	Liu et al. (2018a)
	hsa-miR-103a-3p(▲), hsa-miR-127-3p(▲), hsa-miR-151a-5p(▲), hsa-miR-17-5p(▲),	Tumor tissue Plasma	Early detection	Zhang et al. (2019a)

(continued)

Table 1 (continued)

	hsa-miR-129-1-3p(▲), hsa-miR-566(▲)	Urine	Early detection	Iwasaki et al. (2022)
	hsa-miR-135b-5p(▲)	Feces Plasma	Advanced stage cancer detection	Li et al. (2020a)
	hsa-miR-29a(▲), hsa-miR-92a(▲)	Plasma	Early detection	Huang et al. (2010)
	hsa-miR-409-3p(▲), hsa-miR-7(▼), hsa-miR-93(▼)	Plasma	Early detection	Wang et al. (2015)
	hsa-miR-23a-3p(▲), hsa-miR-27a-3p(▲), hsa-miR-142-5p(▲), hsa-miR-376c-3p(▲)	Serum	Early detection	Vychytilova-Faltejskova et al. (2016)
	hsa-miR-193a-3p(▲), hsa-miR-23a(▲), hsa-miR-338-5p(▲)	Tumor tissue/whole blood	Early detection	Yong et al. (2013)
	hsa-miR-1290(▲), hsa-miR-320d(▼)	Plasma	Early detection	Liu et al. (2019)
	hsa-miR-592(▲)	Serum	Early detection	Pan and Miao (2020)
Gastric cancer	hsa-miR-627(▲), hsa-miR-629(▲), hsa-miR-652(▲)	Plasma	Early detection	Qiu et al. (2016)
	hsa-miR10b-5p(▲), hsa-miR132-3p(▲), hsa-miR185-5p(▲), hsa-miR195-5p(▲), hsa-miR-20a3p(▲), hsa-miR296-5p(▲)	Serum	Cancer detection	Huang et al. (2017)
	hsa-miR-6807-5p(▲), hsa-miR-6856-5p(▲)	Urine	Early detection	Iwasaki et al. (2019)
	hsa-miR-425-5p(▼), hsa-miR-1180-3p(▼), hsa-miR-122-5p(▼), hsa-miR-24-3p(▲), hsa-miR-4632-5p(▼)	Plasma	Early detection	Zhu et al. (2019)
	hsa-miR-1246(▲)	Serum	Early detection	Shi et al. (2020)
	hsa-miR-92b-3p(▲), hsa-miR-146b-5p(▲) hsa-miR-9-5p(▼), hsa-let-7g-5p(▼)	Tumor tissue/serum	Early detection	Tang et al. (2020)
	hsa-miR-16(▲), hsa-miR-25(▲), hsa-miR-92a(▲), hsa-miR-451(▲), hsa-miR-486-5p(▲)	Plasma	Early detection	Zhu et al. (2014)
	hsa-miR-106b(▲), hsa-miR-20a(▲), hsa-miR-221(▲)	Plasma	Early detection	Cai et al. (2013)
	hsa-miR-378(▲)	Serum	Early detection	Liu et al. (2012)
	hsa-miR-196a(▲), hsa-miR-148a(▼)	Gastric mucosa	Early detection	Chen et al. (2012)
Prostate cancer	hsa-miR-4289(▲), hsa-miR-326(▲), hsa-miR-152-3p(▲), hsa-miR-98-5p(▲)	Plasma	Early detection	Matin et al. (2018)
	hsa-miR-141(▲), hsa-miR-21(▲), hsa-miR-375(▲)	Serum	Early detection	Porzycki et al. (2018)
	hsa-miR-375(▼), hsa-miR-451a(▲), hsa-miR-486-3p(▲), hsa-miR-486-5p(▲)	Urine	Cancer detection	Li et al. (2021)
	hsa-let-7c(▼), hsa-let-7e(▼), hsa-miR-30c(▼), hsa-miR-622(▲), hsa-miR-1285(▲)	Plasma	Cancer detection	Zheng et al. (2014)
	hsa-miR-146a-5p(▲), hsa-miR-24-3p(▲), hsa-miR-93-5p(▲)	Serum	Cancer detection	Zhang et al. (2021)
	hsa-miR-205(▼), hsa-miR-214(▼)	**Tumor tissue**	**Cancer detection**	**Srivastava et al. (2013)**
	hsa-miR-142-3p(▲), hsa-miR-142-5p(▲), hsa-miR-223-3p(▲)	Semen	Cancer detection	Barcelo et al. (2019)

hsa-miR-205(▼), hsa-miR-214(▼)	**Tumor tissue**	**Cancer detection**	**Srivastava et al. (2013)**
hsa-miR-142-3p(▲), hsa-miR-142-5p(▲), hsa-miR-223-3p(▲)	Semen	Cancer detection	Barcelo et al. (2019)

proliferation and invasion. In short, researchers found that has-miR-433 is downregulated in colon cancer and inhibits colon cancer cell proliferation and invasion by directly targeting HOXA1 (Li et al. 2018a). Another study examined the effect of hsa-miR-17-5p and vimentin on colon cancer. Previous studies have shown that hsa-miR-17-5p is effective in breast and lung cancer (Jiang et al. 2014; Wang et al. 2019a). Vimentin, a protein encoded by VIM, plays a role in cellular structure and integrity as well as affecting cell shape and motility in the process of epithelial-mesenchymal transition (EMT) during metastasis (Mendez et al. 2010). In this context, hsa-miR-17-5p precursors and inhibitors were transformed into four CRC cells to see if hsa-miR-17-5p regulates EMT. Argonaute 2 (Ago2) immunoprecipitation (IP) and luciferase assay were used to evaluate the regulatory mechanism. Results indicate that overexpression of hsa-miR-17-5p inhibited vimentin expression and decreased cell migration and invasion, while inhibition of hsa-miR-17-5p had the opposite effect (Kim et al. 2020). A similar study looked at the link between hsa-miR-708 and zinc finger E-box binding homeobox 1 (ZEB1), which belongs to the ZEB transcription factor family and has an important role in normal embryonic development, including the induction of EMT to depolarize epithelial cells to become mesenchymal cells. During this process, its primary function is to bind to the promoter region, suppressing E-cadherin expression to cause cells to lose their epithelial properties (Vandewalle et al. 2009). This transcription factor also regulates invasive carcinomas by triggering the epithelial-mesenchymal transition (EMT) (Zhang et al. 2015a). To summarize, miR-708 is significantly downregulated in cancerous tissues, while overexpression of hsa-miR-708 and silencing of ZEB1 could inhibit CRC cell proliferation, migration and invasion, as well as promote cell apoptosis (Sun et al. 2019).

Another study examined the relationship between hsa-miR-18a-5p and interferon regulatory factor 2 (IRF2). IRF2 belongs to the IRF protein family and plays an important role in cellular response during adaptive immunity and tumorigenesis (Taniguchi et al. 2001). The study used CCK-8 assay to measure cell proliferation, luciferase reporter assay to assess the relationship between the target and miRNA, and western blot to understand gene expression. As a result, hsa-miR-18a-5p was shown to support cell proliferation and reduce migration (Liang et al. 2017). Another example study examined the relationship between has-miR-16 and MAPK kinase 1 (MEK1) in the ERK/MAPK signaling pathway. ERK/MAPK signaling pathways play an important role in cell proliferation and invasion (Guo et al. 2020). MAPK kinase 1 (MEK1) is a protein kinase that can activate this signaling pathway. A luciferase reporter assay was used to determine the regulatory relationship between hsa-miR-16 and MEK1. Plate cloning, flow cytometry and Transwell assay demonstrated the proliferation and invasion of cells transfected with wild-type and mutant MEK1. This study showed that hsa-miR-16 expression is downregulated in cancerous cells, indicating that its overexpression downregulates the ERK/MAPK pathway via MEK1 expression inhibition and may prevent cell invasion and proliferation (Chen et al. 2019).

Another study on lung cancer focused on the relationship between Fas Associated Factor 1 (FAF1) and hsa-miR-26a-5p. FAF1 is a member of the death-inducing

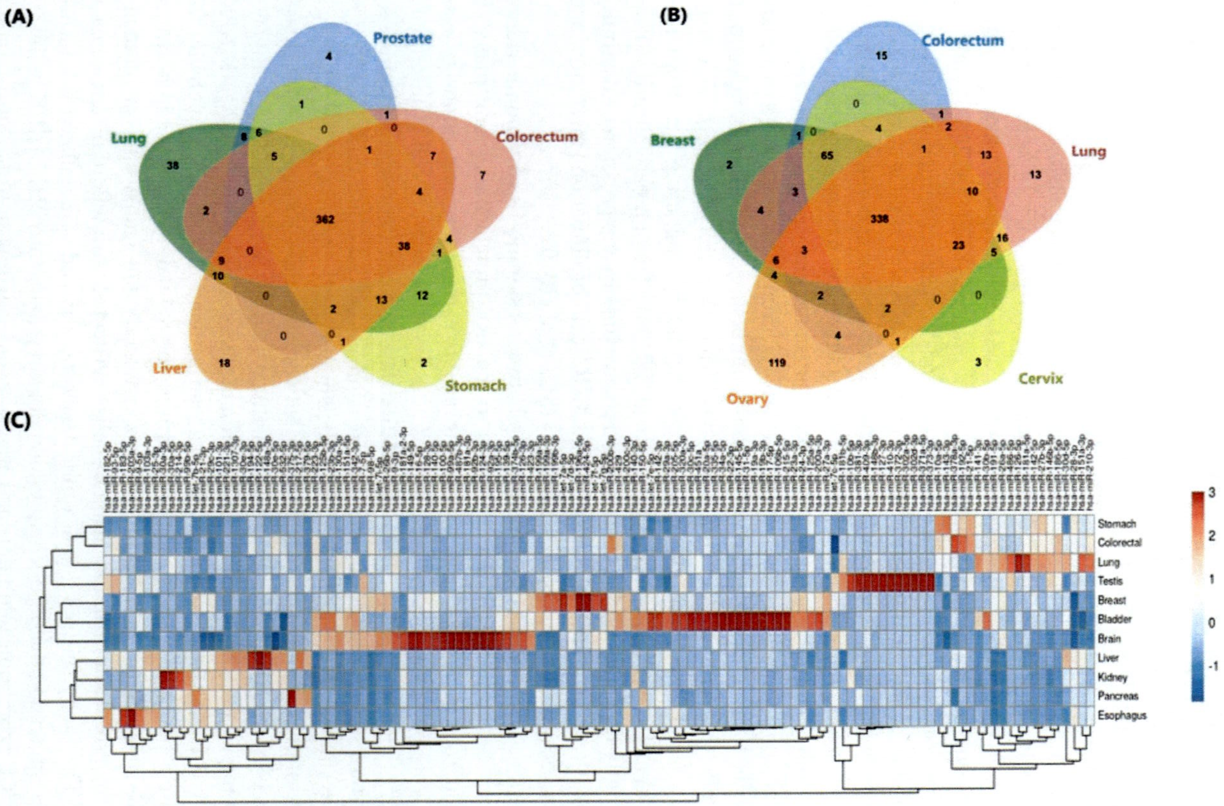

Fig. 1 (**a**) The number of common highly expressed miRNAs in most frequent cancer types in males (lung, prostate, colorectal, liver, gastric) and (**b**) in females (breast, ovarian, colorectal, lung, cervical). (**c**) Heatmap indicating the normalized expression of top differentially expressed human miRNA genes across eleven major cancer types

signaling complex and is expressed by the FAF1 gene. Interaction between the fas ligand and corresponding antigen can regulate cell death in many organs (Ryu et al. 2003). MMT assay was used to target hsa-miR-26a-3p for FAS1 investigation by dual-luciferase reporter assay to evaluate cell viability, cell cycle assay and EdU proliferation assay were used to understand the effect of hsa-miR26-5p on cell replication and the cell cycle, and a Transwell assay was used for metastasis. Results concluded that hsa-miR-26a-3p directly targets and downregulates the expression of FAF1. Inhibiting hsa-miR-26a-5p-promoted apoptosis and inhibited cell growth, migration and invasion (Ye et al. 2020). The table below summarizes examples of miRNA targets and their effects, according to five main types of cancer (Table 2).

4 Functional Role and Emerging Therapeutic Target of Long Non-coding RNAs in Oncogenesis

Despite their initial perception as "transcriptional noise," lncRNAs play a role in cell development, diseases, and essential pathways, and regulate a wide range of cellular functions (Wang et al. 2018b, 2019c). These RNAs of more than 200 nucleotides engage in nearly all biological processes, including epigenetics, transcription and the post-transcriptional regulation of gene signature, contributing to various physiological and pathological processes in various different cell compartments. Numerous studies have shown that lncRNAs can alter the gene expression network through multiple mechanisms, including miRNA sponging, chromatin and histone modifications, and translation regulation (Ghafouri-Fard et al. 2021; Zhou et al. 2020a). In addition, because of their regulation of cell proliferation, apoptosis, metastasis, stemness and therapeutic resistance, lncRNAs are believed to play a crucial role in many aspects of cell biology and may contribute to the occurrence and development of tumors. However, only a small proportion of lncRNAs have been identified and their myriad roles are still not entirely understood. Here, we compile a detailed summary of several well-known cancer-related lncRNAs or "oncogenic lncRNAs" to explore their roles in cancer.

HOTAIR—HOX Antisense Intergenic RNA: HOX Transcript Antisense Intergenic RNA (HOTAIR) is a well-studied lncRNA that plays an important role in the tumorigenesis of various human cancers. Expression is upregulated in both primary and metastatic breast cancer, and an increase of up to 2000-fold has been demonstrated in the latter. High expression of HOTAIR in primary breast cancer also correlates with metastasis and poor survival. Increased expression is seen in patients with lymph node metastases in hepatocellular cancer (Ghafouri-Fard and Taheri 2019a), hepatocellular carcinoma (HCC) and HCC patients with liver transplantation, compared with normal liver tissue (Rajagopal et al. 2020).

ANRIL—Antisense ncRNA in the INK4 locus: Antisense transcription (transcription from the opposite strand to a protein-coding or sense strand) is thought to play a role in gene regulation involving the degradation of the corresponding sense

Table 2 Highly altered miRNA expression, target genes and mode of action in the most prevalent human cancers

Cancer type	Candidate miRNA	Target	Mode of action	References
Lung cancer	hsa-miR-146a(▼)	COX-2, IRS2, Cyclin D1, Cyclin D2	Suppress	Cornett and Lutz (2014), Park et al. (2015), Li et al. (2016a)
	hsa-miR-21(▲)	TPM1, TIMP3, PTEN	Enhance	Zhu et al. (2007), Qi et al. (2003), Shen et al. (2014a)
	hsa-miR-365(▼)	TRIM25 NKX2-1	Suppress	Han et al. (2019), Kang et al. (2013)
	hsa-miR-361-5p(▼)	FOXM1, SMAD2	Suppress	Hou et al. (2017), Othman and Nagoor (2019)
	hsa-miR-221(▲)	PUMA	Enhance	Zhang et al. (2010a)
	hsa-miR-17-5p(▼)	TBC1D2, PTEN, TBP2	Both	Serva et al. (2012), Wang et al. (2021a)
	hsa-miR-18a-5p(▲)	IRF2, ATM, HIF-1α	Enhance	Liang et al. (2017), Chen et al. (2018a)
	hsa-miR-661(▲)	RB1, RUNX3	Enhance	Liu et al. (2017), Wang et al. (2017)
	hsa-miR-378(▲)	RBX1	Enhance	Ho et al. (2018)
	hsa-miR-93(▲)	DAB2, TBP2	Enhance	Du et al. (2014), Li et al. (2020b)
	hsa-miR-7(▼)	KCNJ2	Suppress	Liu et al. (2015a)
	hsa-miR-22(▼)	ErbB3, SIRT	Suppress	Ling et al. (2012), Gu et al. (2021a)
	hsa-miR-19a/b(▲)	c-Met, PP2A, BIM	Enhance	Cao et al. (2017), Baumgartner et al. (2018)
	hsa-miR-150(▲)	HMGA2, FOXO4	Enhance	Dai et al. (2019), Li et al. (2016b)
	hsa-miR135b(▲)	LZTS1, LATS1, MOB-1A, NDR2, βTrCP	Enhance	Lin et al. (2013)
	hsa-miR-145(▼)	OCT4, SOX2, Fascin1	Suppress	Chiou et al. (2012)
	hsa-miR-584-5p(▼)	YKT6, MMP-14	Suppress	Lee et al. (2021), Guo et al. (2019)
	hsa-miR-18a-5p(▲)	IRF2	Enhance	Liang et al. (2017)
	hsa-miR-16(▼)	MEK1	Suppress	Chen et al. (2019)
	hsa-miR-26a-5p(▼)	FAF1	Suppress	Ye et al. (2020)
	hsa-miR-708-5p(▼)	DNMT3A	Suppress	Liu et al. (2018b)
	hsa-miR-25-3p(▼)	Cyclin E2	Suppress	Zhao et al. (2014)
	hsa-miR-138(▲)(▼)	cyclin D3, H2AX	Both	Han et al. (2016), Yang et al. (2015)
	hsa-miR-1827(▲)	CRKL	Enhance	Ho et al. (2018)
	hsa-miR-17, 20a, 20b(▼)	TGFβR2	Suppress	Wang et al. (2019a)
	hsa-miR-19a-3p(▲)	Fra-1	Enhance	Yang et al. (2014)
	hsa-miR-34a(▼)	E2F3, CD44, SIRT1	Suppress	Li et al. (2012)
	hsa-miR-10b(▲)	Hoxd10	Enhance	Ma et al. (2010)
	hsa-miR-497(▼)	HIF-1α	Suppress	Wu et al. (2016)
	hsa-miR-182(▲)	FBXW7	Enhance	Chiang et al. (2016)
	hsa-miR-200c/hsa-miR147(▲)	SerpinB2, FOXP3	Enhance	Jin et al. (2017)
	hsa-miR-200b(▼)	ERM	Suppress	Zhang et al. (2017b), Hong et al. (2016)

(continued)

Table 2 (continued)

	hsa-miR-26a-5p(▼)	RNF6	Suppress	Huang et al. (2019)
	hsa-miR-29b(▼)	Akt3	Suppress	Li et al. (2017)
	hsa-miR-138(▼)	Vimentin	Suppress	Zhang et al. (2016a)
	hsa-miR-153(▼)	MTDH	Suppress	Li et al. (2015a)
	hsa-miR-221/222(▲)	Notch3	Enhance	Liang et al. (2018)
	hsa-miR-4472(▲)	RGMA	Enhance	Li et al. (2020c)
	Hsa-let-7d/hsa-miR-185(▼)	Rab25	Suppress	Shahabi et al. (2021)
	hsa-miR-423(▲)	TNIP2	Enhance	Dai et al. (2020)
Breast cancer	hsa-miR-20b(▲)	PTEN	Enhance	Zhou et al. (2014)
	hsa-miR-660-5p(▲)	TET2	Enhance	Peng et al. (2020)
	hsa-miR-3178(▼)	Notch1	Suppress	Kong et al. (2018)
	hsa-miR-216b(▼)	SDCBP	Suppress	Jana et al. (2017)
	hsa-miR-424-5p(▼)	PD-L1	Suppress	Dastmalchi et al. (2020)
	hsa-miR-125a/hsa-let-7e(▲)	IL-6	Enhance	Park and Kim (2019)
	hsa-miR-204(▼)	PI3K-α, c-SRC, VEGF, FAK, RAF1, MAPK, ANGPT1, TGFBR2	Suppress	Salinas-Vera et al. (2018)
	hsa-miR-100(▼)	VEGF, HIF-1α	Suppress	Flores-Perez et al. (2016), Pakravan et al. (2017)
	hsa-miR-193a(▼)	WT1	Suppress	Xie et al. (2017)
	hsa-miR-143-5p(▼)	HIF-1α, GLUT1	Suppress	Xu et al. (2022)
	hsa-miR-17-5p	netrin 4	Enhance	Jiang et al. (2014)
	hsa-miR-433(▼)	HOXA1	Suppress	Li et al. (2018a)
	hsa-miR-17-5p(▼)	Vimentin	Suppress	Kim et al. (2020)
	hsa-miR-708(▼)	ZEB1	Suppress	Sun et al. (2019)
	hsa-miR-181a(▲)	SRCIN1	Enhance	Sun et al. (2018a)
	hsa-miR-10b(▲)	KLF4	Enhance	Xie et al. (2019)
	hsa-miR-143(▼)	DNMT3A	Suppress	Ng et al. (2009)
	hsa-miR-22(▲)	BTG1	Enhance	Zhang et al. (2015b)
	hsa-miR-21-5p(▲)	KRIT1	Enhance	He et al. (2021)
	hsa-miR-27a-3p(▲)	BTG1	Enhance	Su et al. (2019)
Colorectal cancer	hsa-let-7e(▼)	DCLK1	Suppress	Kohlan et al. (2019)
	hsa-miR-30a(▲)	Beclin-1	Enhance	Liu et al. (2016)
	hsa-miR-200b-3p	Wnt1	Suppress	Chen et al. (2018b)
	hsa-miR-203a-3p(▲)	PDE4D	Enhance	Chen et al. (2018c)
	hsa-miR-22(▼)	HDAC1	Suppress	Hu et al. (2019)
	hsa-miR-216a(▲)	MAP1S	Enhance	Wang et al. (2019b)
	hsa-miR-98(▼)	HK2	Suppress	Zhu et al. (2017)
	hsa-miR-140(▼)	Smad3	Suppress	Li et al. (2018b)
	hsa-miR-28-5p(▼)	SSRP1	Suppress	Wu et al. (2019a)

(continued)

Table 2 (continued)

	hsa-miR-196a-5p(▲)	IκBα	Enhance	Xin et al. (2019)
	hsa-miR-23a(▲)	PDK4	Enhance	Deng et al. (2018)
	hsa-miR-130(▲)	TGFβR2	Enhance	Duan et al. (2016)
	hsa-miR-1207-5p/hsa-miR-1266, hsa-miR-1182(▼)	hTERT	Suppress	Chen et al. (2014b)
	hsa-miR-24(▲)	BCL2L11	Enhance	Zhang et al. (2016b)
	hsa-miR-100(▼)	ZBTB7A	Suppress	Shi et al. (2015), Wu et al. (2015)
	hsa-miR-125b(▲)	PPP1CA	Enhance	Zheng et al. (2013)
	hsa-miR-145(▼)	ETS1	Suppress	Zheng et al. (2011)
	hsa-miR-148a(▼)	ROCK1	Suppress	Liu et al. (2015b)
	hsa-miR-940(▲)	ZNF24	Enhance	Zhang et al. (2015c)
	hsa-miR-339-5p	ALKBH1	Suppress	Wang et al. (2020a)
	hsa-miR-135b-5p(▲)	KLF4	Enhance	Chen et al. (2020a)
Gastric cancer	hsa-miR-181a(▲)	Caprin-1	Enhance	Lu et al. (2019)
	hsa-miR-96-5p(▲)	FOXO3	Enhance	He and Zou (2020)
	hsa-miR-598(▼)	RRS1	Suppress	Ma et al. (2019)
	hsa-miR-103a-3p(▲)	ATF7	Enhance	Hu et al. (2018)
	hsa-miR-26a(▼)	FGF9	Suppress	Deng et al. (2013)
	hsa-miR-375(▼)	JAK2, ERBB2	Suppress	Ding et al. (2010)
	hsa-miR-29a(▼)	VEGF-A	Suppress	Chen et al. (2014c)
	hsa-miR-107(▲)	NF1	Enhance	Wang et al. (2016), Shen et al. (2014b)
	hsa-miR-214-5p(▼)	SOX4	Suppress	Xu et al. (2021a)
	hsa-miR-136(▼)	MKK4	Suppress	Zhu et al. (2018)
	hsa-miR-143(▼)	Bcl-2	Suppress	Ma et al. (2017)
	hsa-miR-200b(▼)	Bmi-1	Suppress	Yu et al. (2014)
	hsa-miR-144(▼)	EZH2	Suppress	Sun et al. (2021)
	hsa-miR-31-3p(▼)	GABBR2	Suppress	Choi et al. (2022)
	hsa-miR-1207-5p(▼)	PYCR1	Suppress	Xu et al. (2021b)
	hsa-miR-33b-3p(▼)	DOCK4	Suppress	Mei et al. (2021)
	hsa-miR-378a(▼)	GLUT1	Suppress	Cannistraci et al. (2022)
	hsa-miR-101-3p(▲)	CUL4B	Enhance	Gu et al. (2021b)
	hsa-miR-489-3p(▼)	DLX1	Suppress	Bai et al. (2020)
	hsa-miR-219-5p(▼)	HMGA2	Suppress	Huang et al. (2020)
Prostate cancer	hsa-miR-185(▼)	ALK4	Suppress	Li et al. (2020d)
	hsa-miR-15a/hsa-miR-16-1(▼)	BCL2, CCND1, WNT3A	Suppress	Bonci et al. (2009)
	hsa-miR-203(▼)	CKAP2, LASP1, BIRC5, WASF1, ASAP1	Suppress	Viticchie et al. (2011)
	hsa-miR-515-5p(▼)	TRIP13	Suppress	Zhang et al. (2019b)
	hsa-miR-1-3p(▼)	E2F5, PFTK1	Suppress	Li et al. (2018c)
	hsa-miR-338-3p(▼)	RAB23	Suppress	Wang and Qin (2018)
	hsa-miR-148-3p/hsa-miR-152-3p(▼)	KLF4	Suppress	Feng et al. (2019)
	hsa-miR-653-5p(▲)	SOX30	Enhance	Fu et al. (2019)
	hsa-miR-139-5p(▼)	Notch1	Suppress	Sun et al. (2018b)

transcripts (RNA interference) and gene silencing at the chromatin level. Global transcriptome analyses demonstrate that a significant fraction of the genome can produce transcripts from both strands and that antisense transcripts often link adjacent "genes" in complex loci to form chains of interconnected transcription units (Lou et al. 2020). ANRIL activates two Polycomb repressor complexes, PRC1 and PRC2, resulting in chromatin refolding that silences the INK4b-ARF-INK4a locus, which encodes the tumor suppressors p15INK4b, p14ARF, and p16INK4a, which are involved in cell cycle inhibition. It also activates senescence and stress-induced apoptosis. The over-expression of ANRIL in prostate cancer has been shown to silence INK4b-ARF-INK4a and p15/CDKN2B by heterochromatin reformation (Yu et al. 2020a). However, suppression is mediated by direct binding to combox 7 (CBX 7) and SUZ12, which are members of PRC1 and PRC2, respectively (Yap et al. 2010).

MALAT1—Metastasis-Associated Lung Adenocarcinoma Transcript 1: Several chromosomal translocation breakpoints have been identified at 11q13.1 at the MALAT1 locus (Li et al. 2018d). Several types of cancer are associated with MALAT1 activation, including hepatocellular carcinoma, breast cancer, pancreatic cancer, osteosarcoma, colon cancer and prostate cancer (Goyal et al. 2021). MALAT1 is a novel biomarker for predicting tumor recurrence after liver transplantation that could also serve as a potential therapeutic target. In addition, it has been shown to play an essential role in tumor progression (Arun and Spector 2019). The findings of one study suggest that MALAT1 is a novel class of non-coding RNAs that promote cell motility by regulating expression of related genes through transcription and post-transcriptional mechanisms. Furthermore, short-interfering RNA-mediated MALAT1 silencing has been shown to affect lung cancer cell motility and gene expression *in vitro* (Zhang et al. 2019c).

H19: Despite high-level expression in embryos, this imprinted ncRNA did not cause any apparent phenotype change following target deletion. However, the fact that it has been conserved throughout evolution suggests that this is a functional RNA. This locus contains imprinted genes for H19 and insulin-like growth factor 2 (IGF2). These genes are expressed differently due to H19 being a maternal allele and IGF2 a paternal allele (Yu et al. 2020b). H19 is misexpressed in many tumor types, including hepatocellular and bladder cancers, due to the loss of imprinting. Oncogenic, tumor-suppressive and apoptotic properties have been observed in this lncRNA. H19 enhances tumorigenesis as a result of cMYC expression on various cell types. As well as downregulating IGF2 expression, c-MYC blocks the expression of the transcription factor.

MEG3-Maternally expressed gene 3: MEG3 encodes a noncoding RNA with an antiproliferative function. Arachnoid cells exhibit high levels of MEG3 mRNA, which is encoded by an imprinted gene located at 14q32. The IOMM-Lee and CH157-MN cell lines and the majority of human meningiomas do not express MEG3. A strong correlation has been found between tumor grade and loss of MEG3 expression; as tumor grade increases, so does the frequency of MEG3 loss in meningiomas. Moreover, MEG3 gene promoters and imprinting control regions are methylated more extensively in meningiomas. It has been demonstrated that

MEG3 suppresses DNA synthesis functionally in both IOMM-Lee and CH157-MN cells by 60%. Furthermore, based on the results of colony-forming efficiency assays, MEG3 inhibits the growth of CH157-MN cells by 80%. Additionally, MEG3 activates p53-mediated transactivation in these cell lines. Consequently, these data support the hypothesis that MEG3, which encodes a noncoding RNA, is a tumor suppressor gene on chromosome 14q32 that plays a role in meningioma progression in a novel manner (Reyes-Gonzalez and Vivas-Mejia 2021).

GAS5-Growth Arrest-Specific 5: The steroid receptors bind to specific sequences in the promoters of target genes, called 'response elements' to activate transcription. One study showed that non-coding RNA can also act on steroid receptors, inhibiting their binding to DNA response elements and inhibiting steroid receptor function. According to the authors, Gas5, which accumulates in cells in the absence of nutrients and growth factors, binds to glucocorticoid receptors and inhibits glucocorticoid-mediated transcription. Glucocorticoids prevented the transcription of antiapoptotic genes in cells under stress. Likewise, GAS5 inhibited the transcriptional activity of other steroid receptors, suggesting that decoy RNA response elements similarly regulate transcription factor activity (Zhang et al. 2010b). Specifically, GAS5 induces apoptosis directly or indirectly in prostate and breast cancer cell lines, and breast cancers have been found to express GAS5 significantly less than normal breast epithelial cells (Kino et al. 2010).

LincRNA-p21: A direct effect of the p53 signaling pathway is the induction of expression. As an essential component of cellular apoptosis regulation, it represses genes involved in p53 function. By physically interacting with RNA-binding protein hnRNP K, LincRNA-p21 mediates gene repression by suppressing the promoters of genes in a p53-dependent manner (Wang et al. 2008).

Many more studies are needed to gain a deeper understanding of the regulatory processes of lncRNAs. The dysregulation of lncRNAs in complex cellular processes leads to human disease. Studying the mechanisms by which lncRNAs are involved in oncogenic and tumor suppressive pathways will allow us to identify possible new cancer diagnostic markers and therapeutic targets (Table 3).

Uncovering the lncRNA-miRNA-mRNA relationship is a rapidly advancing field. These ncRNAs have a two-way effect in target-directed regulation. One is the regulatory role of miRNAs on lncRNAs, and the other is the role of lncRNA-controlled miRNAs in changing expression levels of the mRNA target (Sun et al. 2020). In addition to the effects of individual miRNAs and lncRNAs on mRNA, they direct miRNA-lncRNA interactions in transcriptome regulation. These interactions lead to miRNA-dependent RNA degradation, competition of miRNAs and lncRNAs for the same mRNA targets to act as miRNA sponges (Paraskevopoulou and Hatzigeorgiou 2016). The lncRNA DDX11-AS1-miR-873-5p-SPC18 interaction was observed to have regulatory effects on gastric cancer, having a sponge effect on hsa-miR-873-5p and increasing the expression of SPC18 and DDX11-AS1 (Saito et al. 2015). In another study, the relationship between lncRNA MALAT1, hsa-miR-140 and VEGF-A was examined. MALAT1 and miR-140 appeared to be negatively correlated, and inhibition of hsa-miR-140 increased VEGF-A expression, resulting in increased angiogenesis. Targeting MALAT1 or hsa-miR-140 for liver cancer may

Table 3 Oncogenic lncRNAs and their therapeutic potential in human cancer

lncRNA	Cancer type	Role in cancer	References
AK023948	Papillary thyroid carcinoma	Downregulated in most PTC tumors, is a candidate susceptibility gene for PTC	Samson et al. (2018)
ANRIL	Prostate, leukemia	Activating two polycomb repressor complexes, PRC1 and PRC2, resulting in chromatin refolding that silences the INK4b-ARF-INK4a locus, which encodes the tumor suppressors p15INK4b, p14ARF and p16INK4a	Yang et al. (2011)
BC200	Breast, cervix, esophagus, lung, ovary, parotid, tongue	Expressed in carcinomas of the breast, cervix, oesophagus, lung, ovary, parotid and tongue, but not in normal tissues	Chen et al. (1997)
PRNCR1	Prostate	Upregulated in PC cells and precursor lesion prostatic intraepithelial neoplasia involved in prostate carcinogenesis through androgen receptor activity	Chung et al. (2011)
H19	Bladder, lung, liver, breast, esophagus, choriocarcinoma, colon	Downregulating IGF2 expression, c-MYC blocks the expression of a precursor to miR-675, which suppresses a tumor suppressor gene	Shermane Lim et al. (2021)
HOTAIR	Breast, hepatocellular	Strongly upregulated in breast and hepatocelluar cancer, used as a prognostic marker	Chen et al. (2021a)
HULC	Hepatocellular	Highly upregulated in liver cancer transcribing to an mRNA-like ncRNA by the RNA polymerase II and processed by capping, splicing and polyadenylation	Abbastabar et al. (2018)
LincRNA-p21	Cancer cells	Represses p53 pathway, induces apoptosis	Huarte et al. (2010)
Loc285194	Osteosarcoma	Focal copy-number changes and loss of heterozygosity implicate two ncRNAs and one tumor suppressor gene at this locus	Xu et al. (2019)
Malat1	Breast, prostate, colon, liver, uterus	Promoting cell motility by regulating gene expression related to motility through transcription and post-transcriptional mechanisms.	Guffanti et al. (2009)
MEG3	Brain	Suppressing gene plays a role in meningioma progression and DNA synthesis in both IOMM-Lee and CH157-MN cells, inhibiting the growth of CH157-MN cells and activating p53-mediated transactivation	Wu et al. (2019b)

(continued)

Table 3 (continued)

lncRNA	Cancer type	Role in cancer	References
PTNEP1	Prostate	PTENP1 (pseudogene of PTEN) is strongly upregulated in prostate cancer patients	Lou et al. (2019)
Spry4-it1	Melanoma	Differentially expressed in melanoma cell lines in comparison to melanocytes and keratinocyte controls	Li et al. (2019b)
SRA	Breast, uterus, ovary	Activating RNA as a biological marker for steroid-dependent tumors	Lanz et al. (2003)
UCA1/ CUDR	Bladder, colon, cervix, lung, thyroid, liver, breast, esophagus, stomach	Functioning as a ncRNA in the regulation of embryonic development and cancer invasion and progression	Ghafouri-Fard and Taheri (2019b)
Wt1-as	Acute myeloid leukemia	Encoding antisense RNAs have a regulatory role in WT1 expression via RNA:RNA interactions, and become deregulated by mechanisms in cancer	Cui et al. (2019b)
PCA3	Prostate	Very sensitive and specific marker for the detection of prostate tumor cells in a high background of normal (prostate) cells	Gunelli et al. (2021)
GAS5	Breast	Acting on steroid receptors, inhibiting their binding to DNA response elements and inhibiting steroid receptor function, inducing apoptosis in breast cancer cell lines	Filippova et al. (2021)

have a therapeutic effect to inhibit the development of the disease. The Table 4 summarizes current knowledge on lncRNA-miRNA-mRNA target interactions (Li et al. 2015b).

5 Conclusion

During cancer development, miRNAs and lncRNAs play roles in tumor proliferation, migration, invasion, and cellular and metabolic reprogramming. Accumulating evidence shows that harnessing and targetting the various different classes of ncRNAs, including miRNAs or lncRNAs, in order to eliminate viable tumor cells is a crucial therapeutic approach to managing and treating cancer. However, research on the actionable and druggable potentials of miRNAs and lncRNAs remains insufficient. Thus, integrative approaches such as multi-omics are necessary to determine the roles played by miRNAs and lncRNAs in both cancerous and normal tissues.

Table 4 The lncRNA-miRNA-mRNA relationship in various human cancer types

Cancer type	lncRNA	miRNA	mRNA target	References
Bladder cancer	UCA1	hsa-miR-145	ZEB1, ZEB2	Xue et al. (2016)
	linc00511	hsa-miR-15a-3p	D1, c-myc	Li et al. (2018e)
	CCAT1	hsa-miR-218-5p	Bmi1	Chen et al. (2015)
	lncRNA-LET	hsa-miR-145	TGFβ1	Zhuang et al. (2017)
	PVT1	hsa-miR-194-5p	BCLAF1	Chen et al. (2020b)
Brain cancer	HOTAIR	hsa-miR-141	SKA2	Zhou et al. (2020b)
	HOXA-AS2	hsa-miR-373	EGFR	Gao et al. (2018)
	TUG1	hsa-miR-299	VEGFA	Cai et al. (2017)
Breast cancer	H19	hsa-miR-200b/c, let-7b	Git2, Cyth3	Zhou et al. (2017)
	linc-ROR	hsa-miR-205	ZEB2	Hou et al. (2014)
	DANCR	hsa-miR-758-3p	PAX6	Zhang et al. (2020)
Colorectal cancer	ZFAS1	hsa-miR-200	Vimentin	O'Brien et al. (2021)
	SNHG6	hsa-miR-26a-5p	ULK1	Wang et al. (2019d)
	EGOT	miR-33b-5p	CROT	Ni et al. (2020)
	SOX21-AS1	hsa-miR-145	MYO6	Wei and Li (2017)
Gastric cancer	ATB	hsa-miR-200c	ZEB1	Saito et al. (2015)
	ASNR	hsa-miR-519e-5p	FGFR2	Chen et al. (2021b)
	DDX11-AS1	hsa-miR-873-5p	SPC18	Ren et al. (2020)
	BC032469	hsa-miR-1207-5p	hTERT	Lu et al. (2016)
Kidney cancer	UCA1	hsa-miR-182-5p	DLL4	Wang et al. (2020b)
Liver cancer	PTENP1	hsa-miR-17, hsa-miR-19b, hsa-miR-20a	PTEN, PHLPP	Ma et al. (2015)
	ZFAS1	hsa-miR-150	ZEB1, MMP14, MMP16	Li et al. (2015b)
	MALAT1	hsa-miR-140	VEGF-A	Hou et al. (2020)

(continued)

Table 4 (continued)

Cancer type	lncRNA	miRNA	mRNA target	References
Lung cancer	MALAT1	hsa-miR-217	EZH2	(Lu et al. (2015)
	MCM3AP-AS1	hsa-miR-340-5p	KPNA4	Li et al. (2020e)
	XIST	hsa-miR-449a	Bcl-2	Zhang et al. (2017c)
Ovarian cancer	SNHG12	hsa-miR-129	SOX4	Sun and Fan (2019)
	PART1	hsa-miR-512-3p	CHRAC1	Yang et al. (2021)
Pancreatic cancer	H19	let-7	HMGA2	Ma et al. (2014)
	PVT1	hsa-miR-619-5p	Pygo2, ATG14	Bian et al. (2016)
Prostate cancer	UCA1	hsa-miR-204	CXCR4	He et al. (2019b)
Throat cancer	GAS5	hsa-miR-26a-5p	ULK2	Wang et al. (2021b)

Acknowledgements This study was funded by Scientific Research Projects Coordination Unit of Istanbul University. Project number: 37223. We would like to thank Rachel Mollman, a native English-speaker, for language editing and proofreading.

References

Abbastabar M, Sarfi M, Golestani A, Khalili E (2018) lncRNA involvement in hepatocellular carcinoma metastasis and prognosis. EXCLI J 17:900–913

Adam-Artigues A, Garrido-Cano I, Carbonell-Asins JA, Lameirinhas A, Simon S, Ortega-Morillo-B, Martinez MT, Hernando C, Constancio V, Burgues O, Bermejo B, Henrique R, Lluch A, Jeronimo C, Eroles P, Cejalvo JM (2021a) Identification of a two-microRNA signature in plasma as a novel biomarker for very early diagnosis of breast cancer. Cancer 13:2848

Adam-Artigues A, Garrido-Cano I, Simon S, Ortega B, Moragon S, Lameirinhas A, Constancio V, Salta S, Burgues O, Bermejo B, Henrique R, Lluch A, Jeronimo C, Eroles P, Cejalvo JM (2021b) Circulating miR-30b-5p levels in plasma as a novel potential biomarker for early detection of breast cancer. ESMO Open 6:100039

Anastasiadou E, Jacob LS, Slack FJ (2018) Non-coding RNA networks in cancer. Nat Rev Cancer 18:5–18

Arab A, Karimipoor M, Irani S, Kiani A, Zeinali S, Tafsiri E, Sheikhy K (2017) Potential circulating miRNA signature for early detection of NSCLC. Cancer Genet 216–217:150–158

Arun G, Spector DL (2019) MALAT1 long non-coding RNA and breast cancer. RNA Biol 16:860–863

Asakura K, Kadota T, Matsuzaki J, Yoshida Y, Yamamoto Y, Nakagawa K, Takizawa S, Aoki Y, Nakamura E, Miura J, Sakamoto H, Kato K, Watanabe SI, Ochiya T (2020) A miRNA-based diagnostic model predicts resectable lung cancer in humans with high accuracy. Commun Biol 3:134

Bai P, Li W, Wan Z (2020) miR-489-3p inhibits prostate cancer progression by targeting DLX1. Cancer Manag Res 12:12581–12581
Barcelo M, Castells M, Bassas L, Vigues F, Larriba S (2019) Semen miRNAs contained in exosomes as non-invasive biomarkers for prostate cancer diagnosis. Sci Rep 9:13772
Baumgartner U, Berger F, Gheinani AH, Burgener SS, Monastyrskaya K, Vassella E (2018) miR-19b enhances proliferation and apoptosis resistance via the EGFR signaling pathway by targeting PP2A and BIM in non-small cell lung cancer. Mol Cancer 17:44
Bian EB, Ma CC, He XJ, Wang C, Zong G, Wang HL, Zhao B (2016) Epigenetic modification of miR-141 regulates SKA2 by an endogenous 'sponge' HOTAIR in glioma. Oncotarget 7:30610–30625
Bitaraf A, Babashah S, Garshasbi M (2020) Aberrant expression of a five-microRNA signature in breast carcinoma as a promising biomarker for diagnosis. J Clin Lab Anal 34:e23063
Bonci D, Coppola V, Musumeci M, Addario A, D'urso L, Collura D, Peschle C, de Maria R, Muto G (2009) The Mir-15a/Mir-16-1 cluster controls prostate cancer progression control by targeting of multiple oncogenic activities. J Urol 181:188–188
Brummer A, Hausser J (2014) MicroRNA binding sites in the coding region of mRNAs: extending the repertoire of post-transcriptional gene regulation. Bioessays 36:617–626
Cai H, Yuan Y, Hao YF, Guo TK, Wei X, Zhang YM (2013) Plasma microRNAs serve as novel potential biomarkers for early detection of gastric cancer. Med Oncol 30:452
Cai H, Liu X, Zheng J, Xue Y, Ma J, Li Z, Xi Z, Li Z, Bao M, Liu Y (2017) Long non-coding RNA taurine upregulated 1 enhances tumor-induced angiogenesis through inhibiting microRNA-299 in human glioblastoma. Oncogene 36:318–331
Calin GA, Croce CM (2006) MicroRNA signatures in human cancers. Nat Rev Cancer 6:857–866
Calin GA, Sevignani C, Dumitru CD, Hyslop T, Noch E, Yendamuri S, Shimizu M, Rattan S, Bullrich F, Negrini M, Croce CM (2004) Human microRNA genes are frequently located at fragile sites and genomic regions involved in cancers. Proc Natl Acad Sci U S A 101:2999–3004
Cannistraci A, Hascoet P, Ali A, Mundra P, Clarke NW, Pavet V, Marais R (2022) MiR-378a inhibits glucose metabolism by suppressing GLUT1 in prostate cancer. Oncogene 41:1445–1455
Cao XN, Lai SY, Hu FY, Li GD, Wang GH, Luo XL, Fu XN, Hu JB (2017) miR-19a contributes to gefitinib resistance and epithelial mesenchymal transition in non-small cell lung cancer cells by targeting c-Met. Sci Rep 7:2939
Cao W, Cheng W, Wu W (2018) MicroRNAs reprogram tumor immune response. Methods Mol Biol 1699:67–74
Carninci P, Kasukawa T, Katayama S, Gough J, Frith MC, Maeda N, Oyama R, Ravasi T, Lenhard B, Wells C, Kodzius R, Shimokawa K, Bajic VB, Brenner SE, Batalov S, Forrest ARR, Zavolan M, Davis MJ, Wilming LG, Aidinis V, Allen JE, Ambesi-Impiombato X, Apweiler R, Aturaliya RN, Bailey TL, Bansal M, Baxter L, Beisel KW, Bersano T, Bono H, Chalk AM, Chiu KP, Choudhary V, Christoffels A, Clutterbuck DR, Crowe ML, Dalla E, Dalrymple BP, de Bono B, Della Gatta G, Di Bernardo D, Down T, Engstrom P, Fagiolini M, Faulkner G, Fletcher CF, Fukushima T, Furuno M, Futaki S, Gariboldi M, Georgii-Hemming P, Gingeras TR, Gojobori T, Green RE, Gustincich S, Harbers M, Hayashi Y, Hensch TK, Hirokawa N, Hill D, Huminiecki L, Iacono M, Ikeo K, Iwama A, Ishikawa T, Jakt M, Kanapin A, Katoh M, Kawasawa Y, Kelso J, Kitamura H, Kitano H, Kollias G, Krishnan SPT, Kruger A, Kummerfeld SK, Kurochkin IV, Lareau LF, Lazarevic D, Lipovich L, Liu J, Liuni S, McWilliam S, Babu MM, Madera M, Marchionni L, Matsuda H, Matsuzawa S, Miki H, Mignone F, Miyake S, Morris K, Mottagui-Tabar S, Mulder N, Nakano N, Nakauchi H, Ng P, Nilsson R, Nishiguchi S, Nishikawa S et al (2005) The transcriptional landscape of the mammalian genome. Science 309:1559–1563
Chen W, Bocker W, Brosius J, Tiedge H (1997) Expression of neural BC200 RNA in human tumours. J Pathol 183:345–351
Chen ZH, Zhang GL, Li HR, Luo JD, Li ZX, Chen GM, Yang J (2012) A panel of five circulating microRNAs as potential biomarkers for prostate cancer. Prostate 72:1443–1452

Chen W, Wang H, Chen H, Liu S, Lu H, Kong D, Huang X, Kong Q, Lu Z (2014a) Clinical significance and detection of microRNA-21 in serum of patients with diffuse large B-cell lymphoma in Chinese population. Eur J Haematol 92:407–412
Chen L, Lu MH, Zhang D, Hao NB, Fan YH, Wu YY, Wang SM, Xie R, Fang DC, Zhang H, Hu CJ, Yang SM (2014b) miR-1207-5p and miR-1266 suppress gastric cancer growth and invasion by targeting telomerase reverse transcriptase. Cell Death Dis 5:e1034
Chen L, Xiao H, Wang ZH, Huang Y, Liu ZP, Ren H, Song H (2014c) miR-29a suppresses growth and invasion of gastric cancer cells in vitro by targeting VEGF-A. BMB Rep 47:39–44
Chen CL, Tseng YW, Wu JC, Chen GY, Lin KC, Hwang SM, Hu YC (2015) Suppression of hepatocellular carcinoma by baculovirus-mediated expression of long non-coding RNA PTENP1 and MicroRNA regulation. Biomaterials 44:71–81
Chen X, Wu L, Li DZ, Xu YM, Zhang LP, Niu K, Kong R, Gu JY, Xu ZH, Chen ZT, Sun JG (2018a) Radiosensitizing effects of miR-18a-5p on lung cancer stem-like cells via downregulating both ATM and HIF-1 alpha. Cancer Med 7:3834–3847
Chen LJ, Wang XQ, Zhu YH, Zhu J, Lai QZ (2018b) miR-200b-3p inhibits proliferation and induces apoptosis in colorectal cancer by targeting Wnt1. Mol Med Rep 18:2571–2580
Chen L, Gao HB, Liang JT, Qiao JJ, Duan J, Shi HJ, Zhen TT, Li H, Zhang FF, Zhu ZW, Han AJ (2018c) miR-203a-3p promotes colorectal cancer proliferation and migration by targeting PDE4D. Am J Cancer Res 8:2387–2401
Chen TM, Xiao Q, Wang XJ, Wang ZQ, Hu JW, Zhang Z, Gong ZN, Chen SL (2019) miR-16 regulates proliferation and invasion of lung cancer cells via the ERK/MAPK signaling pathway by targeted inhibition of MAPK kinase 1 (MEK1). J Int Med Res 47:5194–5204
Chen Z, Gao Y, Gao S, Song D, Feng Y (2020a) MiR-135b-5p promotes viability, proliferation, migration and invasion of gastric cancer cells by targeting Kruppel-like factor 4 (KLF4). Arch Med Sci 16:167–176
Chen M, Zhang R, Lu L, Du J, Chen C, Ding K, Wei X, Zhang G, Huang Y, Hou J (2020b) LncRNA PVT1 accelerates malignant phenotypes of bladder cancer cells by modulating miR-194-5p/BCLAF1 axis as a ceRNA. Aging (Albany NY) 12:22291–22312
Chen L, Qian X, Wang Z, Zhou X (2021a) The HOTAIR lncRNA: a remarkable oncogenic promoter in human cancer metastasis. Oncol Lett 21:302
Chen ZH, Li Y, Tan BB, Li F, Zhao Q, Fan LQ, Zhang ZD, Zhao XF, Liu Y, Wang D (2021b) Long non-coding RNA ASNR targeting miR-519e-5p promotes gastric cancer development by regulating FGFR2. Front Cell Dev Biol 9:679176
Chiang CH, Chu PY, Hou MF, Hung WC (2016) MiR-182 promotes proliferation and invasion and elevates the HIF-1 alpha-VEGF-A axis in breast cancer cells by targeting FBXW7. American. J Cancer Res 6:1785
Chiou GY, Cherng JY, Hsu HS, Wang ML, Tsai CM, Lu KH, Chien Y, Hung SC, Chen YW, Wong CI, Tseng LM, Huang PI, Yu CC, Hsu WH, Chiou SH (2012) Cationic polyurethanes-short branch PEI-mediated delivery of Mir145 inhibited epithelial-mesenchymal transdifferentiation and cancer stem-like properties and in lung adenocarcinoma. J Control Release 159:240–250
Choi S, Lee S, Han YH, Choi J, Kim I, Lee J, An HJ (2022) miR-31-3p functions as a tumor suppressor by directly targeting GABBR2 in prostate cancer. Front Oncol 12:945057
Chung S, Nakagawa H, Uemura M, Piao L, Ashikawa K, Hosono N, Takata R, Akamatsu S, Kawaguchi T, Morizono T, Tsunoda T, Daigo Y, Matsuda K, Kamatani N, Nakamura Y, Kubo M (2011) Association of a novel long non-coding RNA in 8q24 with prostate cancer susceptibility. Cancer Sci 102:245–252
Cieslik M, Chinnaiyan AM (2018) Cancer transcriptome profiling at the juncture of clinical translation. Nat Rev Genet 19:93–109
Cornett AL, Lutz CS (2014) Regulation of COX-2 expression by miR-146a in lung cancer cells. RNA 20:1419–1430
Cortez MA, Bueso-Ramos C, Ferdin J, Lopez-Berestein G, Sood AK, Calin GA (2011) MicroRNAs in body fluids--the mix of hormones and biomarkers. Nat Rev Clin Oncol 8:467–477

Cory S, Adams JM (2019) A joint odyssey into cancer genetics. Annu Rev Cancer Biol 3(3):1–19
Cui M, Wang H, Yao X, Zhang D, Xie Y, Cui R, Zhang X (2019a) Circulating microRNAs in cancer: potential and challenge. Front Genet 10:626
Cui L, Nai M, Zhang K, Li L, Li R (2019b) lncRNA WT1-AS inhibits the aggressiveness of cervical cancer cell via regulating p53 expression via sponging miR-330-5p. Cancer Manag Res 11:651–667
Dai FQ, Li CR, Fan XQ, Tan L, Wang RT, Jin H (2019) miR-150-5p Inhibits non-small-cell lung cancer metastasis and recurrence by targeting HMGA2 and beta-catenin signaling. Mol Ther Nucleic Acids 16:675–685
Dai T, Zhao X, Li Y, Yu L, Li Y, Zhou X, Gong Q (2020) miR-423 promotes breast cancer invasion by activating NF-kappaB signaling. Onco Targets Ther 13:5467–5478
Dastmalchi N, Hosseinpourfeizi MA, Khojasteh SMB, Baradaran B, Safaralizadeh R (2020) Tumor suppressive activity of miR-424-5p in breast cancer cells through targeting PD-L1 and modulating PTEN/PI3K/AKT/mTOR signaling pathway. Life Sci 259:118239
Deng M, Tang HL, Lu XH, Liu MY, Lu XM, Gu YX, Liu JF, He ZM (2013) miR-26a suppresses tumor growth and metastasis by targeting FGF9 in gastric cancer. PLoS One 8:e72662
Deng YH, Deng ZH, Hao H, Wu XL, Gao H, Tang SH, Tang H (2018) MicroRNA-23a promotes colorectal cancer cell survival by targeting PDK4. Exp Cell Res 373:171–179
Ding L, Xu YJ, Zhang W, Deng YJ, Si MS, Du Y, Yao HM, Liu XY, Ke YH, Si JM, Zhou TH (2010) MiR-375 frequently downregulated in gastric cancer inhibits cell proliferation by targeting JAK2. Cell Res 20:784–793
Du L, Zhao Z, Ma X, Hsiao TH, Chen Y, Young E, Suraokar M, Wistuba I, Minna JD, Pertsemlidis A (2014) miR-93-directed downregulation of DAB2 defines a novel oncogenic pathway in lung cancer. Oncogene 33:4307–4315
Duan J, Zhang H, Qu Y, Deng T, Huang D, Liu R, Zhang L, Bai M, Zhou L, Ying G, Ba Y (2016) Onco-miR-130 promotes cell proliferation and migration by targeting TGFbetaR2 in gastric cancer. Oncotarget 7:44522–44533
Duan X, Qiao S, Li D, Li S, Zheng Z, Wang Q, Zhu X (2021) Circulating miRNAs in serum as biomarkers for early diagnosis of non-small cell lung cancer. Front Genet 12:673926
Esteller M (2011) Non-coding RNAs in human disease. Nat Rev Genet 12:861–874
Feng F, Liu H, Chen AP, Xia QH, Zhao Y, Jin XB, Huang JJ (2019) miR-148-3p and miR-152-3p synergistically regulate prostate cancer progression via repressing KLF4. J Cell Biochem 120: 17228–17239
Filippova EA, Fridman MV, Burdennyy AM, Loginov VI, Pronina IV, Lukina SS, Dmitriev AA, Braga EA (2021) Long noncoding RNA GAS5 in breast cancer: epigenetic mechanisms and biological functions. Int J Mol Sci 22:6810
Flores-Perez A, Marchat LA, Rodriguez-Cuevas S, Bautista-Pina V, Hidalgo-Miranda A, Ocampo EA, Martinez MS, Palma-Flores C, Fonseca-Sanchez MA, Astudillo-De La Vega H, Ruiz-Garcia E, Gonzalez-Barrios JA, Perez-Plasencia C, Streber ML, Lopez-Camarillo C (2016) Dual targeting of ANGPT1 and TGFBR2 genes by miR-204 controls angiogenesis in breast cancer. Sci Rep 6:34504
Friedman RC, Farh KK, Burge CB, Bartel DP (2009) Most mammalian mRNAs are conserved targets of microRNAs. Genome Res 19:92–105
Fu Q, Sun ZY, Yang F, Mao TC, Gao YY, Wang H (2019) SOX30, a target gene of miR-653-5p, represses the proliferation and invasion of prostate cancer cells through inhibition of Wnt/beta-catenin signaling. Cell Mol Biol Lett 24:71
Gao Y, Yu H, Liu Y, Liu X, Zheng J, Ma J, Gong W, Chen J, Zhao L, Tian Y, Xue Y (2018) Long non-coding RNA HOXA-AS2 regulates malignant glioma behaviors and vasculogenic mimicry formation via the MiR-373/EGFR axis. Cell Physiol Biochem 45:131–147
Gehring WJ, Hiromi Y (1986) Homeotic genes and the homeobox. Annu Rev Genet 20:147–173
Ghafouri-Fard S, Taheri M (2019a) Maternally expressed gene 3 (MEG3): a tumor suppressor long non coding RNA. Biomed Pharmacother 118:109129

Ghafouri-Fard S, Taheri M (2019b) UCA1 long non-coding RNA: an update on its roles in malignant behavior of cancers. Biomed Pharmacother 120:109459
Ghafouri-Fard S, Dashti S, Hussen BM, Farsi M, Taheri M (2021) BCYRN1: an oncogenic lncRNA in diverse cancers. Pathol Res Pract 220:153385
Gil N, Ulitsky I (2020) Regulation of gene expression by cis-acting long non-coding RNAs. Nat Rev Genet 21:102–117
Goyal B, Yadav SRM, Awasthee N, Gupta S, Kunnumakkara AB, Gupta SC (2021) Diagnostic, prognostic, and therapeutic significance of long non-coding RNA MALAT1 in cancer. Biochim Biophys Acta Rev Cancer 1875:188502
Grunnet M, Sorensen JB (2012) Carcinoembryonic antigen (CEA) as tumor marker in lung cancer. Lung Cancer 76:138–143
Gu Y, Pais G, Becker V, Korbel C, Ampofo E, Ebert E, Hohneck J, Ludwig N, Meese E, Bohle RM, Zhao YJ, Menger MD, Laschke MW (2021a) Suppression of endothelial miR-22 mediates non-small cell lung cancer cell-induced angiogenesis. Mol Ther Nucleic Acids 26:849–864
Gu ZH, You ZX, Yang YC, Ding R, Wang ML, Pu JM, Chen J (2021b) Inhibition of MicroRNA miR-101-3p on prostate cancer progression by regulating Cullin 4B (CUL4B) and PI3K/AKT/mTOR signaling pathways. Bioengineered 12:4719–4735
Guffanti A, Iacono M, Pelucchi P, Kim N, Solda G, Croft LJ, Taft RJ, Rizzi E, Askarian-Amiri M, Bonnal RJ, Callari M, Mignone F, Pesole G, Bertalot G, Bernardi LR, Albertini A, Lee C, Mattick JS, Zucchi I, de Bellis G (2009) A transcriptional sketch of a primary human breast cancer by 454 deep sequencing. BMC Genomics 10:163
Gunelli R, Fragala E, Fiori M (2021) PCA3 in prostate cancer. Methods Mol Biol 2292:105–113
Guo TM, Zheng CZ, Wang ZG, Zheng XF (2019) miR-584-5p regulates migration and invasion in non-small cell lung cancer cell lines through regulation of MMP-14. Mol Med Rep 19:1747–1752
Guo YJ, Pan WW, Liu SB, Shen ZF, Xu Y, Hu LL (2020) ERK/MAPK signalling pathway and tumorigenesis. Exp Ther Med 19:1997–2007
Hamamoto J, Soejima K, Yoda S, Naoki K, Nakayama S, Satomi R, Terai H, Ikemura S, Sato T, Yasuda H, Hayashi Y, Sakamoto M, Takebayashi T, Betsuyaku T (2013) Identification of microRNAs differentially expressed between lung squamous cell carcinoma and lung adenocarcinoma. Mol Med Rep 8:456–462
Hamfjord J, Stangeland AM, Hughes T, Skrede ML, Tveit KM, Ikdahl T, Kure EH (2012) Differential expression of miRNAs in colorectal cancer: comparison of paired tumor tissue and adjacent normal mucosa using high-throughput sequencing. PLoS One 7:e34150
Han LP, Fu T, Lin Y, Miao JL, Jiang QF (2016) MicroRNA-138 negatively regulates non-small cell lung cancer cells through the interaction with cyclin D3. Tumor Biol 37:291–298
Han Q, Cheng P, Yang HJ, Liang HP, Lin FC (2019) Altered expression of microRNA-365 is related to the occurrence and development of non-small-cell lung cancer by inhibiting TRIM25 expression. J Cell Physiol 234:22321–22330
Hanahan D, Weinberg RA (2011) Hallmarks of cancer: the next generation. Cell 144:646–674
Hawkes N (2019) Cancer survival data emphasise importance of early diagnosis. BMJ 364:l408
He XH, Zou KJ (2020) MiRNA-96-5p contributed to the proliferation of gastric cancer cells by targeting FOXO3. J Biochem 167:101–108
He Q, Fang Y, Lu F, Pan J, Wang L, Gong W, Fei F, Cui J, Zhong J, Hu R, Liang M, Fang L, Wang H, Yu M, Zhang ZF (2019a) Analysis of differential expression profile of miRNA in peripheral blood of patients with lung cancer. J Clin Lab Anal 33:e23003
He C, Lu XW, Yang F, Qin L, Guo ZF, Sun Y, Wu JW (2019b) LncRNA UCA1 acts as a sponge of miR-204 to up-regulate CXCR4 expression and promote prostate cancer progression. Biosci Rep 39:BSR20181465
He QL, Ye AH, Ye WBA, Liao XM, Qin GQ, Xu YQ, Yin YT, Luo HQ, Yi MH, Xian LY, Zhang SA, Qin XY, Zhu W, Li YL (2021) Cancer-secreted exosomal miR-21-5p induces angiogenesis and vascular permeability by targeting KRIT1. Cell Death Dis 12:576

Hill M, Tran N (2021) miRNA interplay: mechanisms and consequences in cancer. Dis Model Mech 14:dmm047662

Hirschfeld M, Rucker G, Weiss D, Berner K, Ritter A, Jager M, Erbes T (2020) Urinary exosomal microRNAs as potential non-invasive biomarkers in breast cancer detection. Mol Diagn Ther 24:215–232

Ho CS, Noor SM, Nagoor NH (2018) MiR-378 and MiR-1827 regulate tumor invasion, migration and angiogenesis in human lung adenocarcinoma by targeting RBX1 and CRKL, respectively. J Cancer 9:331–345

Hong H, Yu HZ, Yuan JF, Guo CY, Cao HY, Li WB, Xiao CH (2016) MicroRNA-200b impacts breast cancer cell migration and invasion by regulating ezrin-radixin-moesin. Med Sci Monit 22: 1946–1952

Hosono Y, Niknafs YS, Prensner JR, Iyer MK, Dhanasekaran SM, Mehra R, Pitchiaya S, Tien J, Escara-Wilke J, Poliakov A, Chu SC, Saleh S, Sankar K, Su FY, Guo SL, Qiao YY, Freier SM, Bui HH, Cao XH, Malik R, Johnson TM, Beer DG, Feng FY, Zhou WB, Chinnaiyan AM (2017) Oncogenic role of THOR, a conserved cancer/testis long non-coding RNA. Cell 171: 1559

Hou P, Zhao Y, Li Z, Yao R, Ma M, Gao Y, Zhao L, Zhang Y, Huang B, Lu J (2014) LincRNA-ROR induces epithelial-to-mesenchymal transition and contributes to breast cancer tumorigenesis and metastasis. Cell Death Dis 5:e1287

Hou XW, Sun X, Yu Y, Zhao HM, Yang ZJ, Wang X, Cao XC (2017) miR-361-5p suppresses lung cancer cell lines progression by targeting FOXM1. Neoplasma 64:526–534

Hou ZH, Xu XW, Fu XY, Zhou LD, Liu SP, Tan DM (2020) Long non-coding RNA MALAT1 promotes angiogenesis and immunosuppressive properties of HCC cells by sponging miR-140. Am J Physiol Cell Physiol 318:C649–C663

Hu XY, Miao JY, Zhang M, Wang XF, Wang ZZ, Han J, Tong DD, Huang C (2018) miRNA-103a-3p promotes human gastric cancer cell proliferation by targeting and suppressing ATF7 in vitro. Mol Cells 41:390–400

Hu Y, French SW, Chau T, Liu HX, Sheng L, Wei F, Stondell J, Garcia JC, Du Y, Bowlus CL, Wan YY (2019) RARbeta acts as both an upstream regulator and downstream effector of miR-22, which epigenetically regulates NUR77 to induce apoptosis of colon cancer cells. FASEB J 33: 2314–2326

Huang Z, Huang D, Ni S, Peng Z, Sheng W, Du X (2010) Plasma microRNAs are promising novel biomarkers for early detection of colorectal cancer. Int J Cancer 127:118–126

Huang Z, Zhu D, Wu L, He M, Zhou X, Zhang L, Zhang H, Wang W, Zhu J, Cheng W, Chen Y, Fan Y, Qi L, Yin Y, Zhu W, Shu Y, Liu P (2017) Six serum-based miRNAs as potential diagnostic biomarkers for gastric cancer. Cancer Epidemiol Biomarkers Prev 26:188–196

Huang ZM, Ge HF, Yang CC, Cai Y, Chen Z, Tian WZ, Tao JL (2019) MicroRNA-26a-5p inhibits breast cancer cell growth by suppressing RNF6 expression. Kaohsiung J Med Sci 35:467–473

Huang WT, Zhang H, Jin Z, Li K, Hu C, Li ML, Situ J (2020) MiR-219-5p inhibits prostate cancer cell growth and metastasis by targeting HMGA2. Eur Rev Med Pharmacol Sci 24:4710–4718

Huarte M, Guttman M, Feldser D, Garber M, Koziol MJ, Kenzelmann-Broz D, Khalil AM, Zuk O, Amit I, Rabani M, Attardi LD, Regev A, Lander ES, Jacks T, Rinn JL (2010) A large intergenic noncoding RNA induced by p53 mediates global gene repression in the p53 response. Cell 142: 409–419

Itani MM, Nassar FJ, Tfayli AH, Talhouk RS, Chamandi GK, Itani ARS, Makoukji J, Boustany RN, Hou L, Zgheib NK, Nasr RR (2021) A signature of four circulating microRNAs as potential biomarkers for diagnosing early-stage breast cancer. Int J Mol Sci 22:6121

Iwasaki H, Shimura T, Yamada T, Okuda Y, Natsume M, Kitagawa M, Horike SI, Kataoka H (2019) A novel urinary microRNA biomarker panel for detecting gastric cancer. J Gastroenterol 54:1061–1069

Iwasaki H, Shimura T, Kitagawa M, Yamada T, Nishigaki R, Fukusada S, Okuda Y, Katano T, Horike SI, Kataoka H (2022) A novel urinary miRNA biomarker for early detection of colorectal cancer. Cancer 14:461

Iyer MK, Niknafs YS, Malik R, Singhal U, Sahu A, Hosono Y, Barrette TR, Prensner JR, Evans JR, Zhao S, Poliakov A, Cao XH, Dhanasekaran SM, Wu YM, Robinson DR, Beer DG, Feng FY, Iyer HK, Chinnaiyan AM (2015) The landscape of long noncoding RNAs in the human transcriptome. Nat Genet 47:199

Jana S, Sengupta S, Biswas S, Chatterjee A, Roy H, Bhattacharyya A (2017) miR-216b suppresses breast cancer growth and metastasis by targeting SDCBP. Biochem Biophys Res Commun 482: 126–133

Jang JY, Kim YS, Kang KN, Kim KH, Park YJ, Kim CW (2021) Multiple microRNAs as biomarkers for early breast cancer diagnosis. Mol Clin Oncol 14:31

Jiang Z, Yin J, Fu W, Mo Y, Pan Y, Dai L, Huang H, Li S, Zhao J (2014) MiRNA 17 family regulates cisplatin-resistant and metastasis by targeting TGFbetaR2 in NSCLC. PloS One 9: e94639

Jin TF, Kim HS, Choi SK, Hwang EH, Woo J, Ryu HS, Kim K, Moon A, Moon WK (2017) microRNA-200c/141 upregulates SerpinB2 to promote breast cancer cell metastasis and reduce patient survival. Oncotarget 8:32769–32782

Jusoh AR, Mohan SV, Lu Ping T, Tengku Din T, Haron J, Romli RC, Jaafar H, Nafi SN, Tuan Salwani TI, Yahya MM (2021) Plasma circulating mirnas profiling for identification of potential breast cancer early detection biomarkers. Asian Pac J Cancer Prev 22:1375–1381

Kang SM, Lee HJ, Cho JY (2013) MicroRNA-365 regulates NKX2-1, a key mediator of lung cancer. Cancer Lett 335:487–494

Kavakiotis I, Alexiou A, Tastsoglou S, Vlachos IS, Hatzigeorgiou AG (2022) DIANA-miTED: a microRNA tissue expression database. Nucleic Acids Res 50:D1055–D1061

Kim TW, Lee YS, Yun NH, Shin CH, Hong HK, Kim HH, Cho YB (2020) MicroRNA-17-5p regulates EMT by targeting vimentin in colorectal cancer. Br J Cancer 123:1123–1130

Kino T, Hurt DE, Ichijo T, Nader N, Chrousos GP (2010) Noncoding RNA Gas5 is a growth arrest- and starvation-associated repressor of the glucocorticoid receptor. Sci Signal 3:ra8

Kohlan AK, Saidijam M, Amini R, Samadi P, Najafi R (2019) Induction of let-7e gene expression attenuates oncogenic phenotype in HCT-116 colorectal cancer cells through targeting of DCLK1 regulation. Life Sci 228:221–227

Kong P, Chen L, Yu M, Tao J, Liu J, Wang Y, Pan H, Zhou W, Wang S (2018) miR-3178 inhibits cell proliferation and metastasis by targeting Notch1 in triple-negative breast cancer. Cell Death Dis 9:1059

Lanz RB, Chua SS, Barron N, Soder BM, Demayo F, O'Malley BW (2003) Steroid receptor RNA activator stimulates proliferation as well as apoptosis in vivo. Mol Cell Biol 23:7163–7176

Lee TJ, Yuan X, Kerr K, Yoo JY, Kim DH, Kaur B, Eltzschig HK (2020) Strategies to modulate microRNA functions for the treatment of cancer or organ injury. Pharmacol Rev 72:639–667

Lee SB, Park YS, Sung JS, Lee JW, Kim B, Kim YH (2021) Tumor suppressor miR-584-5p inhibits migration and invasion in smoking related non-small cell lung cancer cells by targeting YKT6. Cancer 13:1159

Li LS, Xie XH, Luo JM, Liu M, Xi SY, Guo JL, Kong YN, Wu MQ, Gao J, Xie ZM, Tang J, Wang X, Wei WD, Yang MT, Hung MC, Xie XM (2012) Targeted expression of miR-34a using the T-VISA system suppresses breast cancer cell growth and invasion. Mol Ther 20:2326–2334

Li W, Zhai L, Zhao C, Lv S (2015a) MiR-153 inhibits epithelial-mesenchymal transition by targeting metadherin in human breast cancer. Breast Cancer Res Treat 150:501–509

Li T, Xie J, Shen C, Cheng D, Shi Y, Wu Z, Deng X, Chen H, Shen B, Peng C, Li H, Zhan Q, Zhu Z (2015b) Amplification of long noncoding RNA ZFAS1 promotes metastasis in hepatocellular carcinoma. Cancer Res 75:3181–3191

Li YL, Wang J, Zhang CY, Shen YQ, Wang HM, Ding L, Gu YC, Lou JT, Zhao XT, Ma ZL, Jin YX (2016a) MiR-146a-5p inhibits cell proliferation and cell cycle progression in NSCLC cell lines by targeting CCND1 and CCND2. Oncotarget 7:59287–59298

Li H, Ouyang RY, Wang Z, Zhou WH, Chen HY, Jiang YW, Zhang YB, Li H, Liao MT, Wang WW, Ye M, Ding ZG, Feng XP, Liu J, Zhang B (2016b) MiR-150 promotes cellular metastasis in non-small cell lung cancer by targeting FOXO4. Sci Rep 6:39001

Li Y, Cai B, Shen L, Dong Y, Lu Q, Sun S, Liu S, Ma S, Ma PX, Chen J (2017) MiRNA-29b suppresses tumor growth through simultaneously inhibiting angiogenesis and tumorigenesis by targeting Akt3. Cancer Lett 397:111–119
Li HM, Li JF, Yang TS, Lin SW, Li H (2018a) MicroRNA-433 represses proliferation and invasion of colon cancer cells by targeting homeobox A1. Oncol Res 26:315–322
Li J, Zou K, Yu L, Zhao W, Lu Y, Mao J, Wang B, Wang L, Fan S, Song B, Li L (2018b) MicroRNA-140 inhibits the epithelial-mesenchymal transition and metastasis in colorectal cancer. Mol Ther Nucleic Acids 10:426–437
Li SM, Wu HL, Yu X, Tang K, Wang SG, Ye ZQ, Hu J (2018c) The putative tumour suppressor miR-1-3p modulates prostate cancer cell aggressiveness by repressing E2F5 and PFTK1. J Exp Clin Cancer Res 37:219
Li ZX, Zhu QN, Zhang HB, Hu Y, Wang G, Zhu YS (2018d) MALAT1: a potential biomarker in cancer. Cancer Manag Res 10:6757–6768
Li J, Li Y, Meng F, Fu L, Kong C (2018e) Knockdown of long non-coding RNA linc00511 suppresses proliferation and promotes apoptosis of bladder cancer cells via suppressing Wnt/ beta-catenin signaling pathway. Biosci Rep 38:BSR20171701
Li M, Zou X, Xia T, Wang T, Liu P, Zhou X, Wang S, Zhu W (2019a) A five-miRNA panel in plasma was identified for breast cancer diagnosis. Cancer Med 8:7006–7017
Li Z, Tang X, Duan S (2019b) Interference from LncRNA SPRY4-IT1 restrains the proliferation, migration, and invasion of melanoma cells through inactivating MAPK pathway by up-regulating miR-22-3p. Int J Clin Exp Pathol 12:477–487
Li L, Wang A, Cai M, Tong M, Chen F, Huang L (2020a) Identification of stool miR-135b-5p as a non-invasive diaognostic biomarker in later tumor stage of colorectal cancer. Life Sci 260: 118417
Li Y, Liang M, Zhang YH, Yuan B, Gao WC, Shi ZZ, Bai J (2020b) miR-93, miR-373, and miR-17-5p negatively regulate the expression of TBP2 in lung cancer. Front Oncol 10:526
Li Y, Wang YW, Chen X, Ma RR, Guo XY, Liu HT, Jiang SJ, Wei JM, Gao P (2020c) MicroRNA-4472 promotes tumor proliferation and aggressiveness in breast cancer by targeting RGMA and inducing EMT. Clin Breast Cancer 20:e113–e126
Li YK, Zhong W, Zhu M, Li MB, Yang ZW (2020d) miR-185 inhibits prostate cancer angiogenesis induced by the nodal/ALK4 pathway. BMC Urol 20:49
Li XM, Yu M, Yang CY (2020e) YY1-mediated overexpression of long noncoding RNA MCM3AP-AS1 accelerates angiogenesis and progression in lung cancer by targeting miR-340-5p/KPNA4 axis. J Cell Biochem 121:2258–2267
Li Z, Li LX, Diao YJ, Wang J, Ye Y, Hao XK (2021) Identification of urinary exosomal miRNAs for the non-invasive diagnosis of prostate cancer. Cancer Manag Res 13:25–35
Li RD, Qu H, Wang SB, Chater JM, Wang XS, Cui YR, Yu L, Zhou R, Jia Q, Traband R, Wang MY, Xie WB, Yuan DB, Zhu JG, Zhong WD, Jia ZY (2022) CancerMIRNome: an interactive analysis and visualization database for miRNome profiles of human cancer. Nucleic Acids Res 50:D1139–D1146
Liang C, Zhang X, Wang HM, Liu XM, Zhang XJ, Zheng B, Qian GR, Ma ZL (2017) MicroRNA-18a-5p functions as an oncogene by directly targeting IRF2 in lung cancer. Cell Death Dis 8: e2764
Liang YK, Lin HY, Dou XW, Chen M, Wei XL, Zhang YQ, Wu Y, Chen CF, Bai JW, Xiao YS, Qi YZ, Kruyt FAE, Zhang GJ (2018) MiR-221/222 promote epithelial-mesenchymal transition by targeting Notch3 in breast cancer cell lines. NPJ Breast Cancer 4:20
Lin CP, He L (2017) Noncoding RNAs in cancer development. Annu Rev Cancer Biol 1(1): 163–184
Lin JT, Tsai KW (2021) Circulating miRNAs act as diagnostic biomarkers for bladder cancer in urine. Int J Mol Sci 22:4278
Lin CW, Chang YL, Chang YC, Lin JC, Chen CC, Pan SH, Wu CT, Chen HY, Yang SC, Hong TM, Yang PC (2013) MicroRNA-135b promotes lung cancer metastasis by regulating multiple targets in the Hippo pathway and LZTS1. Nat Commun 4:1877

Ling B, Wang GX, Long G, Qiu JH, Hu ZL (2012) Tumor suppressor miR-22 suppresses lung cancer cell progression through post-transcriptional regulation of ErbB3. J Cancer Res Clin Oncol 138:1355–1361

Liu H, Zhu L, Liu B, Yang L, Meng X, Zhang W, Ma Y, Xiao H (2012) Genome-wide microRNA profiles identify miR-378 as a serum biomarker for early detection of gastric cancer. Cancer Lett 316:196–203

Liu HX, Huang J, Peng J, Wu XX, Zhang Y, Zhu WL, Guo LL (2015a) Upregulation of the inwardly rectifying potassium channel Kir2.1 (KCNJ2) modulates multidrug resistance of small-cell lung cancer under the regulation of miR-7 and the Ras/MAPK pathway. Mol Cancer 14:59

Liu X, Ge X, Zhang Z, Zhang X, Chang J, Wu Z, Tang W, Gan L, Sun M, Li J (2015b) MicroRNA-940 promotes tumor cell invasion and metastasis by downregulating ZNF24 in gastric cancer. Oncotarget 6:25418–25428

Liu L, Meng T, Wang QS, Jin HZ, Sun ZQ, Jin B, Fang F, Wang HJ (2016) Association of Beclin-1 and microRNA-30a expression with the severity and treatment response of colorectal cancer. Genet Mol Res 15. https://doi.org/10.4238/gmr.15027704

Liu FY, Cai YJ, Rong XX, Chen JZ, Zheng DY, Chen L, Zhang JY, Luo RC, Zhao P, Ruan J (2017) MiR-661 promotes tumor invasion and metastasis by directly inhibiting RB1 in non small cell lung cancer. Mol Cancer 16:122

Liu X, Pan B, Sun L, Chen X, Zeng K, Hu X, Xu T, Xu M, Wang S (2018a) Circulating exosomal miR-27a and miR-130a act as novel diagnostic and prognostic biomarkers of colorectal cancer. Cancer Epidemiol Biomarkers Prev 27:746–754

Liu TC, Wu XP, Chen T, Luo ZW, Hu XH (2018b) Downregulation of DNMT3A by miR-708-5p inhibits lung cancer stem cell-like phenotypes through repressing Wnt/beta-catenin signaling. Clin Cancer Res 24:1748–1760

Liu X, Xu X, Pan B, He B, Chen X, Zeng K, Xu M, Pan Y, Sun H, Xu T, Hu X, Wang S (2019) Circulating miR-1290 and miR-320d as novel diagnostic biomarkers of human colorectal cancer. J Cancer 10:43–50

Lou W, Ding B, Zhong G, Du C, Fan W, Fu P (2019) Dysregulation of pseudogene/lncRNA-hsa-miR-363-3p-SPOCK2 pathway fuels stage progression of ovarian cancer. Aging (Albany NY) 11:11416–11439

Lou N, Liu G, Pan Y (2020) Long noncoding RNA ANRIL as a novel biomarker in human cancer. Future Oncol 16:2981–2995

Lu L, Luo F, Liu Y, Liu X, Shi L, Lu X, Liu Q (2015) Posttranscriptional silencing of the lncRNA MALAT1 by miR-217 inhibits the epithelial-mesenchymal transition via enhancer of zeste homolog 2 in the malignant transformation of HBE cells induced by cigarette smoke extract. Toxicol Appl Pharmacol 289:276–285

Lu MH, Tang B, Zeng S, Hu CJ, Xie R, Wu YY, Wang SM, He FT, Yang SM (2016) Long noncoding RNA BC032469, a novel competing endogenous RNA, upregulates hTERT expression by sponging miR-1207-5p and promotes proliferation in gastric cancer. Oncogene 35: 3524–3534

Lu S, Kong H, Hou Y, Ge D, Huang W, Ou J, Yang D, Zhang L, Wu G, Song Y, Zhang X, Zhai C, Wang Q, Zhu H, Wu Y, Bai C (2018) Two plasma microRNA panels for diagnosis and subtype discrimination of lung cancer. Lung Cancer 123:44–51

Lu Q, Chen YC, Sun D, Wang SK, Ding K, Liu MY, Zhang Y, Miao YJ, Liu HC, Zhou FH (2019) MicroRNA-181a functions as an oncogene in gastric cancer by targeting caprin-1. Front Pharmacol 9:1565

Ma L, Reinhardt F, Pan E, Soutschek J, Bhat B, Marcusson EG, Teruya-Feldstein J, Bell GW, Weinberg RA (2010) Therapeutic silencing of miR-10b inhibits metastasis in a mouse mammary tumor model. Nat Biotechnol 28:341–U67

Ma C, Nong K, Zhu H, Wang W, Huang X, Yuan Z, Ai K (2014) H19 promotes pancreatic cancer metastasis by derepressing let-7′s suppression on its target HMGA2-mediated EMT. Tumour Biol 35:9163–9169

Ma MZ, Chu BF, Zhang Y, Weng MZ, Qin YY, Gong W, Quan ZW (2015) Long non-coding RNA CCAT1 promotes gallbladder cancer development via negative modulation of miRNA-218-5p. Cell Death Dis 6:e1583

Ma ZW, Luo YZ, Qiu MX (2017) miR-143 induces the apoptosis of prostate cancer LNCap cells by suppressing Bcl-2 expression. Med Sci Monit 23:359–365

Ma YL, Yan F, Wei WJ, Deng J, Li L, Liu L, Sun JH (2019) MicroRNA-598 inhibits the growth and maintenance of gastric cancer stem-like cells by down-regulating RRS1. Cell Cycle 18: 2757–2769

Marrugo-Ramirez J, Mir M, Samitier J (2018) Blood-based cancer biomarkers in liquid biopsy: a promising non-invasive alternative to tissue biopsy. Int J Mol Sci 19:2877

Matin F, Jeet V, Moya L, Selth LA, Chambers S, Australian Prostate Cancer BioResource, Clements JA, Batra J (2018) A plasma biomarker panel of four microRNAs for the diagnosis of prostate cancer. Sci Rep 8:6653

Mei Y, Li K, Zhang ZC, Li MM, Yang H, Wang H, Huang XM, Li XY, Shi SH, Yang HJ (2021) miR-33b-3p acts as a tumor suppressor by targeting DOCK4 in prostate cancer. Front Oncol 11: 740452

Mendez MG, Kojima S, Goldman RD (2010) Vimentin induces changes in cell shape, motility, and adhesion during the epithelial to mesenchymal transition. FASEB J 24:1838–1851

Meng W, McElroy JP, Volinia S, Palatini J, Warner S, Ayers LW, Palanichamy K, Chakravarti A, Lautenschlaeger T (2013) Comparison of microRNA deep sequencing of matched formalin-fixed paraffin-embedded and fresh frozen cancer tissues. PLoS One 8:e64393

Menon A, Abd-Aziz N, Khalid K, Poh CL, Naidu R (2022) miRNA: a promising therapeutic target in cancer. Int J Mol Sci 23:11502

Mitchell PS, Parkin RK, Kroh EM, Fritz BR, Wyman SK, Pogosova-Agadjanyan EL, Peterson A, Noteboom J, O'Briant KC, Allen A, Lin DW, Urban N, Drescher CW, Knudsen BS, Stirewalt DL, Gentleman R, Vessella RL, Nelson PS, Martin DB, Tewari M (2008) Circulating microRNAs as stable blood-based markers for cancer detection. Proc Natl Acad Sci U S A 105:10513–10518

Ng EKO, Tsang WP, Ng SSM, Jin HC, Yu J, Li JJ, Rocken C, Ebert MPA, Kwok TT, Sung JJY (2009) MicroRNA-143 targets DNA methyltransferases 3A in colorectal cancer. Br J Cancer 101:699–706

Ni Y, Li C, Bo C, Zhang B, Liu Y, Bai X, Cui B, Han P (2020) LncRNA EGOT regulates the proliferation and apoptosis of colorectal cancer by miR-33b-5p/CROT axis. Biosci Rep BSR20193893

Niveditha D, Jasoria M, Narayan J, Majumder S, Mukherjee S, Chowdhury R, Chowdhury S (2020) Common and unique microRNAs in multiple carcinomas regulate similar network of pathways to mediate cancer progression. Sci Rep 10:2331

O'Brien J, Hayder H, Zayed Y, Peng C (2018) Overview of microRNA biogenesis, mechanisms of actions, and circulation. Front Endocrinol (Lausanne) 9:402

O'Brien SJ, Fiechter C, Burton J, Hallion J, Paas M, Patel A, Patel A, Rochet A, Scheurlen K, Gardner S, Eichenberger M, Sarojini H, Srivastava S, Rai S, Kalbfleisch T, Polk HC Jr, Galandiuk S (2021) Long non-coding RNA ZFAS1 is a major regulator of epithelial-mesenchymal transition through miR-200/ZEB1/E-cadherin, vimentin signaling in colon adenocarcinoma. Cell Death Dis 7:61

Orom UA, Nielsen FC, Lund AH (2008) MicroRNA-10a binds the 5′UTR of ribosomal protein mRNAs and enhances their translation. Mol Cell 30:460–471

Othman N, Nagoor NH (2019) Overexpression of miR-361-5p plays an oncogenic role in human lung adenocarcinoma through the regulation of SMAD2. Int J Oncol 54:306–314

Otmani K, Lewalle P (2021) Tumor suppressor miRNA in cancer cells and the tumor microenvironment: mechanism of deregulation and clinical implications. Front Oncol 11:708765

Pakravan K, Babashah S, Sadeghizadeh M, Mowla SJ, Mossahebi-Mohammadi M, Ataei F, Dana N, Javan M (2017) MicroRNA-100 shuttled by mesenchymal stem cell-derived exosomes suppresses in vitro angiogenesis through modulating the mTOR/HIF-1alpha/VEGF signaling axis in breast cancer cells. Cell Oncol (Dordr) 40:457–470

Pan Z, Miao L (2020) Serum microRNA-592 serves as a novel potential biomarker for early diagnosis of colorectal cancer. Oncol Lett 20:1119–1126

Pan J, Zhou C, Zhao X, He J, Tian H, Shen W, Han Y, Chen J, Fang S, Meng X, Jin X, Gong Z (2018) A two-miRNA signature (miR-33a-5p and miR-128-3p) in whole blood as potential biomarker for early diagnosis of lung cancer. Sci Rep 8:16699

Paraskevopoulou MD, Hatzigeorgiou AG (2016) Analyzing MiRNA-LncRNA interactions. Methods Mol Biol 1402:271–286

Park Y, Kim J (2019) Regulation of IL-6 signaling by miR-125a and let-7e in endothelial cells controls vasculogenic mimicry formation of breast cancer cells. BMB Rep 52:214–219

Park DH, Jeon HS, Lee SY, Choi YY, Lee HW, Yoon S, Lee JC, Yoon YS, Kim DS, Na MJ, Kwon SJ, Kim DS, Kang J, Park JY, Son JW (2015) MicroRNA-146a inhibits epithelial mesenchymal transition in non-small cell lung cancer by targeting insulin receptor substrate 2. Int J Oncol 47: 1545–1553

Peng B, Li C, He L, Tian M, Li X (2020) miR-660-5p promotes breast cancer progression through down-regulating TET2 and activating PI3K/AKT/mTOR signaling. Braz J Med Biol Res 53: e9740

Porzycki P, Ciszkowicz E, Semik M, Tyrka M (2018) Combination of three miRNA (hsa-miR-141, hsa-miR-21, and hsa-miR-375) as potential diagnostic tool for prostate cancer recognition. Int Urol Nephrol 50:1619–1626

Qi JH, Ebrahem Q, Moore N, Murphy G, Claesson-Welsh L, Bond M, Baker A, Anand-Apte B (2003) A novel function for tissue inhibitor of metalloproteinases-3 (TIMP3): inhibition of angiogenesis by blockage of VEGF binding to VEGF receptor-2. Nat Med 9:407–415

Qiu X, Zhang J, Shi W, Liu S, Kang M, Chu H, Wu D, Tong N, Gong W, Tao G, Zhao Q, Qiang F, Zhu H, Wu Q, Wang M, Zhang Z (2016) Circulating microRNA-26a in plasma and its potential diagnostic value in gastric cancer. PLoS One 11:e0151345

Rajagopal T, Talluri S, Akshaya RL, Dunna NR (2020) HOTAIR LncRNA: a novel oncogenic propellant in human cancer. Clin Chim Acta 503:1–18

Ren Y, Zhang H, Jiang P (2018) MicroRNA-382 inhibits cell growth and migration in colorectal cancer by targeting SP1. Biol Res 51:51

Ren Z, Liu XC, Si YR, Yang DS (2020) Long non-coding RNA DDX11-AS1 facilitates gastric cancer progression by regulating miR-873-5p/SPC18 axis. Artif Cells Nanomed Biotechnol 48: 572–583

Reyes-Gonzalez JM, Vivas-Mejia PE (2021) c-MYC and epithelial ovarian cancer. Front Oncol 11: 601512

Rosenfeld N, Aharonov R, Meiri E, Rosenwald S, Spector Y, Zepeniuk M, Benjamin H, Shabes N, Tabak S, Levy A, Lebanony D, Goren Y, Silberschein E, Targan N, Ben-Ari A, Gilad S, Sion-Vardy N, Tobar A, Feinmesser M, Kharenko O, Nativ O, Nass D, Perelman M, Yosepovich A, Shalmon B, Polak-Charcon S, Fridman E, Avniel A, Bentwich I, Bentwich Z, Cohen D, Chajut A, Barshack I (2008) MicroRNAs accurately identify cancer tissue origin. Nat Biotechnol 26:462–469

Ryu SW, Lee SJ, Park MY, Jun JI, Jung YK, Kim E (2003) Fas-associated factor 1, FAF1, is a member of Fas death-inducing signaling complex. J Biol Chem 278:24003–24010

Saito T, Kurashige J, Nambara S, Komatsu H, Hirata H, Ueda M, Sakimura S, Uchi R, Takano Y, Shinden Y, Iguchi T, Eguchi H, Ehata S, Murakami K, Sugimachi K, Mimori K (2015) A long non-coding RNA activated by transforming growth factor-beta is an independent prognostic marker of gastric cancer. Ann Surg Oncol 22(suppl 3):S915–S922

Salinas-Vera YM, Marchat LA, Garcia-Vazquez R, Gonzalez De La Rosa CH, Castaneda-Saucedo-E, Tito NN, Flores CP, Perez-Plasencia C, Cruz-Colin JL, Carlos-Reyes A, Lopez-Gonzalez JS, Alvarez-Sanchez ME, Lopez-Camarillo C (2018) Cooperative multi-targeting of signaling networks by angiomiR-204 inhibits vasculogenic mimicry in breast cancer cells. Cancer Lett 432:17–27

Samson J, Cronin S, Dean K (2018) BC200 (BCYRN1) - the shortest, long, non-coding RNA associated with cancer. Noncoding RNA Res 3:131–143

Santiago K, Chen Wongworawat Y, Khan S (2020) Differential microRNA-signatures in thyroid cancer subtypes. J Oncol 2020:2052396
Serva A, Knapp B, Tsai YT, Claas C, Lisauskas T, Matula P, Harder N, Kaderali L, Rohr K, Erfle H, Eils R, Braga V, Starkuviene V (2012) miR-17-5p regulates endocytic trafficking through targeting TBC1D2/Armus. PloS One 7:e52555
Shahabi A, Naghili B, Ansarin K, Montazeri M, Dadashpour M, Zarghami N (2021) Let-7d and miR-185 impede epithelial-mesenchymal transition by downregulating Rab25 in breast cancer. Asian Pac J Cancer Prev 22:305–313
Shen H, Zhu F, Liu JY, Xu TP, Pei D, Wang R, Qian YY, Li Q, Wang L, Shi ZM, Zheng JT, Chen QD, Jiang BH, Shu YQ (2014a) Alteration in Mir-21/PTEN expression modulates gefitinib resistance in non-small cell lung cancer. PloS One 9:e103305
Shen ZY, Zhang ZZ, Liu H, Zhao EH, Cao H (2014b) miR-375 inhibits the proliferation of gastric cancer cells by repressing ERBB2 expression. Exp Ther Med 7:1757–1761
Shermane Lim YW, Xiang X, Garg M, Le MT, Li-Ann Wong A, Wang L, Goh BC (2021) The double-edged sword of H19 lncRNA: insights into cancer therapy. Cancer Lett 500:253–262
Shi DB, Wang YW, Xing AY, Gao JW, Zhang H, Guo XY, Gao P (2015) C/EBPalpha-induced miR-100 expression suppresses tumor metastasis and growth by targeting ZBTB7A in gastric cancer. Cancer Lett 369:376–385
Shi Y, Wang Z, Zhu X, Chen L, Ma Y, Wang J, Yang X, Liu Z (2020) Exosomal miR-1246 in serum as a potential biomarker for early diagnosis of gastric cancer. Int J Clin Oncol 25:89–99
Shimomura A, Shiino S, Kawauchi J, Takizawa S, Sakamoto H, Matsuzaki J, Ono M, Takeshita F, Niida S, Shimizu C, Fujiwara Y, Kinoshita T, Tamura K, Ochiya T (2016) Novel combination of serum microRNA for detecting breast cancer in the early stage. Cancer Sci 107:326–334
Shin VY, Siu JM, Cheuk I, Ng EK, Kwong A (2015) Circulating cell-free miRNAs as biomarker for triple-negative breast cancer. Br J Cancer 112:1751–1759
Sohel MMH (2020) Circulating microRNAs as biomarkers in cancer diagnosis. Life Sci 248: 117473
Sokilde R, Vincent M, Moller AK, Hansen A, Hoiby PE, Blondal T, Nielsen BS, Daugaard G, Moller S, Litman T (2014) Efficient identification of miRNAs for classification of tumor origin. J Mol Diagn 16:106–115
Srivastava A, Goldberger H, Dimtchev A, Ramalinga M, Chijioke J, Marian C, Oermann EK, Uhm S, Kim JS, Chen LN, Li X, Berry DL, Kallakury BV, Chauhan SC, Collins SP, Suy S, Kumar D (2013) MicroRNA profiling in prostate cancer--the diagnostic potential of urinary miR-205 and miR-214. PLoS One 8:e76994
Su C, Huang DP, Liu JW, Liu WY, Cao YO (2019) miR-27a-3p regulates proliferation and apoptosis of colon cancer cells by potentially targeting BTG1. Oncol Lett 18:2825–2834
Sun D, Fan XH (2019) LncRNA SNHG12 accelerates the progression of ovarian cancer via absorbing miRNA-129 to upregulate SOX4. Eur Rev Med Pharmacol Sci 23:2345–2352
Sun W, Wang XJ, Li JL, You CY, Lu P, Feng HJ, Kong Y, Zhang HY, Liu YQ, Jiao RH, Chen X, Ba Y (2018a) MicroRNA-181a promotes angiogenesis in colorectal cancer by targeting SRCIN1 to promote the SRC/VEGF signaling pathway. Cell Death Dis 9:438
Sun Q, Weng DH, Li KZ, Li S, Bai XY, Fang C, Luo DF, Wu P, Chen G, Wei JC (2018b) MicroRNA-139-5P inhibits human prostate cancer cell proliferation by targeting Notch1. Oncol Lett 16:793–800
Sun SN, Hang TY, Zhang BY, Zhu L, Wu Y, Lv XW, Huang Q, Yao HH (2019) miRNA-708 functions as a tumor suppressor in colorectal cancer by targeting ZEB1 through Akt/mTOR signaling pathway. Am J Transl Res 11:5338–5356
Sun B, Liu C, Li H, Zhang L, Luo G, Liang S, Lu M (2020) Research progress on the interactions between long non-coding RNAs and microRNAs in human cancer. Oncol Lett 19:595–605
Sun XB, Chen YW, Yao QS, Chen XH, He M, Chen CB, Yang Y, Gong XX, Huang L (2021) MicroRNA-144 suppresses prostate cancer growth and metastasis by targeting EZH2. Technol Cancer Res Treat 20:1533033821989817

Switlik WZ, Karbownik MS, Suwalski M, Kozak J, Szemraj J (2019) Serum miR-210-3p as a potential noninvasive biomarker of lung adenocarcinoma: a preliminary study. Genet Test Mol Biomarkers 23:353–358

Tang S, Cheng J, Yao Y, Lou C, Wang L, Huang X, Zhang Y (2020) Combination of four serum exosomal MiRNAs as novel diagnostic biomarkers for early-stage gastric cancer. Front Genet 11:237

Taniguchi T, Ogasawara K, Takaoka A, Tanaka N (2001) IRF family of transcription factors as regulators of host defense. Annu Rev Immunol 19:623–655

Turnbull C, Sud A, Houlston RS (2018) Cancer genetics, precision prevention and a call to action. Nat Genet 50:1212–1218

Tutar L, Ozgur A, Tutar Y (2018) Involvement of miRNAs and pseudogenes in cancer. Methods Mol Biol 1699:45–66

Uhlen M, Zhang C, Lee S, Sjostedt E, Fagerberg L, Bidkhori G, Benfeitas R, Arif M, Liu ZT, Edfors F, Sanli K, von Feilitzen K, Oksvold P, Lundberg E, Hober S, Nilsson P, Mattsson J, Schwenk JM, Brunnstrom H, Glimelius B, Sjoblom T, Edqvist PH, Djureinovic D, Micke P, Lindskog C, Mardinoglu A, Ponten F (2017) A pathology atlas of the human cancer transcriptome. Science 357:660

Vandewalle C, van Roy F, Berx G (2009) The role of the ZEB family of transcription factors in development and disease. Cell Mol Life Sci 66:773–787

Viticchie G, Lena AM, Latina A, Formosa A, Gregersen LH, Lund AH, Bernardini S, Mauriello A, Miano R, Spagnoli LG, Knight RA, Candi E, Melino G (2011) MiR-203 controls proliferation, migration and invasive potential of prostate cancer cell lines. Cell Cycle 10:1121–1131

Vo JN, Cieslik M, Zhang YJ, Shukla S, Xiao LB, Zhang YP, Wu YM, Dhanasekaran SM, Engelke CG, Cao XH, Robinson DR, Nesvizhskii AI, Chinnaiyan AM (2019) The landscape of circular RNA in cancer. Cell 176:869

Vychytilova-Faltejskova P, Radova L, Sachlova M, Kosarova Z, Slaba K, Fabian P, Grolich T, Prochazka V, Kala Z, Svoboda M, Kiss I, Vyzula R, Slaby O (2016) Serum-based microRNA signatures in early diagnosis and prognosis prediction of colon cancer. Carcinogenesis 37:941–950

Wang YX, Qin HY (2018) miR-338-3p targets RAB23 and suppresses tumorigenicity of prostate cancer cells. Am J Cancer Res 8:2564–2574

Wang XT, Arai S, Song XY, Reichart D, Du K, Pascual G, Tempst P, Rosenfeld MG, Glass CK, Kurokawa R (2008) Induced ncRNAs allosterically modify RNA-binding proteins in cis to inhibit transcription. Nature 454:126–130

Wang K, Zhang S, Weber J, Baxter D, Galas DJ (2010) Export of microRNAs and microRNA-protective protein by mammalian cells. Nucleic Acids Res 38:7248–7259

Wang S, Xiang J, Li Z, Lu S, Hu J, Gao X, Yu L, Wang L, Wang J, Wu Y, Chen Z, Zhu H (2015) A plasma microRNA panel for early detection of colorectal cancer. Int J Cancer 136:152–161

Wang SZ, Ma GX, Zhu HX, Lv CY, Chu HY, Tong N, Wu DM, Qiang FL, Gong WD, Zhao QH, Tao GQ, Zhou JW, Zhang ZD, Wang ML (2016) miR-107 regulates tumor progression by targeting NF1 in gastric cancer. Sci Rep 6:36531

Wang Y, Li YQ, Wu B, Shi C, Li C (2017) MicroRNA-661 promotes non-small cell lung cancer progression by directly targeting RUNX3. Mol Med Rep 16:2113–2120

Wang H, Peng R, Wang J, Qin Z, Xue L (2018a) Circulating microRNAs as potential cancer biomarkers: the advantage and disadvantage. Clin Epigenetics 10:59

Wang ZH, Yang B, Zhang M, Guo WW, Wu ZY, Wang Y, Jia L, Li S, Xie W, Yang D, Cancer Genome Atlas Research Network (2018b) lncRNA epigenetic landscape analysis identifies EPIC1 as an oncogenic lncRNA that interacts with MYC and promotes cell-cycle progression in cancer. Cancer Cell 33:706

Wang Y, Xu W, Wang Y, Xu X, Lv S, Dong X (2019a) miR-17-5p promotes migration and invasion in breast cancer cells by repressing netrin 4. Int J Clin Exp Pathol 12:1649–1657

Wang Y, Zhang S, Dang S, Fang X, Liu M (2019b) Overexpression of microRNA-216a inhibits autophagy by targeting regulated MAP1S in colorectal cancer. Onco Targets Ther 12:4621–4629

Wang Y, Li WJ, Chen XY, Li Y, Wen PH, Xu F (2019c) MIR210HG predicts poor prognosis and functions as an oncogenic lncRNA in hepatocellular carcinoma. Biomed Pharmacother 111:1297–1301

Wang XK, Lan ZX, He J, Lai QH, Yao X, Li QY, Liu YF, Lai HS, Gu CC, Yan Q, Fang YX, Zhang Y, Li AM, Liu SD (2019d) LncRNA SNHG6 promotes chemoresistance through ULK1-induced autophagy by sponging miR-26a-5p in colorectal cancer cells. Cancer Cell Int 19:234

Wang C, Huang Y, Zhang J, Fang Y (2020a) MiRNA-339-5p suppresses the malignant development of gastric cancer via targeting ALKBH1. Exp Mol Pathol 115:104449

Wang W, Hu WT, Wang Y, An Y, Song L, Shang PF, Yue ZJ (2020b) Long non-coding RNA UCA1 promotes malignant phenotypes of renal cancer cells by modulating the miR-182-5p/DLL4 axis as a ceRNA. Mol Cancer 19:18

Wang MY, Zhao MN, Guo QM, Lou JT, Wang L (2021a) Non-small cell lung cancer cell-derived exosomal miR-17-5p promotes osteoclast differentiation by targeting PTEN. Exp Cell Res 408:112834

Wang J, Zhu YM, Ni S, Liu SY (2021b) LncRNA GAS5 suppressed proliferation and promoted apoptosis in laryngeal squamous cell carcinoma by targeting MiR-26a-5p and modifying ULK2. Cancer Manage Res 13:871–887

Weber JA, Baxter DH, Zhang S, Huang DY, Huang KH, Lee MJ, Galas DJ, Wang K (2010) The microRNA spectrum in 12 body fluids. Clin Chem 56:1733–1741

Wei AW, Li LF (2017) Long non-coding RNA SOX21-AS1 sponges miR-145 to promote the tumorigenesis of colorectal cancer by targeting MYO6. Biomed Pharmacother 96:953–959

Wiemer EA (2007) The role of microRNAs in cancer: no small matter. Eur J Cancer 43:1529–1544

Wu JG, Wang JJ, Jiang X, Lan JP, He XJ, Wang HJ, Ma YY, Xia YJ, Ru GQ, Ma J, Zhao ZS, Zhou R (2015) MiR-125b promotes cell migration and invasion by targeting PPP1CA-Rb signal pathways in gastric cancer, resulting in a poor prognosis. Gastric Cancer 18:729–739

Wu ZH, Cai XH, Huang CG, Xu J, Liu A (2016) miR-497 suppresses angiogenesis in breast carcinoma by targeting HIF-1 alpha. Oncol Rep 35:1696–1702

Wu W, He K, Guo Q, Chen J, Zhang M, Huang K, Yang D, Wu L, Deng Y, Luo X, Yu H, Ding Q, Xiang G (2019a) SSRP1 promotes colorectal cancer progression and is negatively regulated by miR-28-5p. J Cell Mol Med 23:3118–3129

Wu M, Huang Y, Chen T, Wang W, Yang S, Ye Z, Xi X (2019b) LncRNA MEG3 inhibits the progression of prostate cancer by modulating miR-9-5p/QKI-5 axis. J Cell Mol Med 23:29–38

Xie F, Hosany S, Zhong S, Jiang Y, Zhang F, Lin L, Wang X, Gao S, Hu X (2017) MicroRNA-193a inhibits breast cancer proliferation and metastasis by downregulating WT1. PLoS One 12:e0185565

Xie Y, Zhao J, Liang YL, Chen M, Luo YH, Cui XB, Jiang B, Peng L, Wang XY (2019) MicroRNA-10b controls the metastasis and proliferation of colorectal cancer cells by regulating Kruppel-like factor 4. Artif Cells Nanomed Biotechnol 47:1722–1729

Xin H, Wang C, Liu Z (2019) miR-196a-5p promotes metastasis of colorectal cancer via targeting IkappaBalpha. BMC Cancer 19:30

Xu Y, Li J, Wang P, Zhang Z, Wang X (2019) LncRNA HULC promotes lung squamous cell carcinoma by regulating PTPRO via NF-kappaB. J Cell Biochem 120:19415–19421

Xu GC, Meng Y, Wang LH, Dong B, Peng FF, Liu ST, Li SK, Liu T (2021a) miRNA-214-5p inhibits prostate cancer cell proliferation by targeting SOX4. World J Surg Oncol 19:338

Xu H, He Y, Lin L, Li M, Zhou Z, Yang Y (2021b) MiR-1207-5p targets PYCR1 to inhibit the progression of prostate cancer. Biochem Biophys Res Commun 575:56–64

Xu J, Li X, Zhang PR, Luo J, Mou EX, Liu SW (2022) miR-143-5p suppresses breast cancer progression by targeting the HIF-1 alpha-related GLUT1 pathway. Oncol Lett 23:147

Xue M, Pang H, Li X, Li H, Pan J, Chen W (2016) Long non-coding RNA urothelial cancer-associated 1 promotes bladder cancer cell migration and invasion by way of the hsa-miR-145-ZEB1/2-FSCN1 pathway. Cancer Sci 107:18–27

Yang Z, Zhou L, Wu LM, Lai MC, Xie HY, Zhang F, Zheng SS (2011) Overexpression of long non-coding RNA HOTAIR predicts tumor recurrence in hepatocellular carcinoma patients following liver transplantation. Ann Surg Oncol 18:1243–1250

Yang J, Zhang Z, Chen C, Liu Y, Si Q, Chuang TH, Li N, Gomez-Cabrero A, Reisfeld RA, Xiang R, Luo Y (2014) MicroRNA-19a-3p inhibits breast cancer progression and metastasis by inducing macrophage polarization through downregulated expression of Fra-1 proto-oncogene. Oncogene 33:3014–3023

Yang H, Luo JW, Liu ZG, Zhou R, Luo H (2015) MicroRNA-138 regulates DNA damage response in small cell lung cancer cells by directly targeting H2AX. Cancer Invest 33:126–136

Yang X, Zhang Q, Zhang M, Su W, Wang Z, Li Y, Zhang J, Beer DG, Yang S, Chen G (2019) Serum microRNA signature is capable of early diagnosis for non-small cell lung cancer. Int J Biol Sci 15:1712–1722

Yang HB, Zhang XG, Zhu L, Yang Y, Yin XM (2021) YY1-induced lncRNA PART1 enhanced resistance of ovarian cancer cells to cisplatin by regulating miR-512-3p/CHRAC1 axis. DNA Cell Biol 40:821–832

Yao Q, Chen Y, Zhou X (2019) The roles of microRNAs in epigenetic regulation. Curr Opin Chem Biol 51:11–17

Yap KL, Li SD, Munoz-Cabello AM, Raguz S, Zeng L, Mujtaba S, Gil J, Walsh MJ, Zhou MM (2010) Molecular interplay of the noncoding RNA ANRIL and methylated histone H3 lysine 27 by polycomb CBX7 in transcriptional silencing of INK4a. Mol Cell 38:662–674

Ye MF, Lin D, Li WJ, Xu HP, Zhang J (2020) MiR-26a-5p serves as an oncogenic microRNA in non-small cell lung cancer by targeting FAF1. Cancer Manage Res 12:7131–7142

Yerukala Sathipati S, Ho SY (2018) Identifying a miRNA signature for predicting the stage of breast cancer. Sci Rep 8:16138

Yi Z, Fu Y, Zhao S, Zhang X, Ma C (2010) Differential expression of miRNA patterns in renal cell carcinoma and nontumorous tissues. J Cancer Res Clin Oncol 136:855–862

Yong FL, Law CW, Wang CW (2013) Potentiality of a triple microRNA classifier: miR-193a-3p, miR-23a and miR-338-5p for early detection of colorectal cancer. BMC Cancer 13:280

Yu X, Zhang X, Bi T, Ding Y, Zhao J, Wang C, Jia T, Han D, Guo G, Wang B, Jiang J, Cui S (2013) MiRNA expression signature for potentially predicting the prognosis of ovarian serous carcinoma. Tumour Biol 34:3501–3508

Yu JJ, Lu YY, Cui D, Li EH, Zhu YP, Zhao YY, Zhao FJ, Xia SJ (2014) miR-200b suppresses cell proliferation, migration and enhances chemosensitivity in prostate cancer by regulating Bmi-1. Oncol Rep 31:910–918

Yu Y, Chen Q, Zhang X, Yang J, Lin K, Ji C, Xu A, Yang L, Miao L (2020a) Long noncoding RNA ANRIL promotes the malignant progression of cholangiocarcinoma by epigenetically repressing ERRFI1 expression. Cancer Sci 111:2297–2309

Yu L, Shao X, Huo L, Zhang T (2020b) Long non-coding RNA (lncRNA) metastasis-associated lung adenocarcinoma transcript 1 (MALAT1) promotes cell proliferation and migration by regulating miR-143-3p and MAGE Family Member A9 (MAGEA9) in oral squamous cell carcinoma. Med Sci Monit 26:e924187

Yuan C, Zhu X, Han Y, Song C, Liu C, Lu S, Zhang M, Yu F, Peng Z, Zhou C (2016) Elevated HOXA1 expression correlates with accelerated tumor cell proliferation and poor prognosis in gastric cancer partly via cyclin D1. J Exp Clin Cancer Res 35:15

Yuan C, Zhang Y, Tu W, Guo Y (2019) Integrated miRNA profiling and bioinformatics analyses reveal upregulated miRNAs in gastric cancer. Oncol Lett 18:1979–1988

Zha TZ, Hu BS, Yu HF, Tan YF, Zhang Y, Zhang K (2012) Overexpression of HOXA1 correlates with poor prognosis in patients with hepatocellular carcinoma. Tumour Biol 33:2125–2134

Zhang B, Pan X, Cobb GP, Anderson TA (2007) microRNAs as oncogenes and tumor suppressors. Dev Biol 302:1–12

Zhang CZ, Zhang JX, Zhang AL, Wang YY, Han L, You YP, Pu PY, Kang CS (2010a) PUMA is a novel target of miR-221/222 in human epithelial cancers. Int J Oncol 37:1621–1626
Zhang X, Gejman R, Mahta A, Zhong Y, Rice KA, Zhou YL, Cheunsuchon P, Louis DN, Klibanski A (2010b) Maternally expressed gene 3, an imprinted noncoding RNA gene, is associated with meningioma pathogenesis and progression. Cancer Res 70:2350–2358
Zhang P, Sun Y, Ma L (2015a) ZEB1: at the crossroads of epithelial-mesenchymal transition, metastasis and therapy resistance. Cell Cycle 14:481–487
Zhang HH, Tang JL, Li C, Kong JL, Wang JY, Wu YH, Xu EP, Lai MD (2015b) MiR-22 regulates 5-FU sensitivity by inhibiting autophagy and promoting apoptosis in colorectal cancer cells. Cancer Lett 356:781–790
Zhang D, Xiao YF, Zhang JW, Xie R, Hu CJ, Tang B, Wang SM, Wu YY, Hao NB, Yang SM (2015c) miR-1182 attenuates gastric cancer proliferation and metastasis by targeting the open reading frame of hTERT. Cancer Lett 360:151–159
Zhang J, Liu D, Feng Z, Mao J, Zhang C, Lu Y, Li J, Zhang Q, Li Q, Li L (2016a) MicroRNA-138 modulates metastasis and EMT in breast cancer cells by targeting vimentin. Biomed Pharmacother 77:135–141
Zhang H, Duan J, Qu Y, Deng T, Liu R, Zhang L, Bai M, Li J, Ning T, Ge S, Wang X, Wang Z, Fan Q, Li H, Ying G, Huang D, Ba Y (2016b) Onco-miR-24 regulates cell growth and apoptosis by targeting BCL2L11 in gastric cancer. Protein Cell 7:141–151
Zhang L, Shan X, Wang J, Zhu J, Huang Z, Zhang H, Zhou X, Cheng W, Shu Y, Zhu W, Liu P (2017a) A three-microRNA signature for lung squamous cell carcinoma diagnosis in Chinese male patients. Oncotarget 8:86897–86907
Zhang GX, Zhang W, Li BJ, Stringer-Reasor E, Chu CJ, Sun LY, Bae S, Chen DQ, Wei S, Jiao K, Yang WH, Cui RJ, Liu RH, Wang LZ (2017b) MicroRNA-200c and microRNA-141 are regulated by a FOXP3-KAT2B axis and associated with tumor metastasis in breast cancer. Breast Cancer Res 19:73
Zhang YL, Li XB, Hou YX, Fang NZ, You JC, Zhou QH (2017c) The lncRNA XIST exhibits oncogenic properties via regulation of miR-449a and Bcl-2 in human non-small cell lung cancer. Acta Pharmacol Sin 38:371–381
Zhang H, Zhu M, Shan X, Zhou X, Wang T, Zhang J, Tao J, Cheng W, Chen G, Li J, Liu P, Wang Q, Zhu W (2019a) A panel of seven-miRNA signature in plasma as potential biomarker for colorectal cancer diagnosis. Gene 687:246–254
Zhang XY, Zhou J, Xue DW, Li Z, Liu YL, Dong LM (2019b) MiR-515-5p acts as a tumor suppressor via targeting TRIP13 in prostate cancer. Int J Biol Macromol 129:227–232
Zhang Y, Zhang H, Zhang Z, Li S, Jiang W, Li X, Lv J (2019c) LncRNA MALAT1 cessation antagonizes hypoxia/reoxygenation injury in hepatocytes by inhibiting apoptosis and inflammation via the HMGB1-TLR4 axis. Mol Immunol 112:22–29
Zhang XH, Li BF, Ding J, Shi L, Ren HM, Liu K, Huang CC, Ma FX, Wu XY (2020) LncRNA DANCR-miR-758-3p-PAX6 molecular network regulates apoptosis and autophagy of breast cancer cells. Cancer Manage Res 12:4073–4084
Zhang S, Liu C, Zou X, Geng X, Zhou X, Fan X, Zhu D, Zhang H, Zhu W (2021) MicroRNA panel in serum reveals novel diagnostic biomarkers for prostate cancer. PeerJ 9:e11441
Zhao ZY, Liu JT, Wang CL, Wang Y, Jiang YG, Guo M (2014) MicroRNA-25 regulates small cell lung cancer cell development and cell cycle through cyclin E2. Int J Clin Exp Pathol 7:7726–7734
Zheng B, Liang L, Wang C, Huang S, Cao X, Zha R, Liu L, Jia D, Tian Q, Wu J, Ye Y, Wang Q, Long Z, Zhou Y, Du C, He X, Shi Y (2011) MicroRNA-148a suppresses tumor cell invasion and metastasis by downregulating ROCK1 in gastric cancer. Clin Cancer Res 17:7574–7583
Zheng L, Pu J, Qi T, Qi M, Li D, Xiang X, Huang K, Tong Q (2013) miRNA-145 targets v-ets erythroblastosis virus E26 oncogene homolog 1 to suppress the invasion, metastasis, and angiogenesis of gastric cancer cells. Mol Cancer Res 11:182–193
Zheng G, Xiong Y, Xu W, Wang Y, Chen F, Wang Z, Yan Z (2014) A two-microRNA signature as a potential biomarker for early gastric cancer. Oncol Lett 7:679–684

Zhou W, Shi G, Zhang Q, Wu Q, Li B, Zhang Z (2014) MicroRNA-20b promotes cell growth of breast cancer cells partly via targeting phosphatase and tensin homologue (PTEN). Cell Biosci 4:62

Zhou W, Ye XL, Xu J, Cao MG, Fang ZY, Li LY, Guan GH, Liu Q, Qian YH, Xie D (2017) The lncRNA H19 mediates breast cancer cell plasticity during EMT and MET plasticity by differentially sponging miR-200b/c and let-7b. Sci Signal 10:eaak9557

Zhou Y, Tian B, Tang JM, Wu J, Wang H, Wu ZN, Li X, Yang DS, Zhang BH, Xiao YH, Wang Y, Ma JL, Wang WX, Su M (2020a) SNHG7: a novel vital oncogenic lncRNA in human cancers. Biomed Pharmacother 124:109921

Zhou C, Yi C, Yi Y, Qin W, Yan Y, Dong X, Zhang X, Huang Y, Zhang R, Wei J, Ali DW, Michalak M, Chen XZ, Tang J (2020b) LncRNA PVT1 promotes gemcitabine resistance of pancreatic cancer via activating Wnt/beta-catenin and autophagy pathway through modulating the miR-619-5p/Pygo2 and miR-619-5p/ATG14 axes. Mol Cancer 19:118

Zhu SM, Si ML, Wu HL, Mo YY (2007) MicroRNA-21 targets the tumor suppressor gene tropomyosin 1 (TPM1). J Biol Chem 282:14328–14336

Zhu C, Ren C, Han J, Ding Y, Du J, Dai N, Dai J, Ma H, Hu Z, Shen H, Xu Y, Jin G (2014) A five-microRNA panel in plasma was identified as potential biomarker for early detection of gastric cancer. Br J Cancer 110:2291–2299

Zhu W, Zhou K, Zha Y, Chen D, He J, Ma H, Liu X, Le H, Zhang Y (2016) Diagnostic value of serum miR-182, miR-183, miR-210, and miR-126 levels in patients with early-stage non-small cell lung cancer. PloS One 11:e0153046

Zhu W, Huang Y, Pan Q, Xiang P, Xie N, Yu H (2017) MicroRNA-98 suppress warburg effect by targeting HK2 in colon cancer cells. Dig Dis Sci 62:660–668

Zhu YD, Shao SL, Pan HF, Cheng ZL, Rui X (2018) MicroRNA-136 inhibits prostate cancer cell proliferation and invasion by directly targeting mitogen-activated protein kinase kinase 4. Mol Med Rep 17:4803–4810

Zhu XL, Ren LF, Wang HP, Bai ZT, Zhang L, Meng WB, Zhu KX, Ding FH, Miao L, Yan J, Wang YP, Liu YQ, Zhou WC, Li X (2019) Plasma microRNAs as potential new biomarkers for early detection of early gastric cancer. World J Gastroenterol 25:1580–1591

Zhuang J, Shen L, Yang L, Huang X, Lu Q, Cui Y, Zheng X, Zhao X, Zhang D, Huang R, Guo H, Yan J (2017) TGFbeta1 promotes gemcitabine resistance through regulating the LncRNA-LET/NF90/miR-145 signaling axis in bladder cancer. Theranostics 7:3053–3067

Precision Medicine: Translation of Individualistic Treatment in Oncology

Hemant Joshi

Abbreviations

ADC	Antibody-drug conjugate
ADCC	Antibody dependent cell mediated cytotoxicity
ADCP	Antibody-dependent cellular phagocytosis
BCMA	B-cell maturation antigen
CDK	Cyclin dependent kinase
CSF1R	Colony stimulating factor 1 receptor
FGFR	Fibroblast growth factor receptor
FLT3	FMS-like tyrosine kinase 3
GD2	Glycolipid disialoganglioside
HGFR	Hepatocyte growth factor receptor
Hh	Hedgehog pathway
HIF-2α	Hypoxia-inducible factor 2 alpha
HONEY	Hyperbaric OxygeN therapy on brEast cancer patients with late radiation toxicity
HSCT	Hematopoietic stem cell transplantation
IGF-1R	Insulin-like growth factor-1 receptor
IMMUNOGAST	IMMUNOtherapy-based Regimens in Recurrent GASTric Adenocarcinoma
Itk	Interleukin-2 receptor-inducible T-cell kinase
Lck	Lymphocyte-specific protein tyrosine kinase
MMAE	Monomethyl auristatin E
mTOR	Mammalian target of rapamycin
N^2M^2	NCT neuro master match

H. Joshi (✉)
School of Biotechnology, Jawaharlal Nehru University, New Delhi, Delhi, India
e-mail: hemant17_sbt@jnu.ac.in

H. S. Tuli, M. B. Yerer Aycan (eds.), *Oncology: Genomics, Precision Medicine and Therapeutic Targets*, https://doi.org/10.1007/978-981-99-1529-3_4

NSCLC	Non-small cell lung cancer
PDGFR	Platelet-derived growth factor receptor
PD-L1	Programmed cell death ligand 1
PSMA	Prostate-specific membrane antigen
RAINBO	Refining adjuvant treatment in endometrial cancer based on molecular features
RANKL	Receptor activator of nuclear factor kappa-B ligand
RON	Recepteur d'Origine Nantais
TRKA	Tropomyosin receptor tyrosine kinases
VEGF	Vascular endothelial growth factor

1 Significance of Precision Medicine in Oncology

Precision medicine entails tailoring specific drugs or treatments to specific people, in accordance with the genetic or molecular characteristics of their diseases. It is a patient care approach based on the idea that one patient's disease might not be the same as another who seems to have the same disease. In addition to the genes you inherit from your parents, the environment in which you reside can have an impact on your health, manifestations of your symptoms, and treatment effectiveness. Precision medicine aims to uncover a deeper understanding of variations at the genetic and proteomic level and how these factors affect health. The field of precision medicine that deals with therapies for different tumors is known as precision oncology. Tumors have genetic variations that result in tumor proliferation and dissemination. The molecular alterations that are seen in patients with tumors differ from those seen in other patients with the same tumor. Cancers differ from one another in the number and type of gene mutations they contain, as well as other cellular changes they undergo (Le Tourneau et al. 2019; Joshi et al. 2023). Precision oncology aims to comprehend the underlying causes of tumor heterogeneity and complexity as well as the processes that drive tumor heterogeneity by extracting information from various approaches, including big data, proteomics, transcriptomics, single cell analysis, epigenetics, immunophenotyping, molecular imaging, etc. as shown in Fig. 1.

The precision oncology strategy takes genetic consideration to create more refined and efficient treatment options designed specifically for patients with similar molecular characteristics known as precision medicine or personalized medicine (Le Tourneau et al. 2019). The process may require analyzing biological fluids and tumor tissue using liquid biopsies (LB) and targeted therapies, in which small drug molecules or monoclonal antibodies (mAb) that target specific cells to prevent tumor proliferation and dissemination. As opposed to chemotherapy, targeted therapies have a lower probability of damaging healthy cells, leading to fewer and milder side effects (Millner and Strotman 2016). With precision medicine, it is becoming possible to treat cancer patients by providing alternatives that could be more

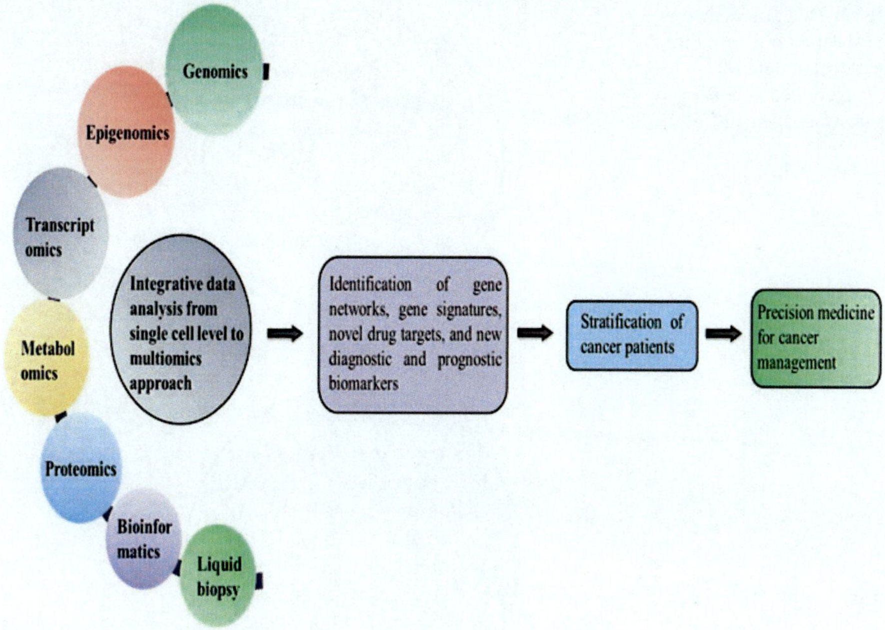

Fig. 1 Schematic representation of precision medicine development using basic to clinic research

effective than ineffective or unnecessary treatments. This chapter discusses the significance of precision medicine and various strategies of precision medicine in the field of oncology. Special attention has been given to the increased demand for LB and targeted therapies for different tumors as precision medicine in the field of oncology. Further, we elaborate on experiences from ongoing clinical trials and challenges associated with precision medicine, which presents possibilities for future tumor therapy and metastasis prevention.

2 Strategies of Precision Medicine in Oncology

The implementation of precision medicine or tailoring treatment for each patient in accordance with tumor biology has been considered a breakthrough era in cancer therapy (Pasic et al. 2013a). A precision medicine approach has a broad range of clinical applications, including screening, early detection, prognosis, predicting treatment effectiveness and drug resistance, detecting tumor metastases, and classifying cancers according to their biological characteristics (Diamandis et al. 2010). Currently, LB and targeted therapies are becoming the main tools for precision medicine or individualised treatment in the field of oncology for cancer management as shown in Fig. 2. LB can provide non-invasive molecular profiling that may better reveal the molecular profiling of subclonal tumor populations existing in a patient, in

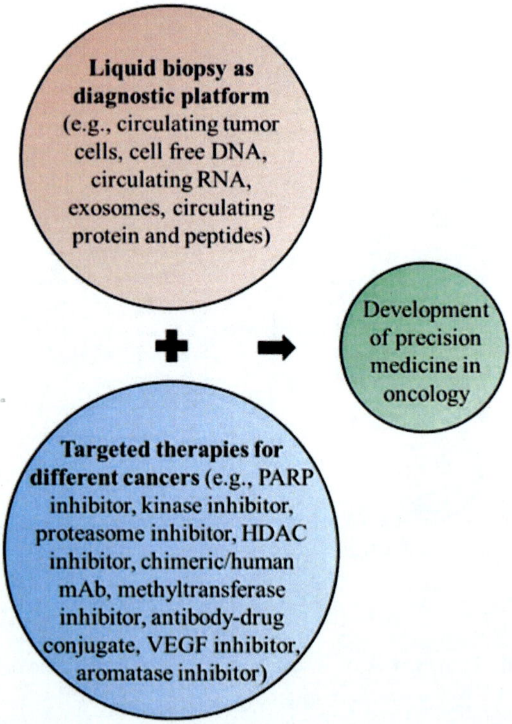

Fig. 2 Diagram depicting the specific tools used for the development of precision medicine for cancer management

contrast to tissue biopsies that only tell about one tumor location (Cheng et al. 2016). This non-invasive tool gains significant attention for early detection of specific cancer through identification of diagnostic biomarkers. On the other hand, targeted cancer therapies are small molecules and therapeutic mAbs specifically designed to inhibit the target molecules indispensable for tumor proliferation and dissemination. Conventional chemotherapies usually interfering with cell division to kill rapidly dividing tumor cells in the body but also affects the normal cell growth. The targeted therapies are come in limelight due to the specific targeting of particular cancer cell type with more precision and possibly less toxicity.

3 Liquid Biopsy

LB is one of the important diagnostic platforms of precision medicine to achieve the clinical management of cancer. It is a real-time based approach for molecular detection of tumor heterogeneity by biomarkers detection in body fluids (Imamura et al. 2016). LB, due to its non-invasive nature, is linked with a substantially low illness rate and can be programmed recurrently to offer detailed information about the disease at consecutive intervals. This strategy has proven quite effective for cancer therapy by determining tumor severity and early detection of drug resistance and tumor metastasis (Bettegowda et al. 2014). Further, it provides several benefits that overcome the potential barriers posed by traditional tissue biopsy such as minimally invasive, inexpensive, easy to obtain from biological fluids, low failure rate, short processing time, evaluation of tumor heterogeneity, high tolerance to treatment, and easy monitoring of therapy response (Heitzer et al. 2019). LB is comprised of different components such as circulating tumor cells (CTC), cell-free nucleic acid (i.e., cfDNA and cfRNA), circulating proteins & peptides, and exosomes. All these circulating molecules are released into the peripheral blood from local and advanced-stage tumors (Di Meo et al. 2017). These circulating molecules are found in a wide range of body samples including plasma, serum, cerebrospinal fluid (CSF), urine, seminal plasma, saliva, and pleural effusion. In this section, we will discuss the role of these circulating molecules in LB as precision medicine for cancer diagnosis.

3.1 *Circulating Tumor Cells*

CTCs are cells that are released from a primary tumor and may be proficient in developing tumor metastases. The latter is the most fatal and terminal aspect of a tumor and is accountable for approximately 90% of tumor-associated deceases (Fidler 2003; Majumder et al. 2020; Kashyap et al. 2017; Tuli et al. 2015a; Kumar et al. 2016). CTCs are the main culprit to induce the metastatic process through a series of events as discussed. Initially, the CTC disengage from the primary tumor and is released into the blood stream. After that, the CTC must circulate in the peripheral blood and trickle into the microvessels of a distant tissue and endure in the microenvironment of the distant tissue. Lastly, the CTC must colonize to produce metastasis (Joshi et al. 2023; Singh et al. 2016; Kashyap et al. 2016; Kumar et al. 2015; Tuli et al. 2015b; De et al. 2019; El-Kenawi et al. 2020).

CTCs are quite rare in terms of frequency approximately 1 in 10^8 circulating cells in the whole blood (Nelson 2010). The potential utility of CTCs as cancer biomarkers are gaining substantial attention despite their rarity. The cytopathological characteristics indicate the CTCs are round and oval in morphology and can be separated from other circulating blood cells through the expression of surface markers. The positive expression of both epithelial cell adhesion molecules (EpCAM) and cytokeratins and negative expression of CD45 on the surface of CTCs especially derived from the epithelial origin (Nagrath et al. 2007; Maltoni et al. 2016). The detection of CTCs has been done in three steps such as CTCs enrichment, isolation, enumeration, and detection. As a result, there is an urgent need for the isolation and molecular assessment of CTCs through several techniques with high detection accuracy and specificity (Alix-Panabières and Pantel 2013). Several strategies are currently being used for the positive selection of CTCs including Herrigbone-chip, CellSearch system, Isoflux, EPHESIA CTC-chip, MagSweeper, CTC-chip, Adnatest system, DEPArray, GEDI microdevice, and Velcro-like device.

The negative selection was accomplished through CTC-iChip, leucocytes depletion kits, and the RosetteSep method (Zhang et al. 2021).

The US FDA only approved the CellSearch® system as a diagnostic platform in January 2004 for CTC detection, mainly for prostate (PCa), breast (BCa), and colorectal cancer (CRC). The CellSearch® system is recognized as the premier option for CTC isolation and detection. In metastatic PCa, BCa, and CRC, if greater than five CTCs are present in 7.5 mL of blood then the chances of overall survival and prognosis are quite low (Cristofanilli et al. 2004; De Bono et al. 2008; Cohen et al. 2009). With the help of CTCs detection, different tumor types can be recognized based on the characteristics of CTCs. For instance, the origin of cancer can be revealed by identifying the molecular profiling of genes, immunohistochemistry, and different phenotypes of CTCs. For example, CDX2 and KRT20 containing CTCs are linked with CRC, TTF-1, and KRT7 containing cells related to lung cancer (LCa), and PSMA and PSA containing CTCs are connected with PCa cancer (Zhang et al. 2021). Other than that, CTC diagnosing platform is also used for the detection of various other tumors, e.g., digestive tract tumor, PCa, small cell lung carcinoma (SCLC), etc. The frequency of CTCs has been quintessential to measuring the severity of diseases and the fluctuation of CTC numbers during cancer treatment serves as a signal of the effectiveness of treatment (Friedlander and Fong 2014).

3.2 Cell-Free DNA

Circulating cfDNA was originally discovered in 1948, however, it has gained attention in recent years as a biomarker of LB for early diagnosis of cancer (Elshimali et al. 2013). CfDNA originated primarily from necrotic and apoptotic tumor cells, while it originated most commonly from apoptotic cells in healthy persons. The main difference between necrotic and apoptotic cfDNA is the DNA size. In apoptotic cells, it fragments heavily and produces small DNA fragments (~166 bp), whereas in necrotic cells, the size of DNA (~1000 bp) is quite large due to low DNA fragmentation (Casadio et al. 2013). The cfDNA fragmentation was found to be increased in renal cell carcinoma, bladder cancer, PCa individuals, last-stage cancer patients, and in individuals that possess large-sized tumors (Gang et al. 2010). This occurs due to the induction of apoptosis by tumor led to the destruction of peripheral non-cancerous tissue. These fragmented DNA are released into the blood circulation and possesses tumor-associated mutations including single nucleotide polymorphism and change in DNA methylation pattern. The genetic change including point mutations and chromosomal aberrations in tumor-associated genes increased the chances of tumor formation. Detection of genetic changes in cfDNA as a non-invasive biomarker present in different biological fluids has increased the chances of early cancer detection. Several studies reported microsatellite DNA mutations in plasma samples of renal cell carcinoma and bladder cancer individuals (Kinde et al. 2013; Goessl et al. 1998). The chromosome 3p mutation and *KRAS2* mutation in plasma cfDNA of renal cell carcinoma and bladder cancer, respectively, for early diagnosis of cancer and *TERT* promoter mutation for relapse of urinary bladder (Kinde et al. 2013; Goessl et al. 1998). The copy number variation was observed in plasma cfDNA of metastatic PCa cancer at different time points (Ulz et al. 2016). The above findings indicate that cfDNA mutations serve as effective biomarkers for early and recurrent cancer detection.

Another important aspect of cfDNA in cancer diagnosis is the change in DNA methylation. The change in DNA methylation pattern of tumor-associated genes substantially affects tumor formation and it can be detected in circulating cfDNA. For example, some tumor-associated genes such as *RARB2*, *GSTP1*, and *RASSF1* were evidenced to be highly methylated in serum derived cfDNA of PCa (Sunami et al. 2009). Likewise, a change in the methylation pattern of *TWIST1*, *PCDH17*, *NID2*, and *POU4F2* genes in urine cfDNA has been linked to bladder cancer (Wang et al. 2016; Renard et al. 2010). These shreds of evidence suggest that changes in DNA methylation pattern serve as an early indicator of disease reactivation. The rate of cfDNA released into the blood relies on various factors such as size, location, and tumor vascularity led to the difference in their abundance in the blood ranging from 0.01 to 90%. The concentration of cfDNA varies around 100 ng/ml in healthy persons and is equal to or greater than 1000 ng/ml in tumor individuals (Cheng et al. 2016; Elazezy and Joosse 2018). All this information is easy to extract from cfDNA because the latter is a reliable and easily accessible biomarker. The major limitation of cfDNA is its short half-life around 16–780 min due to which it is rapidly cleared off during surgery and adversely affects the sensitivity of the detection (Qin et al. 2016). The unfavourable stability of cfDNA could be beneficial for real-time monitoring of disease. Detection of cfDNA demands highly sensitive and accurate methods. Several methods have been developed that meet the requirements of cfDNA analysis including PCR based (mutant allele-specific PCR and nested PCR), Digital PCR (microfluidic type and droplet type), mass spectrometry, targeted sequencing (CAPP-seq, Safe-seqS, and Tam-seq), and whole genome sequencing (karyotyping and personalized analysis of rearranged ends) (Millner and Strotman 2016).

3.3 Circulating RNAs

It is also possible to determine cfRNA in patients' biological fluids, although it is less abundant than cfDNA (Alix-Panabières and Pantel 2013). CfRNA, primarily microRNA (miRNA) and messenger RNA (mRNA), play a major role as non-invasive diagnostic and prognostic biomarkers for cancer detection via monitoring tumor genetics in a biological fluid (Fernandez-Mercado et al. 2015).

3.3.1 Circulating miRNAs

A miRNA is a short non-coding RNA comprised of 20-24 nucleotides, which is responsible for regulating gene expression when it pairs with its target RNA's 3′ untranslated regions (Zhang et al. 2013). Over the last decade, numerous studies have reported establishing a link between oncology and miRNA, showing a substantial role in tumor formation, tumor proliferation, and metastasis (Leong et al. 2018). MiRNAs are either present in complexes with proteins or packaged in exosomes to protect themselves from degradation. The abundance of miRNAs circulating in biological fluids is higher than that of mRNAs, which indicates they could be used as non-invasive biomarkers in cancer diagnosis, prognosis, and therapeutic response prediction (Leong et al. 2018; Otsuji et al. 2017). Various studies have reported elevated expression of urinary miR-146a-5p, miR-15a, and miR-126 in patients with bladder cancer, urothelial carcinoma, and renal cell carcinoma, respectively, compared to healthy individuals (Hanke et al. 2010; von Brandenstein et al. 2012; Sasaki et al. 2016). Circulating plasma levels of miR-145, miR-223, miR-20a, and miR-21 differentiated NSCLC patients from healthy individuals (Zhang et al. 2017). It has been noted that the increased expression of plasma miR-221 and miR-222 have been used to detect renal cell carcinoma, whereas the increased amount of miR-373 and miR-10b derived from plasma have been detected in BCa patients who have lymph node metastasis (Chen et al. 2013; Teixeira et al. 2014). Another study has reported high levels of serum miR-21 in hormone-refractory state tumor individuals, resistant to docetaxel drug treatment (Zhang et al. 2011). Taken together, circulating miRNAs have shown remarkable promise as non-invasive diagnostic and prognostic biomarkers for the detection of multiple tumors.

3.3.2 Circulating mRNAs

Compared to miRNAs, circulating mRNAs are more fragile to RNases (Reddi and Holland 1976) though only a few mRNAs are relatively stable in circulation when they are complex with RNA-binding proteins and lipids (Deligezer et al. 2006). Since circulating mRNAs are involved in cellular protein machinery, circulating mRNAs highlight the progress of cellular pathways and may serve as diagnostic biomarkers for cancer. Increased levels of urinary CK20 mRNA and hTERT mRNA (telomerase reverse transcriptase mRNA) were detected in urothelial carcinoma and bladder cancer patients, respectively compared to the healthy controls, work as diagnostic biomarkers (Guo et al. 2009; Xie et al. 2007). Circulating UBE2C mRNA levels in urine enhanced bladder cancer compared with normal (Kim et al. 2016). Even the CAIX splice variant mRNA in urine has been shown to be diagnostic for different cancer types, such as bladder tumors, PCa, and kidney tumors (Malentacchi et al. 2016). According to a recent study, PLA2G10 mRNA was demonstrated to be a diagnostic and prognostic biomarker for the detection of

NSCLC (Chen et al. 2022). Increased TIMP1 mRNA levels in platelets from CRC patients compared to the platelets from healthy individuals (Yang et al. 2019). Further, B7-H3 mRNA levels were higher in blood samples of metastatic renal cell carcinoma (Qin et al. 2013). The Abiraterone treatment was reported to increase the levels of AR-V7 mRNAs in blood in castration-resistant PCa patients (Antonarakis et al. 2015). Although circulating mRNA has been discovered for decades, clinical applications in this field have not been achieved, possibly because of the low abundance, instability, intracellular mRNA contamination from specimen processing, and inter-individual variability in mRNA levels (Junqueira-Neto et al. 2019).

3.4 Circulating Proteins

Proteomic analyses have opened a new avenue for the prognosis and diagnosis of the tumor via a non-invasive manner by assessing circulating proteins as a biomarker (Di Meo et al. 2016; Pasic et al. 2013b). Several commercial tests are available that measure groups of circulating proteins to detect specific cancer types e.g. 4K score test, ImmunoCyt™, and Aura Tek FDP test. The first one is blood-based test used to detect PCa cancer by measurement of four kallikreins comprising *hKLK2*, intact PSA (prostate-specific antigen), free PSA, and total PSA (Parekh et al. 2015). Second, a urine-based test to measure carcinoembryonic antigen and cytoplasmic mucin to detect urothelial carcinoma (He et al. 2016). The third test diagnoses recurrent bladder cancer patients by assessing the fibrin degradation products (FDPs) in urine samples (Siemens et al. 2003). Several studies reported an increased level of Hsp27 in serum and AQP1 & PLIN2 in urine are screening biomarkers to differentiate clear cells from papillary renal cell carcinoma (White et al. 2014; Morrissey et al. 2015). Another group identified elevated plasma levels of CAV1 and CAV2 assist to distinguish castration resistant PCa relative to non-castration resistant (Sugie et al. 2015). Likewise, a group of four and eight urinary peptides measures from the serum aid to differentiate muscle-invasive bladder cancer from non-muscle invasive and urothelial carcinoma, respectively (Bryan et al. 2011; Schiffer et al. 2009). Taken together, these evidences suggest that proteomics shows a good response in terms of clinical practice. Further, the enhanced specificity, sensitivity, and clinical success of blood and urine-based tests have been achieved by moving forward from single to multiple marker detection for cancer prognosis and diagnosis.

3.5 Exosomes

Exosomes are extracellular vesicles actively secreted as membrane vesicles from all types of cells. It is 40–120 nm in size and exists in most biological fluids including cerebrospinal fluid, urine, blood, serum, and also in breast milk (Yu et al. 2015). It functions as a mediator of intercellular communication through a transfer of different cargo molecules including DNA, RNA, lipids, enzymes, and proteins (Al-Nedawi et al. 2008). The nature of exosomal cargo depends on the cell origin and can be delivered to nearby or distant cells, controlling intracellular communication, gene expression, and phenotype (Thakur et al. 2014). An increased number of exosomes are secreted by the cancerous cells compared to the healthy cells (Rosell et al. 2009). These characteristics of exosomes meet all the specific criteria to be used as non-invasive biomarkers for cancer detection. The conventional method for the isolation and purification of exosomes from a large volume is ultracentrifugation which helps to harvest the exosomes without contamination of cells and other components. Further, the isolation of a specific population of exosomes from a small volume can be achieved using magnetic beads coated with exosomal markers (i.e. CD147, CD9, and CD63) specific antibodies (Jia et al. 2017). Most assays being developed to be used exosomes as a diagnostic biomarker in tumor detection such as ExoScreen, ExoTest, ExoDx, and ExoQuick. ExoScreen is a diagnostic test that can be used to recognize two specific antigens on the surface of circulating exosomes in blood samples using photosensitizer beads. For instance, the presence of CD9 and CD147 antigens on circulating exosomes could help to distinguish CRC individuals from healthy volunteers (Yoshioka et al. 2014). ExoTest is generally used to measure the high amounts of caveolin-1 and CD63 expressing circulating exosomes in plasma sample of melanoma individuals (Fernández-Lázaro et al. 2020). ExoDx is a simple diagnostic test used to evaluate the expression of three genes (i.e., *ERG*, *SPDEF*, and *PCA3*) in circulating exosomes derived from urine samples to detect PCa (McKiernan et al. 2016). ExoQuick is another assay used to harvest a good amount of exosomal RNAs from urine and serum samples that could help to distinguish different cancers (Junqueira-Neto et al. 2019). The detection and characterization of exosomes occur through the use of various biochemical techniques including, western blots, ELISA, flow cytometry (FACS), mass spectrometry, nanoparticle tracking analysis, scanning, and transmission electron microscopy (Lötvall et al. 2014). Recent studies indicate that several miRNAs are predictive biomarkers for guiding cancer diagnosis. For instance, exosomal capture miRNAs may have the ability to distinguish between clear cell renal cell carcinoma and healthy individuals (Butz et al. 2016). High amounts of miRNA21 and EGFRvIII were also found in exosomes harvested from the serum of glioblastoma individuals (Junqueira-Neto et al. 2019). Another study reported a decreased level of exosomal miR34a and miR148a in urine samples of PCa individuals differentiate from benign prostate hyperplasia (Corcoran et al. 2014). Additionally, several InRNAs are also crucial as non-invasive biomarkers for the detection of tumor progression. For e.g., exosomes harvested from the urine samples of the high-grade muscle-invasive condition of urothelial tumor individuals contains increased amount of some InRNAs (i.e. ANRIL, linc-RoR, HOX-AS-2, and HOTAIR) (Berrondo et al. 2016). Further, it was evidenced that the abundance of a few proteins (i.e. HSP90, HSP70, VLA-4, and TYRP-2) increased in exosomes derived from the advanced melanoma patients (Junqueira-Neto et al. 2019). Despite the success, only a few exosomal biomarkers are recognized as potential cancer biomarkers due to the less efficient and sensitive methods for the isolation and purification, and detection of exosomes. The development of more sensitive approaches is likely to translate these exosomal biomarkers into the clinic.

4 Targeted Therapies

Targeted therapies are works as a monomer unit of precision medicine without which the polymer molecule will not be able to form. Targeted therapies are designed in such a way that it targets specifically those molecules that control tumor cell replication, proliferation, progression, and dissemination (Joshi et al. 2023; Lee et al. 2018). Two types of targeted tumor agents are available commercially including therapeutic mAbs and small molecule or low molecular weight drugs. Therapeutic mAbs are designed to recognize, mark, and target those antigens that are expressed on tumor cells to eliminate the cancerous cells by the immune system. Even in a few cases, these antibodies conjugated with radioisotopes and cytotoxic toxins to deliver the complex in specific cancer cell types. For e.g. cetuximab a chimeric mAb that is designed in such a way to specifically targets epidermal growth factor receptor (EGFR); bevacizumab a humanized mAb that targets VEGF-A; and ipilimumab a complete human antibody that targets CTLA-4 (Sharkey and Goldenberg 2006). Whereas, the small drug molecules can easily pass through the cell membrane to make communication with targets present inside a tumor cell. The latter are typically formed to target the specific protein and abolish its enzymatic activity. For instance, seliciclib, imatinib, and bortezomib are small molecules that target and inhibit the enzymatic activity of a cyclin-dependent kinase, tyrosine kinase, and proteasome, respectively (Millner and Strotman 2016). The detailed version of targeted therapies for various cancer types is listed in Table 1. There are numerous types of cancer, and precision medicine has had a profound impact on both diagnosis and therapeutics. As opposed to providing an exhaustive list of all kinds of cancer, we will provide examples of how precision medicine is transforming each field.

4.1 Breast Cancer

Breast cancer is the most commonly occurring cancer-causing an estimated 31% new cancer incidences and 15% cancer mortality in 2022. In the last decade, the occurrence of invasive female BCa has risen by about 0.5% per year (American Cancer Society 2022). Almost two-thirds of women diagnosed with locally advanced BCa (65%) have had 90% and 84% survival rates in the last five and ten years, respectively (American Cancer Society 2022). The clinical manifestations of Bca are breast swelling, long-term alterations in the breast, including skin thickening, skin redness, breast swelling, feeling lumpy in the breast, scaliness or retraction

Table 1 List of FDA approved targeted therapies for various cancers treatment

Drug name	Drug type	Target	Mechanism of action	Caner types
Atezolizumab (Tecentriq)	mAb	PD-L1	Blocks PD-L1 interactions with PD-1 and B7.1 receptors on T lymphocyte and antigen presenting cells (APC)	Bladder cancer
Avelumab (Bavencio)	mAb	PD-L1	Blocks PD-L1 interactions with its receptors on T lymphocyte and APC	Bladder cancer, endocrine and neuroendocrine tumors, kidney cancer, skin cancer
Enfortumab vedotin-ejfv (Padcev)	Nectin-4-targeted mAb and MMAE drug conjugate	Nectin-4, microtubule disrupting agent	The interaction of ADC with Nectin-4-expressing cells, internalization of the complex, and proteolytic release of MMAE. By releasing MMAE, the microtubule network is disrupted within the cell, thereby interrupting cell cycle and triggering apoptosis	Bladder cancer
Erdafitinib (Balversa)	Kinase inhibitor	FGFR enzymes	Cell lines expressing genetic mutations in the FGFR were inhibited by erdafitinib and had decreased viability	Bladder cancer
Nivolumab (Opdivo)	mAb	PD-L1/L2	By inhibiting the interaction of PD-L1 and PD-L2 with PD-1 found on T lymphocytes, it inhibits the proliferation and production of cytokines in T cells	Bladder cancer, liver and bile duct cancer, esophageal cancer, CRC, head and neck cancer (HNC), lymphoma, kidney cancer, LCa, malignant mesothelioma, skin cancer, stomach cancer
Pembrolizumab (Keytruda)	mAb	PD-L1/L2	By inhibiting the interaction of PD-L1 and PD-L2 with PD-1 found on T lymphocytes, it	Bladder cancer, Stomach cancer, BCa, skin cancer cervical cancer, endometrial cancer, CRC,

			inhibits the proliferation and production of cytokines in T cells	esophageal cancer, HNC, LCa, kidney cancer, liver and bile duct cancer, lymphoma
Sacituzumab govitecan-hziy (Trodelvy)	Trop-2-targeted mAb and topoisomerase inhibitor conjugate	Trop-2, topoisomerase	The complex binds to Trop-2-expressing cancer cells and releases SN-38 (topoisomerase inhibitor) after being internalized	Bladder cancer, BCa
Belzutifan (Welireg)	Transcription factor inhibitor	HIF-2α	The drug inhibits the HIF-2-α-HIF-1β interaction, thereby reducing the transcription and expression of HIF-2α regulated genes	Brain cancer, Kidney cancer, Pancreatic cancer
Bevacizumab (Avastin)	Recombinant humanized mAb	VEGF	VEGF is bound by bevacizumab and prevented from interacting with its receptors (Flt-1 and KDR) on the surface of endothelial cells and then inhibits new blood vessel formation	Brain cancer, CRC, kidney cancer, ovarian cancer, liver and bile duct cancer, cervical cancer, LCa
Everolimus (Afinitor)	Macrolide immunosuppressant	FK506 Binding Protein-12 (FKBP-12)	The immunosuppressive complex formed by everolimus and FKBP-12 inhibits the mTOR kinase by binding to it	Brain cancer, pancreatic cancer BCa, kidney cancer
Abemaciclib (Verzenio)	Kinase inhibitor	CDK4 and CDK6	Apoptotic and senescent cells were caused by inhibition of Rb phosphorylation via CDK action and blockage of cell cycle progression from G1 to S phase	BCa
Ado-trastuzumab emtansine (Kadcyla)	Humanized anti-HER2 IgG1 and small molecule cytotoxin DM1	HER2 receptor	The process of internalization of Kadcyla is mediated by the HER2 receptor subdomain IV, followed by lysosomal degradation,	BCa

(continued)

Table 1 (continued)

Drug name	Drug type	Target	Mechanism of action	Caner types
			releasing DM1-containing cytotoxic catabolites into the cell. By binding to tubulin, DM1 disrupts microtubule networks in the cell, leading to cell cycle arrest and apoptosis	
Alpelisib (Piqray)	Kinase inhibitor	Phosphatidylinositol-3-kinase (PI3K)	Alpelisib inhibited the phosphorylation of PI3K downstream targets and hence inhibited the PI3K/Akt signalling pathway	BCa
Anastrozole (Arimidex)	Non-steroidal aromatase inhibitor	Aromatase enzyme	It significantly inhibits estrogen production in normal and tumor tissues through inhibition of the aromatase enzyme	BCa
Exemestane (Aromasin)	Steroidal aromatase inactivator	Aromatase enzyme	It is an androstenedione antagonist interacts with aromatase enzyme and blocks it led to the decreased estrogen synthesis	BCa
Fam-trastuzumab deruxtecan-nxki (Enhertu)	HER2-targeted antibody and topoisomerase inhibitor (DXd) conjugate	HER2 receptor	HER2 binds to this complex, which then undergoes internalization, lysosomal enzyme cleavage, and subsequently releases DXd, causing DNA damage and cell death	BCa, esophageal cancer, stomach cancer
Fulvestrant (Faslodex)	Estrogen receptor antagonist	Estrogen receptor	It interacts with and deactivates the estrogen receptor in breast tumor cells	BCa
Lapatinib ditosylate (Tykerb)	Kinase inhibitors	HER2 (ErbB2) and EGFR (ErbB1) receptor	It inhibits ErbB-mediated tumor cell growth	BCa

Letrozole (Femara)	Non-steroidal aromatase inhibitor	Aromatase enzyme	It significantly inhibits estrogen production in normal and tumor tissues through inhibition of the aromatase enzyme	BCa
Margetuximab-cmkb (Margenza)	HER2/neu receptor antagonist	HER2/neu receptor	Inhibits tumor cell proliferation by binding to the extracellular domain of the HER2 protein	BCa
Neratinib maleate (Nerlynx)	Kinase inhibitor	EGFR, HER2, and HER4	It reduced autophosphorylation and demonstrated antitumor activity by irreversibly binding to EGFR, HER2, and HER4	BCa
Olaparib (Lynparza)	poly (ADP-ribose) polymerase (PARP) inhibitor	PARP enzymes	Increasing PARP-DNA complex formation and inhibition of PARP enzyme activity, causing DNA damage and apoptosis	BCa, ovarian cancer, PCa, pancreatic cancer
Palbociclib (Ibrance)	Kinase inhibitor	CDK4 and CDK6	Apoptotic and senescent cells were caused by inhibition of Rb phosphorylation via CDK action and blockage of cell cycle progression from G1 to S phase	BCa
Pertuzumab (Perjeta)	Recombinant humanized mAb	Extracellular dimerization domain of HER-2 protein	This drug blocked HER2 heterodimerization induced by ligands, which led to the inhibition of MAP kinase and PI3 kinase, resulting in cell cycle arrest and apoptosis	BCa
Ribociclib (Kisqali)	Kinase inhibitor	CDK4 and CDK6	Apoptotic and senescent cells were caused by inhibition of Rb phosphorylation via CDK action and blockage of cell cycle progression from G1 to S phase	BCa

(continued)

Table 1 (continued)

Drug name	Drug type	Target	Mechanism of action	Caner types
Talazoparib tosylate (Talzenna)	PARP inhibitor	PARP enzymes	Increasing PARP-DNA complex formation and inhibition of PARP enzyme activity, causing DNA damage and apoptosis	BCa
Tamoxifen citrate (Soltamox)	Non-steroidal anti-estrogen	Estrogen receptor	By competing with estrogen for binding sites in target tissues, it inhibits the formation of dimethyl benzanthracene-induced tumors	BCa
Toremifene (Fareston)	Estrogen agonist/ antagonist	Estrogen receptor	By competing with estrogen for binding sites in target tissues, it inhibits the formation of dimethyl benzanthracene-induced tumors	BCa
Trastuzumab (Herceptin)	Humanized IgG1 kappa mAb	HER2 receptor	HER2-overexpressing tumor cells are inhibited by it	BCa, esophageal cancer, stomach cancer
Tucatinib (Tukysa)	Kinase inhibitor	HER2 and HER3	The drug inhibits HER2 and HER3 phosphorylation, thereby inhibiting MAPK and AKT signaling, and exhibits anti-tumor activity in tumor cells expressing HER2	Breast cancer
Tisotumab vedotin-tftv (Tivdak)	Human tissue factor (VIIa) targeted mAb and MMAE	Human tissue factor VIIa and microtubules	It interferes with factor X activation by binding factor VIIa. Aside from preventing thrombin formation and cancer-associated venous thromboembolism, this could also inhibit new blood vessel formation. Additionally, MMAE inhibits tubulin	Cervical cancer

			polymerization, causing an arrest in the G2/M phase and cell cycle death	
Cetuximab (Erbitux)	EGFR antagonist	EGFR	It inhibits cell growth, induces apoptosis, and decreases the production of matrix metalloproteinases and vascular endothelial growth factors by binding to EGFR and preventing phosphorylation and activation of receptor-associated kinases	CRC, HNC
Encorafenib (Braftovi)	Kinase inhibitor	BRAF V600E	It inhibits B-raf protein, resulting in inhibition of MAPK and ERK signalling and showed anti-tumor activity	CRC, skin cancer
Ipilimumab (Yervoy)	Cytotoxic T-lymphocyte antigen 4 (CTLA-4)-blocking antibody	CTLA-4	It inhibits binding of CTLA-4 and its ligands, resulting in induction of anti-tumor immune response	CRC, esophageal cancer, kidney cancer, malignant mesothelioma, liver and bile duct cancer, skin cancer, LCa
Panitumumab (Vectibix)	EGFR antagonist	EGFR	It inhibits cell growth, induces apoptosis, and decreases the production of matrix metalloproteinases and vascular endothelial growth factors by binding to EGFR and preventing phosphorylation and activation of receptor-associated kinases	CRC
Ramucirumab (Cyramza)	VEGFR2 antagonist	VEGFR2	Human endothelial cells are inhibited from proliferating and migrating as a result of ligand-	CRC, stomach cancer, liver and bile duct cancer, esophageal cancer, LCa

(continued)

Table 1 (continued)

Drug name	Drug type	Target	Mechanism of action	Caner types
			stimulated inhibition of VEGFR2 by Cyramza	
Regorafenib (Stivarga)	Multikinase inhibitor	Multiple kinases	Inhibiting oncogenesis, tumor angiogenesis, metastasis, and tumor immunity is achieved by binding to multiple membrane-bound and intracellular kinases	CRC, Liver and bile duct cancer
Ziv-aflibercept (Zaltrap)	VEGF inhibitor	VEGF	The inhibition of neovascularization and decreased vascular permeability is caused by the inhibition of their cognate receptors	CRC
Imatinib mesylate (Gleevec)	Kinase inhibitor	BCR-ABL tyrosine kinase	Through its inhibition of BCR-ABL tyrosine kinase, it induces apoptosis and inhibits tumor cell proliferation	Dermatofibrosarcoma protuberans, Leukemia, Myelodysplastic and myeloproliferative disorders
Iobenguane I 131 (Azedra)	Radioactive therapeutic agent	Norephinephrine (NE) agonist	Apoptosis and tumor necrosis are caused by radioactive decay of I131, which is taken up by the NE transporter in adrenergic nerve terminals	Endocrine and neuroendocrine tumors
Lanreotide acetate (Somatuline Depot)	Somatostatin agonist	Somatostatin receptor	It inhibits the amounts of certain natural substances produced by the body	Endocrine and neuroendocrine tumors
Lutetium Lu 177-dotatate (Lutathera)	Radiolabeled somatostatin agonist	Somatostatin receptor subtype 2	Free radicals formed during the radioactive decay of Lu 177 cause cellular damage in cells that express somatostatin receptors and their neighbouring cells	Endocrine and neuroendocrine tumors

Dostarlimab-gxly (Jemperli)	PD-1 blocking IgG4 humanized mAb	PD-1 receptor	Binds to PD-1 receptors and blocks PD-L1 and PD-L2 interactions, releasing immune-suppressive effects mediated by PD-1 pathway	Endometrial cancer
Lenvatinib mesylate (Lenvima)	Kinase inhibitor	VEGF receptors	It inhibits the kinase activities of VEGF receptors known to contribute to new blood vessel formation, tumor proliferation and progression	Endometrial cancer, kidney cancer, thyroid cancer, liver and bile duct cancer
Avapritinib (Ayvakit)	Kinase inhibitor	PDGFRA, KIT D816V, and PDGFRA D842	Tumor cell proliferation is inhibited by inhibiting the autophosphorylation of target proteins	Gastrointestinal stromal tumor; leukemia
Denosumab (Xgeva)	Human IgG2 mAb	RANKL	Osteoclast development, function, and survival are decreased with prevention of the RANKL/RANK interaction, resulting in increases in bone mass and strength in trabecular and cortical bone	Giant cell tumor
Pexidartinib hydrochloride (Turalio)	Kinase inhibitor	CSF1R	Tumor cell proliferation is inhibited by inhibiting the autophosphorylation of target proteins	Giant cell tumor
Axitinib (Inlyta)	Kinase inhibitor	VEGFR-1, -2, and -3	Cancer growth is inhibited by inhibiting the phosphorylation of VEGFR-2 present in tumor cells.	Kidney cancer
Cabozantinib-s-malate (Cabometyx)	Kinase inhibitor	TIE-2, AXL, VEGFR-1, -2, and -3, KIT, RET, ROS1, TYRO3, FLT-3, MER, TRKB, and MET	Tumor formation, dissemination, angiogenesis, antibiotic resistance, and microenvironmental	Kidney cancer, Liver and bile duct cancer

(continued)

Table 1 (continued)

Drug name	Drug type	Target	Mechanism of action	Caner types
			tumor formation are all affected by this molecule that inhibits these receptor tyrosine kinases	
Pazopanib hydrochloride (Votrient)	Kinase inhibitor	VEGFR-1, -2, and -3, c-Fms, PDGFR-α & -β, Lck, FGFR-1 & -3, Kit, Itk	Tumor cell proliferation is inhibited by inhibiting the autophosphorylation of target proteins	Kidney cancer
Sorafenib tosylate (Nexavar)	Kinase inhibitor	VEGFR-1, -2, and -3, BRAF, PDGFR-β, KIT, RET/PTC, FLT-3, c-CRAF, and RET kinases	Tumor cell signalling, angiogenesis, and cell death are inhibited by inhibiting the autophosphorylation of target proteins	Kidney cancer, liver and bile duct cancer, thyroid cancer
Sunitinib malate (Sutent)	Kinase inhibitor	RET, PDGFR-β, FLT-3, VEGFR-2, KIT kinases	Tumor cell signalling, progression, and metastasis are inhibited by inhibiting the autophosphorylation of target proteins	Kidney cancer, pancreatic cancer
Temsirolimus (Torisel)	Antineoplastic agent	mTOR	mTOR controls cell division and is inhibited by this drug when it binds to an intracellular protein (FKBP-12)	Kidney cancer
Tivozanib hydrochloride (Fotivda)	Kinase inhibitor	PDGFR-β, c-kit, VEGFR-1, -2, and -3	Tumor cell signalling, angiogenesis, and cell death are inhibited by inhibiting the autophosphorylation of target proteins	Kidney cancer
Acalabrutinib (Calquence)	Kinase inhibitor	Bruton tyrosine kinase (BTK)	Tumor growth and cancerous B-cell proliferation are inhibited	Leukemia, lymphoma

			by inhibiting the kinase activity of BTK	
Alemtuzumab (Campath)	Humanized mAb	CD52	It interacts with CD52 antigen in different immune cells (i.e. monocytes, NK cells, B-cells, macrophages, and T-cells) led to the ADCC	Leukemia
Asciminib hydrochloride (Scemblix)	Kinase inhibitor	ABL1	It interacts with the myristoyl residue of ABL1 and inhibits its kinase activity in BCR-ABL1 fusion protein	Leukemia
Blinatumomab (Blincyto)	Bi-specific mAb	CD19, CD3	It forms a synapse by binding CD19 antigen on cancerous B-cells and CD3 protein on T-cells and led to the lysis of CD19 positive tumor cells	Leukemia
Bosutinib (Bosulif)	Kinase inhibitor	BCR-ABL kinase, SRC, LYN, and HCK kinases	It prevents the CML by inhibiting kinase activity of BCR-ABL fusion protein and SRC-family kinases	Leukemia
Brexucabtagene autoleucel (Tecartus)	T-cell immunotherapy	CD19	It recognizes and interacts with CD19 antigen present on normal B cells and tumor cells led to the killing of CD19 expressing cancer cells	Leukemia, lymphoma
Dasatinib (Sprycel)	Kinase inhibitor	PDGFR-β, BCR-ABL, c-KIT, SRC family (FYN, LCK, SRC, YES), and EPHA2	Tumor cell signalling, progression, and metastasis are inhibited by inhibiting the autophosphorylation of target proteins	Leukemia

(continued)

Table 1 (continued)

Drug name	Drug type	Target	Mechanism of action	Caner types
Duvelisib (Copiktra)	Small molecule inhibitor	Phosphatidylinositol 3-kinases (PI3K-δ and PI3K-γ)	It inhibits PI3K-δ and PI3K-γ isoform of kinases present on normal and cancerous B-cells	Leukemia, lymphoma
Enasidenib mesylate (Idhifa)	Small molecule inhibitor	Isocitrate dehydrogenase 2 (IDH2) enzyme	The drug reduced 2-hydroxyglutarate (2-HG) levels and induced myeloid differentiation by targeting mutant IDH2 variants R140Q, R172S, and R172K	Leukemia
Gemtuzumab ozogamicin (Mylotarg)	CD33-directed mAb and N-acetyl gamma calicheamicin	CD33	Upon binding to CD33-expressing tumor cells, it internalizes the complex, and the release of the inhibitor molecule leads to double-strand DNA breaks, resulting in cell cycle arrest and apoptosis	Leukemia
Gilteritinib fumarate (Xospata)	Kinase inhibitor	FLT3	It inhibits the signalling of FLT3-ITD to induce apoptosis in leukemic cells expressing FLT3-ITD	Leukemia
Ibrutinib (Imbruvica)	Kinase inhibitor	BTK	Tumor growth and cancerous B-cell proliferation are inhibited by inhibiting the kinase activity of BTK	Leukemia, lymphoma
Idelalisib (Zydelig)	Kinase inhibitor	PI3K-δ	It inhibits PI3K-δ isoform of kinases present on normal and tumor B-cells	Leukemia
Inotuzumab ozogamicin (Besponsa)	CD22-directed mAb and N-acetyl gamma calicheamicin	CD22	Upon binding to CD22-expressing tumor cells, it internalizes the complex, and the	Leukemia

			release of the inhibitor molecule leads to double-strand DNA breaks, resulting in cell cycle arrest and apoptosis	
Ivosidenib (Tibsovo)	Small molecule inhibitor	Isocitrate dehydrogenase 1 (IDH1) enzyme	The drug reduced 2-HG levels and induced myeloid differentiation by targeting mutant IDH1	Leukemia, liver and bile duct cancer
Midostaurin (Rydapt)	Kinase inhibitor	FLT3, KIT, PDGFRα/β, PKC kinase	Inhibit FLT3 receptor signaling and cell proliferation led to the induction of apoptosis in leukemic cells. Additionally, it induces apoptosis in mast cells and inhibits KIT signaling, cell proliferation, and histamine release	Leukemia
Moxetumomab pasudotox-tdfk (Lumoxiti)	CD22-directed cytotoxin	CD22	The drug interacts to CD22 present on B-cells and is internalized, resulting in elongation factor 2 being ADP-ribosylated, thereby inhibiting protein synthesis and activating cell death	Leukemia
Nilotinib (Tasigna)	Kinase inhibitor	BCR-ABL kinase	Through its inhibition of BCR-ABL tyrosine kinase, it induces apoptosis and inhibits tumor cell proliferation	Leukemia
Obinutuzumab (Gazyva)	Humanized anti-CD20 mAb	CD20	It interacts to CD20 antigen present on pre-B and matures B cells, induces the greater ADCC activity in cancer cell lines	Leukemia, lymphoma
Ofatumumab (Arzerra)	IgG1κ human mAb	CD20	It specifically interacts to CD20 antigen present on pre-B and matures B cells, induces the	Leukemia

(continued)

Table 1 (continued)

Drug name	Drug type	Target	Mechanism of action	Caner types
			ADCC mediated cell lysis in cancer cell lines	
Pemigatinib (Pemazyre)	Kinase inhibitor	FGFR-1, -2, -3, and -4	It targets the FGFR kinase and inhibited their phosphorylation and signalling led to the decreased cell viability in cancer cell lines	Leukemia, liver and bile duct cancer, lymphoma, myelodysplastic and myeloproliferative disorders
Ponatinib hydrochloride (Iclusig)	Kinase inhibitor	TIE2, FLT3, VEGFR, EPH, PDGFR, RET, FGFR, KIT, and ABL kinases	It inhibits the tyrosine kinase activity of various kinases results in shrink the size of tumors	Leukemia
Rituximab (Rituxan)	Chimeric murine/human mAb	CD20	It interacts to CD20 antigen present on pre-B and matures B cells, induces the greater ADCC activity in cancer cell lines	Leukemia, lymphoma
Rituximab and hyaluronidase human (Rituxan Hycela)	Rituximab and hyaluronidase human	CD20	It interacts to CD20 antigen present on pre-B and matures B cells, induces the greater ADCC activity in cancer cell lines. Hyaluronidase increases the absorption of rituximab in systemic circulation	Leukemia, lymphoma
Tagraxofusp-erzs (Elzonris)	Recombinant human interleukin-3 and truncated diphtheria toxin	CD123	It interacts to CD123 expressing cells and inhibits protein synthesis and cell death	Leukemia
Tisagenlecleucel (Kymriah)	Anti-CD19 chimeric antigen receptor (CAR)	CD19	In this technique, a patient's own T cells are reprogrammed to express a CAR, which is used to eliminate malignant cells and normal cells that express CD19	Leukemia, lymphoma

Venetoclax (Venclexta)	BCL-2 inhibitor	BCL-2	It shows cell killing activity in BCL-2 expressing tumor cells	Leukemia, lymphoma
Atezolizumab (Tecentriq)	PD-L1 blocking antibody	PD-L1	Blocks PD-L1 interactions with its receptors on T lymphocyte and APC led to the induction of the anti-tumor immune response and reduced cancer cell growth	Liver and bile duct cancer, Skin cancer, LCa
Durvalumab (Imfinzi)	PD-L1 blocking antibody	PD-L1	Blocks PD-L1 interactions with its receptors on T lymphocyte and APC led to the induction of the anti-tumor immune response and reduced cancer cell growth	Liver and bile duct cancer, LCa
Infigratinib phosphate (Truseltiq)	Kinase inhibitor	FGFR-1, -2, -3, and -4	It targets the FGFR kinase and inhibited their phosphorylation and signalling led to the decreased cell viability in cancer cell lines	Liver and bile duct cancer
Afatinib dimaleate (Gilotrif)	Kinase inhibitor	EGFR, HER2, and HER4	Tumor cell proliferation is inhibited by inhibiting the autophosphorylation of target proteins	LCa
Alectinib (Alecensa)	Kinase inhibitor	ALK and RET	STAT3 and AKT are activated by ALK, and the viability of tumor cells is decreased as a result of its inhibition of ALK phosphorylation and ALK-mediated activation	LCa
Amivantamab-vmjw (Rybrevant)	Bispecific antibody	EGFR and MET	It recognizes and interacts to EGFR and MET present on cancerous cells and promote ADCC by immune effector cells	LCa

(continued)

Table 1 (continued)

Drug name	Drug type	Target	Mechanism of action	Caner types
Brigatinib (Alunbrig)	Kinase inhibitor	FLT-3, IGF-1R, ROS1, and ALK	STAT3 and AKT are activated by ALK, and the viability of tumor cells is decreased as a result of its inhibition of ALK phosphorylation and ALK-mediated activation	LCa
Capmatinib hydrochloride (Tabrecta)	Kinase inhibitor	MET	The viability of MET dependent cancerous cells is decreased as a result of its inhibition of MET phosphorylation and MET-mediated activation	LCa
Cemiplimab-rwlc (Libtayo)	PD-L1 blocking antibody	PD-L1	Blocks PD-L1 interactions with its receptors on T lymphocyte and APC led to the induction of the anti-tumor immune response and reduced cancer cell growth	LCa, skin cancer
Ceritinib (Zykadia)	Kinase inhibitor	IGF-1R, insulin receptor (InsR), ROS1, and ALK	STAT3 and AKT are activated by ALK, and the viability of tumor cells is decreased as a result of its inhibition of ALK phosphorylation and ALK-mediated activation	LCa
Crizotinib (Xalkori)	Kinase inhibitor	RON, HGFR, ROS1, c-Met, and ALK	STAT3 and AKT are activated by ALK, and the viability of tumor cells is decreased as a result of its inhibition of ALK phosphorylation and ALK-mediated activation	LCa, lymphoma

Dabrafenib mesylate (Tafinlar)	Kinase inhibitor	BRAF, CRAF, SIK1, NEK11, and LIMK1	Reduces the growth of tumors with BRAF V600 mutations	LCa, Skin cancer, thyroid cancer
Dacomitinib (Vizimpro)	Kinase inhibitor	EGFR, DDR1, EPHA6, LCK, DDR2, and MNK1	Tumor cell proliferation is inhibited by inhibiting the autophosphorylation of EGFR and HER-2	LCa
Entrectinib (Rozlytrek)	Kinase inhibitor	TRKA, TRKB, and TRKC	Inhibits the growth of tumor cell lines originated from different cancerous cells possessing *NTRK*, *ROS1*, and *ALK* fusion genes	LCa
Erlotinib hydrochloride (Tarceva)	Kinase inhibitor	EGFR	Blocking EGFR kinase activity and inhibiting autophosphorylation of receptor tyrosine kinase led to the suppression of further downstream signalling	LCa, Pancreatic cancer
Fam-trastuzumab deruxtecan-nxki (Enhertu)	HER2-targeted antibody and topoisomerase inhibitor conjugate	HER2	HER2 binds to this complex, which then undergoes internalization, lysosomal enzyme cleavage, and subsequently releases DXd, causing DNA damage and cell death	LCa
Gefitinib (Iressa)	Kinase inhibitor	EGFR	Blocking EGFR kinase activity and inhibiting autophosphorylation of receptor tyrosine kinase led to the suppression of further downstream signalling	LCa

(continued)

Table 1 (continued)

Drug name	Drug type	Target	Mechanism of action	Caner types
Lorlatinib (Lorbrena)	Kinase inhibitor	ACK, FAK2, ROS1, FPS, TRKA, TYK1, FER, TRKB, FAK, TRKC, and ALK	STAT3 and AKT are activated by ALK, and the viability of tumor cells is decreased as a result of its inhibition of ALK phosphorylation and ALK-mediated activation	LCa
Mobocertinib succinate (Exkivity)	Kinase inhibitor	EGFR	Blocking EGFR kinase activity and inhibiting autophosphorylation of receptor tyrosine kinase led to the suppression of further downstream signalling	LCa
Necitumumab (Portrazza)	Anti-EGFR recombinant human mAb	EGFR	By binding necitumumab, EGFR is internalized and degraded, which leads to the inhibition of malignant expression, angiogenesis, and apoptosis	LCa
Osimertinib mesylate (Tagrisso)	Kinase inhibitor	EGFR	Blocking EGFR kinase activity and inhibiting autophosphorylation of receptor tyrosine kinase led to the suppression of further downstream signalling	LCa
Pralsetinib (Gavreto)	Kinase inhibitor	VEGFR2, DDR1, RET, PDGFR-β, TRKC, FGFR1, JAK1-2, TRKA, and FLT3	It inhibits the kinase activity of RET, thereby inhibiting further downstream signalling and uncontrolled cell proliferation	LCa, thyroid cancer
Selpercatinib (Retevmo)	Kinase inhibitor	RET, VEGFR1 and VEGFR3	It inhibits the kinase activity of RET, thereby inhibiting further downstream signalling and uncontrolled cell proliferation	LCa, thyroid cancer

Sotorasib (Lumakras)	RAS GTPase inhibitor	KRAS	In tumor cells expressing KRAS, it binds irreversibly to the unique cysteine of KRASG12C, stopping downstream signaling. Thereby, preventing cell growth and induces cell death	LCa
Tepotinib hydrochloride (Tepmetko)	Kinase inhibitor	MET, HGF	Prevents cancerous cell proliferation, anchorage-independent growth, and dissemination in MET- expressing tumor cells	LCa
Trametinib dimethyl sulfoxide (Mekinist)	Kinase inhibitor	MEK1, MEK2	Reduces the growth of tumors with BRAF V600 mutations	LCa, skin cancer, thyroid cancer
Axicabtagene ciloleucel (Yescarta)	T-cell immunotherapy	CD19	T cells are activated, proliferate, acquire effector functions, and inflammatory cytokines and chemokines are secreted when it binds to cancer cells that express CD19, as well as normal B cells	Lymphoma
Belinostat (Beleodaq)	Histone deacetylase (HADC) inhibitor	HDAC	Several tumor cells were arrested in the cell cycle and apoptosis was induced as a result of the accumulation of acetylated histones and other proteins by this inhibitor	Lymphoma
Bexarotene (Targretin)	Retinoid X receptor (RXR) activator	RXR	As a result of its binding and activation of RXR, it controls the expression of genes that regulate cell division and proliferation	Lymphoma
Bortezomib (Velcade)	Antineoplastic agent	26S proteasome	By suppressing the 26S proteasome's chymotrypsin-like	Lymphoma, multiple myeloma

(continued)

Table 1 (continued)

Drug name	Drug type	Target	Mechanism of action	Caner types
			activity, it inhibits targeted proteolysis, thereby affecting different signaling mechanisms in the cell. As a result of disrupting normal homeostatic mechanisms, cancer cells die	
Brentuximab vedotin (Adcetris)	CD30-directed mAb and MMAE conjugate	CD30	Upon binding to CD30-expressing tumor cells, it internalizes the complex, and the release of the inhibitor molecule leads to double-strand DNA breaks, resulting in cell cycle arrest and apoptosis	Lymphoma
Copanlisib hydrochloride (Aliqopa)	Kinase inhibitor	PI3K-α and PI3K-δ	It inhibits PI3K-δ and PI3K-γ isoform of kinases present on normal and cancerous B-cells	Lymphoma
Ibritumomab tiuxetan (Zevalin)	CD20-directed mAb and tiuxetan conjugate	CD20	It binds specifically to CD20 antigen present on precursor B-cells or matures B-cells and on B-cell NHLs	Lymphoma
Lisocabtagene maraleucel (Breyanzi)	T-cell immunotherapy	CD19	T cells are activated, proliferate, acquire effector functions, and inflammatory cytokines and chemokines are secreted when it binds to cancer cells that express CD19, as well as normal B cells	Lymphoma
Loncastuximab tesirine-lpyl (Zynlonta)	CD19-directed antibody and alkylating agent conjugate	CD19	Upon binding to CD19-expressing tumor cells, it internalizes the complex, and the	Lymphoma

			release of the inhibitor molecule leads to apoptosis	
Mogamulizumab-kpkc (Poteligeo)	Recombinant humanized mAb	CCR4	It binds to CCR4 expressing on malignant T-cell and Treg cells lead to the depletion of target cells by ADCC	Lymphoma
Polatuzumab vedotin-piiq (Polivy)	CD79b-directed mAb-MMAE conjugate	CD79b	Upon binding to CD79b, Polivy is internalized followed by release of MMAE via lysosomal proteases inhibiting cell division and inducing apoptosis	Lymphoma
Pralatrexate (Folotyn)	Antineoplastic folate analog	Dihydrofolate reductase	Thymidine and other biological molecules that require single carbon transfer for synthesis were depleted as a result of the competitive inhibition of dihydrofolate reductase by this compound.	Lymphoma
Romidepsin (Istodax)	HDAC inhibitor	HDAC	Several tumor cells were arrested in the cell cycle and apoptosis was induced as a result of the accumulation of acetylated histones and other proteins by this inhibitor	Lymphoma
Selinexor (Xpovio)	Nuclear export inhibitor	Exportin1	Selinexor inhibits exportin1 by causing tumor suppressor proteins to accumulate in the nucleus and some oncoproteins, including c-myc and cyclin D1, to be reduced, resulting in cell cycle arrest and cell death in tumor cells	Lymphoma, multiple myeloma

(continued)

Table 1 (continued)

Drug name	Drug type	Target	Mechanism of action	Caner types
Siltuximab (Sylvant)	Chimeric mAb	IL-6	It interacts with human IL-6 cytokine and inhibits the interaction of IL-6 cytokine to both membrane-bound and soluble IL-6 receptors	Lymphoma
Tafasitamab-cxix (Monjuvi)	CD19-directed cytolytic mAb	CD19	T cells are activated, proliferate, acquire effector functions, and inflammatory cytokines and chemokines are secreted when it binds to cancer cells that express CD19, as well as normal B cells	Lymphoma
Tazemetostat hydrobromide (Tazverik)	Methyltransferase inhibitor	EZH2	It binds and inhibits the methyltransferase activity of EZH2 and suppressed proliferation of B-cell lymphoma	Lymphoma
Vorinostat (Zolinza)	HDAC inhibitor	HDAC	Several tumor cells were arrested in the cell cycle and apoptosis was induced as a result of the accumulation of acetylated histones and other proteins by this inhibitor	Lymphoma
Zanubrutinib (Brukinsa)	Kinase inhibitor	BTK	Tumor growth and cancerous B-cell proliferation are inhibited by inhibiting the kinase activity of BTK	Lymphoma
Belantamab mafodotin-blmf (Blenrep)	BCMA targeted antibody and MMAF conjugate	BCMA	Upon binding to BCMA-expressing tumor cells, it internalizes the complex, and the release of the MMAF molecule	Multiple myeloma

			leads to cell cycle arrest and cell death	
Carfilzomib (Kyprolis)	Proteasome inhibitor	20S proteasome	It impairs tumor growth in models of multiple myeloma, hematologic disorders, and solid tumours by binding irreversibly to the N-terminal threonine-containing active sites of the 20S proteasome	Multiple myeloma
Ciltacabtagene autoleucel (Carvykti)	T-cell immunotherapy	BCMA	The CAR induces T cell activation, proliferation, and elimination of target cells after binding to BCMA-expressing cells	Multiple myeloma
Daratumumab (Darzalex)	CD38-directed human mAb	CD38	In addition to Fc mediated cross linking, it also induces ADCC that inhibits the growth of CD38 expressing tumor cells	Multiple myeloma
Daratumumab and hyaluronidase-fihj (Darzalex Faspro)	CD38-directed human mAb and hyaluronidase	CD38	In addition to Fc mediated cross linking, it also induces ADCC that inhibits the growth of CD38 expressing tumor cells. Hyaluronidase increases the absorption of Daratumumab in systemic circulation	Multiple myeloma
Elotuzumab (Empliciti)	SLAMF7-directed human mAb	SLAMF7	Myeloma cells are targeted by it via SLAMF7, and it engages Fc receptors on effector cells to cause them to die via ADCC and ADCP	Multiple myeloma
	T-cell immunotherapy	BCMA	The CAR induces T cell activation, proliferation, and	Multiple myeloma

(continued)

Table 1 (continued)

Drug name	Drug type	Target	Mechanism of action	Caner types
Idecabtagene vicleucel (Abecma)			elimination of target cells after binding to BCMA-expressing cells	
Isatuximab-irfc (Sarclisa)	CD38-directed cytolytic antibody	CD38	In addition to Fc mediated cross linking, it also induces ADCC that inhibits the growth of CD38 expressing tumor cells	Multiple myeloma
Ixazomib citrate (Ninlaro)	Proteasome inhibitor	20S proteasome	It impairs tumor growth in models of multiple myeloma, hematologic disorders, and solid tumours by binding irreversibly to the N-terminal threonine-containing active sites of the 20S proteasome	Multiple myeloma
Fedratinib hydrochloride (Inrebic)	Kinase inhibitor	JAK2 and FLT3	It inhibits the kinase activity of JAK2, thereby inhibiting further downstream signalling and cell proliferation led to induction of cell death	Myelodysplastic and myeloproliferative disorders
Pacritinib citrate (Vonjo)	Kinase inhibitor	JAK2 and FLT3	It inhibits the kinase activity of JAK2, thereby inhibiting further downstream signalling and cell proliferation led to induction of cell death	Myelodysplastic and myeloproliferative disorders
Ruxolitinib phosphate (Jakafi)	Kinase inhibitor	JAK1 and JAK2	It inhibits the kinase activity of JAK2, thereby inhibiting further downstream signalling and cell proliferation led to induction of cell death	Myelodysplastic and myeloproliferative disorders

Dinutuximab (Unituxin)	GD2-binding chimeric mAb	GD2	It interacts with GD2-expressing cells leading to the induction of cell lysis via ADCC	Neuroblastoma
Naxitamab-gqgk (Danyelza)	GD2-binding chimeric mAb	GD2	It interacts with GD2-expressing cells leading to the induction of cell lysis via ADCC	Neuroblastoma
Selumetinib sulfate (Koselugo)	Kinase inhibitor	MEK1/2	There was a reduction in neurofibroma numbers, volume, and proliferation due to inhibition of ERK phosphorylation	Plexiform neurofibroma
Abiraterone acetate (Zytiga)	Androgen biosynthesis inhibitor	17α-hydroxylase/C17,20-lyase (CYP17)	Inhibition of CYP17 is achieved by its conversion to abiraterone. The enzyme is present in the cells of the testicle, the adrenal gland, and the prostate gland, where it is required for the biosynthesis of androgens	PCa
Apalutamide (Erleada)	Androgen receptor (AR) inhibitor	AR	The inhibitor prevents AR nuclear translocation, prevents DNA binding, and inhibits AR-mediated transcription, resulting in reduced tumor cell growth and enhanced cell death	PCa
Cabazitaxel (Jevtana)	Antineoplastic agent	Microtubules	Disassembly inhibition occurs simultaneously with its binding to tubulin and promoting microtubule assembly. Inhibiting mitotic and interphase cell functions occurs as a result of the stabilization of microtubules.	PCa

(continued)

Table 1 (continued)

Drug name	Drug type	Target	Mechanism of action	Caner types
Darolutamide (Nubeqa)	AR inhibitor	AR	The inhibitor prevents AR nuclear translocation, prevents DNA binding, and inhibits AR-mediated transcription, resulting in reduced tumor cell growth and enhanced cell death	PCa
Enzalutamide (Xtandi)	AR inhibitor	AR	The inhibitor prevents AR nuclear translocation, prevents DNA binding, and inhibits AR-mediated transcription, resulting in reduced tumor cell growth and enhanced cell death	PCa
Lutetium Lu 177 vipivotide tetraxetan (Pluvicto)	Radioligand therapeutic agent	Prostate-specific membrane antigen (PSMA)	During the interaction between Pluvicto and PSMA-expressing cells, the beta-minus radiation emitted by lutetium-177 affects PSMA-expressing cells as well as nearby cells, causing DNA damage that can cause apoptosis	PCa
Radium 223 dichloride (Xofigo)	Radiotherapeutic drug	-	A high frequency of double-strand DNA breaks occurs in neighbouring cells including tumor cells, osteoclasts, and osteoblasts as a result of alpha particles released in areas of increased bone turnover. This results in an anti-tumor effect against bone metastases	PCa

Rucaparib camsylate (Rubraca)	PARP inhibitor	PARP enzymes	Increasing PARP-DNA complex formation and inhibition of PARP enzyme activity, causing DNA damage and apoptosis	PCa, ovarian cancer
Binimetinib (Mektovi)	Kinase inhibitor	MEK1/2	It inhibits the ERK phosphorylation, and reduced tumor growth	Skin cancer
Cobimetinib fumarate (Cotellic)	Kinase inhibitor	MEK1/2	It inhibits the ERK phosphorylation, and reduced tumor growth	Skin cancer
Nivolumab and relatlimab-rmbw (Opdualag)	Two IgG4 kappa mAbs	LAG3 and PD-1	Relatlimab blocks the LAG-3 receptor and enhances the antitumor effects of PD-1 blockade to inhibit tumor growth and promote tumor regression	Skin cancer
Sonidegib (Odomzo)	Hh pathway inhibitor	Hh pathway	Smoothened, which is involved in the transduction of Hh signals, is bound and inhibited by this compound	Skin cancer
Tebentafusp-tebn (Kimmtrak)	Bi-specific gp100 peptide-HLA-targeted T-cell receptor CD3 T-cell engager	Gp100, HLA	Uveal melanoma tumor cells present a gp100 peptide on their cell surfaces bound to human leukocyte antigen-A*02:01 that interacts with this antibody	Skin cancer
Vismodegib (Erivedge)	Hh pathway inhibitor	Hh pathway	Smoothened, which is involved in the transduction of Hh signals, is bound and inhibited by this compound	Skin cancer
Vemurafenib (Zelboraf)	Kinase inhibitor	CRAF, ARAF, BRAF, SRMS, ACK1, MAP4K5, and FGR	Tumor formation, dissemination, angiogenesis, antibiotic resistance, and microenvironmental tumor formation are all affected	Skin cancer

(continued)

Table 1 (continued)

Drug name	Drug type	Target	Mechanism of action	Caner types
			by this molecule that inhibits these receptor tyrosine kinases	
Cabozantinib-s-malate (Cometriq)	Kinase inhibitor	TYRO3, RET, VEGFR-1, -2, -3, AXL, TRKB, ROS1, KIT, MER, TIE-2, FLT-3, and MET	Tumor formation, dissemination, angiogenesis, antibiotic resistance, and microenvironmental tumor formation are all affected by this molecule that inhibits these receptor tyrosine kinases	Thyroid cancer
Vandetanib (Caprelsa)	Kinase inhibitor	SRC, RET, TIE2, VEGFR, EPH, BRK, and EGFR	Tumor formation, dissemination, angiogenesis, antibiotic resistance, and microenvironmental tumor formation are all affected by this molecule that inhibits these receptor tyrosine kinases	Thyroid cancer
Niraparib tosylate monohydrate	PARP inhibitor	PARP enzymes	Increasing PARP-DNA complex formation and inhibition of PARP enzyme activity, causing DNA damage and apoptosis	Ovarian cancer

in the nipple, and spontaneous bloody discharge (Tan 2016). It is important to screen for early-stage BCa since it is often asymptomatic. There are several well-known prognostic and diagnostic tests for the detection of BCa using plasma, blood, and tissue samples (Table 2). For e.g. MammaPrint, Oncotype DX, and Prosigna PAM50. MammaPrint is an FDA-cleared assay that examines 70-genes signature by microarrays to identify early-stage BCa individuals who would profit most from adjuvant chemotherapy if they progress distant metastases (Brandão et al. 2019). Another FDA-cleared test is Prosigna PAM50 which employs reverse transcriptase PCR (RT-PCR) to analyse 50 genes associated with subtypes of BCa. The latter is also providing a risk of relapse score to predict prognosis and supplement substantial evidence to immunohistochemistry tests currently available (Wallden et al. 2015). Further, Oncotype DX, the most common prognostic assay, employs RT-PCR to examine the gene expression of 5 housekeeping genes and 16 cancer-related genes to provide a recurrence score to determine who would most profit from chemo and hormonal therapy (Carlson and Roth 2013). The main predictive biomarker present in the BCa is *ERBB2* mutation in hormonal estrogen receptor (HER2), *PIK3CA* mutation, *Ki67* mutation, *BRAF1*, and *BRAF2* mutation (Table 2). Presently, there are some FDA-approved biomarkers assays available used to identify those biomarkers that are connected with different targeted therapies (Table 2). Several targeted therapies emerged against Bca that comprise recombinant humanized mAbs, steroidal and non-steroidal aromatase inhibitors, antibody-drug conjugate, kinase inhibitors (PI3K, CDK4 & 6, HER2, PARP enzyme), non-steroidal anti-estrogen, estrogen agonist/antagonist, and macrolide immunosuppressant (Table 1).

4.2 *Prostate Cancer*

Cancer of the prostate is the second underlying cause of death in men globally, causing an estimated 27% of new cancer incidences and 11% of deaths in 2022 (American Cancer Society 2022). The early-stage PCa is asymptomatic, while advanced stages show clinical manifestations such as burning or pain sensation with urination, feeble urine flow, recurrent urination, struggle in urination, and presence of blood in the urine. As PCa progresses, it usually affects the bones, resulting in pain in the ribs, spine, hip, and other body regions (Chiong et al. 2005). The prominent diagnostic biomarker for PCa detection is PSA. Though, the latter gain criticism over the years due to the possibility of overdiagnosis, which has led to experiments to find more reliable biomarkers. There is an urgent need to develop some potential biomarkers with better specificity and sensitivity that can differentiate between localized prostatic tumors to advanced-stage cancer. Currently, there are three FDA-approved diagnostic tests for PCa detection such as Prostate health index (PHI), Progensa™, and CellSearch circulating tumor cell kit. PHI is a new promising blood test to detect proPSA, free PSA, and total PSA to predict the PCa progression during active monitoring, and differentiate between overall and high grade tumors with biopsy (Loeb and Catalona 2014). Progensa™ is a molecular-based test to

Table 2 Various biomarkers based companion diagnostic tests approved by FDA for the treatment of cancers

Diagnostic test	Manufacturer	Sample type	Biomarker	Biomarker details	Drug	Cancer type	Approval status
Bond Oracle HER2 IHC system	Leica Biosystems	Tissue	ERBB2 (HER2)	HER-2 protein overexpression	Herceptin	Breast cancer	18 April 2012
BRACAnalysis CDx	Myriad Genetic Laboratories, Inc.	Blood	*BRCA1* and *BRCA2* genes	Gene mutations	Lynparza	Breast cancer	12 Jan 2018
					Talzenna		16 Oct 2018
FoundationOne CDx	Foundation Medicine, Inc.	Tissue	*ERBB2*	*ERBB2* gene amplification	Herceptin	Breast cancer	30 Nov 2017
					Perjeta		
					Kadcyla		
		Tissue	*PIK3CA*	H1047L, C420R, E545D, E542K, E545A, Q546E, E545G, E545K, H1047Y, Q546R, and H1047R mutations	Lynparza	Breast cancer	3 Dec 2019
FoundationOne Liquid CDx		Plasma			Piqray	Breast cancer	26 Oct 2020
HER2 CISH pharmDx kit	Dako Denmark A/S	Tissue	*ERBB2*	*ERBB* gene amplification	Herceptin	Breast cancer	30 Nov 2011
HER2 FISH pharmDx kit	Dako Denmark A/S	Tissue	*ERBB2*	*ERBB* gene amplification	Herceptin	Breast cancer	3 May 2005
					Perjeta		8 June 2012
					Kadcyla		22 Feb 2013
HercepTest	Dako Denmark A/S	Tissue	ERBB2	HER-2 protein overexpression	Herceptin	Breast cancer	25 Sept 1998
					Perjeta		8 June 2012
					Kadcyla		22 Feb 2013

INFORM HER2 Dual ISH DNA probe cocktail	Ventana Medical Systems, Inc.	Tissue	*ERBB2*	*ERBB* gene amplification	Herceptin	Breast cancer	14 June 2011
					Kadcyla		3 May 2019
INFORM HER-2/neu					Herceptin		30 Dec 1997
InSite Her-2/neu (CB11) mAb	Biogenex Laboratories, Inc.	Tissue	*ERBB2*	HER-2 protein overexpression	Herceptin	Breast cancer	22 Dec 2004
Ki-67 IHC MIB-1 pharmDx	Agilent Technologies	Tissue	Ki-67	Ki-67 protein expression	Verzenio	Breast cancer	12 Oct 2021
PathVysion HER-2 DNA probe kit	Abbott Molecular Inc.	Tissue	*ERBB2*	*ERBB* gene amplification	Herceptin	Breast cancer	11 Dec 1998
PATHWAY anti-Her2/neu (4B5) rabbit mAb	Ventana Medical Systems, Inc.	Tissue	*ERBB2*	HER-2 protein overexpression	Herceptin	Breast cancer	28 Nov 2000
			ERBB2	HER-2 protein overexpression	Kadcyla		3 May 2019
PD-L1 IHC 22C3 pharmDx	Dako North America, Inc.	Tissue	PD-L1	PD-L1 protein expression	Keytruda	Triple-negative breast cancer	13 Nov 2020
SPOT-LIGHT HER2 CISH kit	Life Technologies Corporation	Tissue	*ERBB2*	*ERBB* gene amplification	Herceptin	Breast cancer	1 July 2008
therascreen PIK3CA RGQ PCR kit	QIAGEN GmbH	Tissue or plasma	*PIK3CA*	H1047Y, E545D, E542K, E545A, Q546R, E545G, H1047L, E545K, Q546E, H1047R, and C420R mutations	Piqray	Breast cancer	24 May 2019
Ventana HER2 Dual ISH DNA probe cocktail	Ventana Medical Systems, Inc.	Tissue	*ERBB2*	*ERBB2* gene amplification	Herceptin	Breast cancer	28 July 2020
PD-L1 IHC 22C3 pharmDx	Dako North America, Inc.	Tissue	PD-L1	PD-L1 protein expression	Keytruda	Cervical cancer	12 June 2018

(continued)

Table 2 (continued)

Diagnostic test	Manufacturer	Sample type	Biomarker	Biomarker details	Drug	Cancer type	Approval status
cobas KRAS mutation test	Roche Molecular Systems, Inc.	Tissue	*KRAS*	*KRAS* gene mutations at codons 12 and 13	Erbitux	Colorectal cancer	7 May 2015
					Vectibix		
Dako EGFR pharmDx kit	Dako North America, Inc.	Tissue	EGFR (HER1)	HER1 protein expression	Erbitux	Colorectal cancer	12 Feb 2004
					Vectibix		27 Sept 2006
FoundationOne CDx	Foundation Medicine, Inc.	Tissue	*KRAS*	Wild type *KRAS* gene	Erbitux	Colorectal cancer	30 Nov 2017
			KRAS and *NRAS*	Wild type *KRAS* and *NRAS* genes	Vectibix		
ONCO/Reveal Dx Lung & Colon cancer assay (O/RDx-LCCA)	Pillar Biosciences, Inc.	Tissue	*KRAS*	Wild type *KRAS* gene	Erbitux	Colorectal cancer	30 July 2021
ONCO/Reveal Dx Lung & Colon cancer assay (O/RDx-LCCA)	Pillar Biosciences, Inc.	Tissue	*KRAS*	Wild type *KRAS* gene	Vectibix	Colorectal cancer	30 July 2021
Praxis extended RAS panel	Illumina, Inc.	Tissue	*KRAS* and *NRAS*	Wild type *KRAS* and *NRAS* genes	Vectibix	Colorectal cancer	29 June 2017
therascreen KRAS RGQ PCR kit	Qiagen Manchester, Ltd.	Tissue	*KRAS*	G13D, G12C, G12A, G12S, G12D, G12R, and G12V mutations	Vectibix	Colorectal cancer	23 May 2014
					Erbitux		6 July 2012
therascreen BRAF V600E RGQ PCR kit	QIAGEN GmbH	Tissue	*BRAF*	V600E	Braftovi and erbitux	Colorectal cancer	15 April 2020
Ventana MMR RxDx panel		Tissue	deficient mismatch	MSH6, PMS2, MLH1, and MSH2 proteins	Jemperli	Endometrial cancer	22 April 2021

	Ventana Medical Systems, Inc.		repair (dMMR) proteins		Keytruda		16 June 2022
PD-L1 IHC 22C3 pharmDx	Dako North America, Inc.	Tissue	PD-L1	PD-L1 protein expression	Keytruda	Esophageal cancer	30 July 2019
PD-L1 IHC 22C3 pharmDx	Dako North America, Inc.	Tissue	PD-L1	PD-L1 protein expression	Keytruda	Head and neck cancer	10 June 2019
Abbott RealTime IDH1	Abbott Molecular, Inc	Bone marrow/blood	*IDH1*	R132L, R132H, R132S, R132G, and R132C mutations	Tibsovo	Acute myeloid leukemia	20 July 2018
Abbott RealTime IDH2	Abbott Molecular, Inc		*IDH2*	R172W, R140L, R172S, R140G, R172G, R140W, R172M, R172K, and R140Q mutations	Idhifa		01 Aug 2017
LeukoStrat CDx FLT3 mutation assay	Invivoscribe Technologies, Inc.	Bone Marrow/blood	*FLT3*	TKD mutations D835 and I836, ITD mutations	Rydapt	Acute myeloid leukemia	28 April 2017
					Xospata		28 Nov 2018
MRDx BCR-ABL test	MolecularMD Corporation	Peripheral blood	t(9;21) Philadelphia chromosome	*BCR-ABL* fusion	Tasigna	Chronic myeloid leukemia	22 Dec 2017
Vysis CLL FISH probe kit	Abbott Molecular, Inc.	Peripheral blood	*TP53*	17p chromosome deletions	Venclexta	B-cell chronic lymphocytic leukemia	11 April 2016
cobas EGFR mutation test v1	Roche Molecular Systems, Inc.	Tissue	*EGFR (HER1)*	Exon 21 (L858R) substitution and exon 19 deletion	Tarceva	NSCLC	15 July 2013
cobas EGFR mutation test v2			*EGFR*	T790M mutation	Tagrisso	NSCLC	13 Nov 2015
		Tissue or plasma	*EGFR*	Exon 21 (L858R) substitution and exon 19 deletion	Tagrisso	NSCLC	18 April 2018
			EGFR		Iressa	NSCLC	27 Oct 2020

(continued)

Table 2 (continued)

Diagnostic test	Manufacturer	Sample type	Biomarker	Biomarker details	Drug	Cancer type	Approval status
			EGFR		Tarceva	NSCLC	27 Oct 2020
			EGFR		Gilotrif	NSCLC	27 Oct 2020
		Plasma	*EGFR*		Tarceva	NSCLC	01 June 2016
			EGFR)	T790M mutation	Tagrisso	NSCLC	28 Sept 2016
FoundationOne CDx	Foundation Medicine, Inc.	Tissue	*EGFR*	Exon 21 (L858R) substitution and exon 19 deletion	Gilotrif	NSCLC	30 Nov 2017
			EGFR		Iressa	NSCLC	30 Nov 2017
			EGFR		Tarceva	NSCLC	30 Nov 2017
			EGFR	T790M mutation	Tagrisso	NSCLC	30 Nov 2017
			ALK	Genetic rearrangements in *ALK*	Alecensa	NSCLC	30 Nov 2017
			ALK	Genetic rearrangements in *ALK*	Xalkori	NSCLC	30 Nov 2017
			ALK	Genetic rearrangements in *ALK*	Zykadia	NSCLC	30 Nov 2017
			BRAF	V600E mutation	Tafinlar + Mekinist	NSCLC	30 Nov 2017
			EGFR	Exon 21 (L858R) substitution and exon 19 deletion	Tagrisso	NSCLC	1 July 2019
			MET		Tabrecta	NSCLC	

				MET SNPs, insertions, and deletions results in exon 14 skipping			6 May 2020
			ROS1	*ROS1* fusions	Rozlytrek	NSCLC	7 June 2022
FoundationOne Liquid CDx	Foundation Medicine, Inc.	Plasma	*EGFR*	Exon 21 (L858R) substitution and exon 19 deletion	Iressa	NSCLC	26 Aug 2020
					Tagrisso	NSCLC	26 Aug 2020
					Tarceva	NSCLC	26 Aug 2020
			MET	*MET* SNPs, insertions, and deletions results in exon 14 skipping	Tabrecta	NSCLC	15 July 2021
			ALK	*ALK* gene rearrangements	Alecensa	NSCLC	26 Oct 2020
Guardant360 CDx	Guardant Health, Inc.	Plasma	*EGFR*	Exon 21 (L858R) substitution, exon 19 deletion and T790M	Tagrisso	NSCLC	7 Aug 2020
			EGFR	*EGFR* exon 20 insertions	Rybrevant	NSCLC	21 May 2021
			KRAS	G12C mutation	Lumakras	NSCLC	28 May 2021
ONCO/Reveal Dx Lung & Colon cancer assay	Pillar Biosciences, Inc.	Tissue	*EGFR*	Exon 21 (L858R) substitution and exon 19 deletion	Iressa	NSCLC	30 July 2021
					Gilotrif	NSCLC	
					Tarceva	NSCLC	
					Vizimpro	NSCLC	
Oncomine Dx target test	Life Technologies Corporation	Tissue	*BRAF*	V600E mutation	Tafinlar + Mekinist	NSCLC	22 June 2017
			ROS1	*ROS1* fusions	Xalkori	NSCLC	22 June 2017
			EGFR		Iressa	NSCLC	

(continued)

Table 2 (continued)

Diagnostic test	Manufacturer	Sample type	Biomarker	Biomarker details	Drug	Cancer type	Approval status
				Exon 21 (L858R) substitution and exon 19 deletion			22 June 2017
			RET	*RET* fusions	Gavreto	NSCLC	4 Sept 2020
			EGFR	Exon 20 insertion mutations	Rybrevant	NSCLC	1 Dec 2021
			EGFR	Exon 20 insertion mutations	Exkivity	NSCLC	15 Sept 2021
PD-L1 IHC 22C3 pharmDx	Dako North America, Inc.	Tissue	PD-L1	PD-L1 protein expression	Keytruda	NSCLC	2 Oct 2015
					Libtayo	NSCLC	22 Feb 2021
					Opdivo + Yervoy	NSCLC	15 May 2020
therascreen EGFR RGQ PCR kit	Qiagen Manchester, Ltd.	Tissue	*EGFR*	Exon 21 (L858R) substitution and exon 19 deletion	Gilotrif	NSCLC	12 July 2013
			EGFR	Exon 21 (L858R) substitution and exon 19 deletion	Iressa	NSCLC	10 July 2015
			EGFR	L861Q, G719X and S7681 mutations	Gilotrif	NSCLC	12 Jan 2016
			EGFR	Exon 21 (L858R) substitution and exon 19 deletion	Vizimpro	NSCLC	27 Sept 2018
			KRAS	G12C mutation	Lumakras	NSCLC	28 May 2021
Ventana ALK (D5F3) CDx assay	Ventana Medical Systems, Inc.	Tissue	ALK	ALK protein expression	Xalkori	NSCLC	12 June 2015
					Zykadia	NSCLC	

							26 May 2017
					Alecensa	NSCLC	6 Nov 2017
					Lorbrena	NSCLC	3 March 2021
Ventana PD-L1 (SP142) assay	Ventana Medical Systems, Inc.	Tissue	PD-L1	PD-L1 protein expression	Tecentriq	NSCLC	2 July 2018
					Tecentriq	NSCLC	15 Oct 2021
			ALK	Genetic rearrangements in *ALK*	Xalkori	NSCLC	26 Oct 2011
					Alunbrig	NSCLC	22 May 2020
cobas EZH2 mutation test	Roche Molecular Systems, Inc.	Tissue	*EZH2*	A692V, A682G, Y646F or Y646X, and Y646N mutations	Tazverik	Follicular Lymphoma Tumor	18 June 2020
BRACAnalysis CDx	Myriad Genetic Laboratories, Inc.	Blood	*BRCA1* and *BRCA2*	Gene mutations	Lynparza	Ovarian Cancer	19 Dec 2014
					Rubraca	Ovarian Cancer	16 Oct 2018
FoundationFocus CDxBRCA assay	Foundation Medicine, Inc.	Tissue	*BRCA1* and *BRCA2*	Gene mutations	Rubraca	Ovarian Cancer	19 Dec 2016
FoundationOne CDx	Foundation Medicine, Inc.	Tissue	*BRCA1* and *BRCA2*	Gene mutations	Lynparza	Ovarian Cancer	1 July 2019
					Rubraca	Ovarian Cancer	30 Nov 2017
FoundationOne Liquid CDx	Foundation Medicine, Inc.	Plasma	*BRCA1* and *BRCA2*	Gene mutations	Rubraca	Ovarian Cancer	26 Oct 2020

(continued)

Table 2 (continued)

Diagnostic test	Manufacturer	Sample type	Biomarker	Biomarker details	Drug	Cancer type	Approval status
Myriad myChoice CDx	Myriad Genetic Laboratories, Inc.	Tissue	Myriad HRD	Deleterious mutations in *BRCA1* and *BRCA2* genes	Zejula	Ovarian Cancer	23 Oct 2019
					Lynparza	Ovarian Cancer	8 May 2020
BRACAnalysis CDx	Myriad Genetic Laboratories, Inc.	Blood	*BRCA1* and *BRCA2*	Gene mutations	Lynparza	Pancreatic Cancer	27 Dec 2019
BRACAnalysis CDx	Myriad Genetic Laboratories, Inc.	Blood	*BRCA1* and *BRCA2*	Gene mutations	Lynparza	Metastatic castrate resistant prostate cancer (mCRPC)	19 May 2020
FoundationOne CDx	Foundation Medicine, Inc.	Tissue	Homologous recombination repair genes	*RAD54L, BRCA2, RAD51D, ATM, RAD51C, BARD1, RAD51B, BRIP1, PALB2,CDK12, FANCL, CHEK1, BRCA1,* and *CHEK2* mutations	Lynparza		19 May 2020
FoundationOne Liquid CDx	Foundation Medicine, Inc.	Plasma	*BRCA1, ATM, and BRCA2*	*BRCA1, ATM,* and *BRCA2* mutations	Lynparza		26 Oct 2020
			BRCA1 and *BRCA2*	*BRCA1* and *BRCA2* genes mutations	Rubraca		26 Aug 2020
cobas 4800 BRAF V600 mutation test	Roche Molecular Systems, Inc.	Tissue	*BRAF*	V600E mutation	Zelboraf	Melanoma	17 Aug 2011
			BRAF	V600E or V600K mutation	Zelboraf + Cotellic	Melanoma	7 Nov 2016
FoundationOne CDx	Foundation Medicine, Inc.	Tissue	*BRAF*	V600E or V600K mutation	Mekinist	Melanoma	30 Nov 2017
				BRAF V600 mutations	Tecentriq + Cotellic + Zelboraf	Melanoma	19 Jan 2022

THXID BRAF kit	bioMérieux Inc.	Tissue	*BRAF*	V600E or V600K mutation	Mekinist	Melanoma	29 May 2013
			BRAF	V600E mutation	Tafinlar	Melanoma	29 May 2013
			BRAF	V600E or V600K mutation	Braftovi + Mektovi	Melanoma	27 June 2018
HER2 FISH pharmDx kit	Dako Denmark A/S	Tissue	*ERBB2*	*ERBB2* gene amplification	Herceptin	Gastric cancer	20 Oct 2010
HercepTest	Dako Denmark A/S	Tissue	ERBB2	ERBB2 protein overexpression	Herceptin	Gastric cancer	20 Oct 2010
therascreen FGFR RGQ RT-PCR kit	QIAGEN Manchester Ltd.	Tissue	*FGFR3*	Exon 7: R248C, S249C; exon 10: G370C, Y373C	Balversa	Urothelial cancer	12 April 2019
Ventana PD-L1 (SP142) assay	Ventana Medical Systems, Inc.	Tissue	PD-L1	PD-L1 protein expression	Tecentriq	Urothelial cancer	18 May 2016
Ventana MMR RxDx panel	Ventana Medical Systems, Inc.	Tissue	dMMR proteins	MSH6, PMS2, MLH1, and MSH2 proteins	Jemperli	Endometrial cancer	22 April 2021
					Keytruda	Endometrial cancer	16 June 2022

quantitate the expression of PCa gene 3 in urine to differentiate between benign prostatic conditions and PCa (Durand et al. 2011). CellSearch CTC is the only blood test available to enumerate the CTC in blood and helps to diagnose the patient with PCa, CRC, and metastatic BCa (Miller et al. 2010). Using these whole diagnostic tests, it is possible to detect PCa patients in need of targeted therapies. The main targeted therapies against PCa are androgen biosynthesis inhibitors, androgen receptor inhibitors, microtubule inhibitors, PARP inhibitors, and radiotherapeutics drugs (Table 1).

4.3 Lung Cancer

LCa is the main culprit of cancer mortality both in males and females. The new estimated cases of LCa are around 12-13% and estimated death cases are 21% in 2022 (American Cancer Society 2022). The clinical symptoms do not appear till it becomes advanced, such as chest pain, persistent cough, a hoarse voice, sputum streaked with blood, recurrent pneumonia or bronchitis, and worsening shortness of breath (Collins et al. 2007). LCa falls into two broad categories: non-small cell LCa (NSCLC, 82%) and small cell LCa (SCLC, 14%). The NSCLC is further subdivided into three main categories such as large-cell cancer, squamous-cell cancer, and adenocarcinoma. Although the low-dose spiral computed tomography technique is used for LCa screening, it has led to a decrease in LCa deaths to some extent, but efficient and reliable biomarkers are needed for LCa diagnosis (Swensen et al. 2002). The potential predictive biomarkers that can differentiate NSCLC from SCLC include *EGFR* (*HER1*), *ALK* fusion rearrangements, *BRAF* (V600E), *MET* single nucleotide variants, *ROS1* fusion, *KRAS*, PD-L1 and ALK protein expression (Table 2). There are some FDA-approved tests used to diagnose these biomarkers in association with different targeted therapies to treat NSCLC such as Ventana PD-LI (SP263 & SP142) assay, Ventana ALK (D5F3) CDx assay, *therascreen* EGFR RGQ PCR Kit, ONCO/Reveal Dx Lung & Colon Cancer assay, Guardant360 CDx, FoundationOne Liquid CDx, and cobas EGFR Mutation Test v2 (Table 2). Ventana PD-LI (SP263 & SP142) test is an immunohistochemical (IHC) test for qualitative recognition of PD-L1 protein present in formalin-fixed, paraffin-embedded (FFPE) tissue that contains cancerous cells or cancer infiltrating immune cells using rabbit mAb against PD-L1 (Oyan et al. 2018). Ventana ALK (D5F3) CDx test is another qualitative diagnostic assay to detect ALK protein in FFPE NSCLC tissue using rabbit mAb against ALK (Uruga and Mino-Kenudson 2018). The *therascreen* EGFR RGQ PCR Kit and cobas EGFR Mutation Test v2 are RT-PCR based qualitative diagnostic assays for the identification of exon 21 (L858R) substitutions and exon 19 deletions of the *HER-1*gene in DNA derived from FFPE NSCLC tissue samples (Syed 2016). Using amplicon-based target enrichment technology, the O/RDx-LCCA *in vitro* diagnostic test employs next-generation sequencing (NGS) for detecting single nucleotide variations and deletions in two genes in DNA extracted from FFPE NSCLC and CRC tissue samples (Cooper and Chen 2022). Guardant360

CDx is a blood test used for the identification of genetic mutations (*EGFR* and *KRAS* genes) in cfDNA isolated from plasma to identify the NSCLC (Bauml et al. 2022). Using a targeted high throughput hybridization-based capture technique, Foundation One Liquid CDx *in vitro* diagnostic test utilizes an NGS-based qualitative test to recognize insertions, deletions, and substitutions in 311 genes, gene rearrangements in four genes and copy number variations in three genes (Woodhouse et al. 2020). The promising targets of the targeted therapies for NSCLC are PD-L1/L2, VEGF, CTLA-4, VEGFR2, EGFR, and various kinases (Table 1). These targeted therapies are quite valuable as precision medicine specifically targets the molecules and provides effective therapy with few side effects to lessen the chances of LCa death.

4.4 Colorectal Cancer

CRC is the third most prominent cancer-causing death both in males and females. The new estimated incidences and deaths caused by CRC are 8% and 9% respectively in 2022 (American Cancer Society 2022). It is most commonly characterized by rectal bleeding, blood appearance in the stool, stool narrower than normal condition, painful or incomplete bowels, abdominal cramps or pain, weight loss, and reduced appetite. Though, the early stage CRC is difficult to stage because the latter is asymptomatic (Bohorquez et al. 2016). To prevent CRC, it is crucial to identify and remove benign tumor growth during the initial screening process. In addition, it is imperative to diagnose cancer at an earlier stage, which allows for easier treatment and better outcomes. Through routine screenings via fecal immunochemical tests in stool samples, multi-target stool DNA tests, highly sensitive Guaiac-based tests, or diagnostic imaging (computed tomography colonography or colonoscopy), the risk of premature CRC mortality can be reduced (Stracci et al. 2014). In recent years, a lot of effort has been put into developing early detection tests for CRC. There are some FDA-approved diagnostic tests are available to diagnose the CRC at an early stage such as cobas KRAS Mutation Test, Dako EGFR pharmDx Kit, FoundationOne CDx, ONCO/Reveal Dx Lung & Colon Cancer Assay, Praxis Extended RAS Panel, *therascreen* KRAS RGQ PCR Kit, and *therascreen* BRAF V600E RGQ PCR Kit (Table 2). Dako EGFR pharmDx Kit is an IHC-based diagnostic test to ascertain the qualitative expression of EGFR protein present on normal and cancer cells to identify the CRC patients eligible for Erbitux and Vectibix treatment (Buckley and Kakar 2007). Praxis Extended RAS Panel is an *in vitro* diagnostic test using NGS technology for assessing RAS mutations (12 codons in *KRAS* and *NRAS* genes and 56 activating gene mutations) in metastatic CRC individuals eligible for Vectibix treatment (Udar et al. 2020). There are several interesting targeted therapies against CRC including humanized mAbs (nivolumab, pembrolizumab, ipilimumab, and bevacizumab), EGFR antagonist (cetuximab and panitumumab), VEGFR2 antagonist (ramucirumab), and multi-kinase inhibitors (ziv-aflibercept, regorafenib, and encorafenib). The detailed version of these targeted therapies is represented in Table 1.

4.5 Pancreatic Cancer

The pancreatic cancer is the fourth leading cause of cancer deaths in both sexes male and female. The new estimated incidences and deaths responsible for pancreatic cancer are 3% and 8%, respectively in 2022 (American Cancer Society 2022). In the last decade, incidence frequency and death rate of pancreatic cancer has risen by approximately 1% and 0.2% per year, respectively (American Cancer Society 2022). The clinical manifestations of pancreatic cancer are as usual do not appear in the initial stage, but later at advanced-stage shows nausea, weight loss, vomiting, abdominal discomfort, severe abdominal pain, and jaundice (Fernandez-del Castillo et al. 2020). Much effort needs to be done to establish some effective diagnostic biomarkers to identify the pancreatic cancer at an early stage. Liver function test and blood tests is few diagnostic tests that can be used to detect the pancreatic cancer. The initial indication of pancreatic cancer is yellowing of eyes and skin that can be confirmed by measuring the high levels of bilirubin using the liver function test (Kiriyama et al. 2010). Blood test used to measure the concentration of substances released by the tumor cells specifically in case of pancreatic tumor i.e. carbohydrate antigen 19-9 and carcinoembryonic antigen (Kiriyama et al. 2010). Treatment options such as surgery, radiation therapy, and chemotherapy may reduce the risk of recurrence, prolong survival, and relieve symptoms, though cannot produce a cure. Therefore, the immunotherapeutic agents are in demand cure the pancreatic cancer by specifically attacking the target of interest responsible for the genesis of pancreatic cancer. Several targeted therapies are available to cure pancreatic cancer by specifically targeting and destroying the cancerous cells, such as kinase inhibitors (sunitinib malate and erlotinib hydrochloride), HIF-2α transcription factor inhibitors (belzutifan), macrolide immunosuppressant (everolimus), and PARP enzyme inhibitor (olaparib) (Table 1).

4.6 Lymphoma

This cancer is initiated in the immune system cells and then disseminated to other parts of the body. Lymphoma is mainly categorized into Hodgkin and non-Hodgkin lymphoma (NHL) and then further sub-categorized on the basis of cell characteristics, cell-surface markers, and anatomic sites where they present. The NHL is one of the most prominent cancers compared to the Hodgkin lymphoma. The new estimated incidences and deaths will be in Hodgkin lymphoma (8540 new cases and 920 deaths) and NHL (80470 new cases and 20250 deaths), respectively in 2022 (American Cancer Society 2022). In the last decade, the incidence rate of Hodgkin lymphoma and NHL has declined by approximately 1.4% and 0.4% per year, respectively (American Cancer Society 2022). The death rate of Hodgkin lymphoma and NHL has declined by approximately 4.5% and 2.2% per year, respectively (American Cancer Society 2022). The clinical symptoms of lymphoma are appetite loss, chest pain, and breathing of shortness, abdominal fullness, and swollen lymph nodes at different areas of the body (e.g., neck, underarm, or groin) (Brown and Skarin 2004). There is only one FDA-approved lab companion test for the diagnosis of lymphoma is cobas EZH2 Mutation Test (Table 2). The cobas® EZH2 Mutation Test is a RT-PCR test for qualitative detection of allele specific SNPs for A692V, Y646F A682G, or Y646H/S/C, and Y646N of the *EZH2* gene in DNA isolated from FFPE tumor tissue to identify follicular lymphoma in individuals eligible for tazverik treatment (Nagy et al. 2020). The conventional treatment of NHL has done using chemotherapy, radiation therapy, and immunotherapy but seldom to treat the recurrent lymphoma. Several targeted therapies are available that aid to identify the immune system of the body and target cancerous lymphoma cells shown favourable results for the treatment of hard-to-treat or recurrent lymphomas. Various classes of targeted drugs exist to cure lymphoma, such as T-cell immunotherapy (brexucabtagene autoleucel, tisagenlecleucel, axicabtagene ciloleucel, lisocabtagene maraleucel), antibody-drug conjugate (brentuximab vedotin, ibritumomab tiuxetan, loncastuximab tesirine-lpyl, polatuzumab vedotin-piiq), humanized mAb (obinutuzumab, rituximab, mogamulizumab-kpkc, tafasitamab-cxix), histone deacetylase inhibitor (belinostat, romidepsin, vorinostat), antineoplastic folate analog (pralatrexate), kinase inhibitor (acalabrutinib, ibrutinib, pemigatinib, crizotinib, copanlisib hydrochloride, zanubrutinib), nuclear export inhibitor (selinexor), methyltransferase inhibitor (tazemetostat hydrobromide), retinoid X receptor (RXR) activator (bexarotene), and small molecule inhibitor (duvelisib, venetoclax) (Table 1).

4.7 Skin Cancer

The cancerous growth in skin is diagnosed most frequently compared to other cancers in the United States of America. The skin cancer is categorized into two categories melamona and non-melanoma. Non-melanoma is sub-classified as basal cell cancer (BCC) and squamous cell cancer (SCC). Among all skin cancers, melanomas constitute the most serious type of skin cancer, representing roughly one percent of all skin cancer cases, although most deaths are associated with them (American Cancer Society 2022). There will be approximately 99,780 patients of invasive melanoma and 97,920 patients of *in situ* melanoma diagnosed in the USA, though 7650 individuals will succumb to it (American Cancer Society 2022). The clinical manifestation of skin cancers are change in size, morphology, and colour of skin lesion or mole, unhealable sore, and appearance of new skin growth (Gordon 2013). An early diagnosis of skin cancer is important to cure the skin cancer and lessen the chances of becoming an invasive melanoma. There are some FDA-cleared diagnostic tests available that detect skin cancer by identifying the specific biomarker in tissue sample such as cobas 4800 BRAF V600 Mutation Test, FoundationOne CDx, and THXID BRAF Kit (Table 2). cobas 4800 BRAF V600 Mutation Test is an *in vitro* diagnostic test using RT-PCR technology for evaluating the *BRAF* V600E mutation in qualitative manner in genomic DNA isolated from FFPE human melanoma tissue for treatment with vemurafenib (Mourah et al. 2015). *BRAF* gene mutations such as V600E or V600K in DNA isolated from FFPE human melanoma tissue can be detected using the THxID™ BRAF assay using RT-PCR technology. It is anticipated that the diagnostic test will be used to select melanoma patients with the *BRAF* V600E mutation for treatment with dabrafenib. In addition, it will be used as a means of selecting melanoma patients with *BRAF* V600E or V600K gene mutations to provide treatment with trametinib (Cheng et al. 2018). The non-melanoma cancer is treatable with minor surgery but invasive melanoma required targeted therapies to specifically mark those molecules aid to recognize the cancerous cells from normal healthy cells and to destroy them with few side effects. Various biomarkers are exist that can be recognized by targeted therapies to kill skin cancer cells, such as PD-L1/L2, BRAF V600E, CTLA-4, multiple kinases, and hedgehog pathway. The detailed version of targeted therapies that recognize the above biomarkers are listed in Table 1.

4.8 Ovarian Cancer

The ovarian cancer is fifth underlying cause of cancer mortality around 4% in female in 2022. Approximately, 19,880 new estimated incidences of cancer death will be detected in the USA in 2022, while 12,810 women will succumb from this disease (American Cancer Society 2022). The early stage ovarian cancer is asymptomatic, while the late stage shows swelling in abdomen, urinary urgency, abdominal pain, bloating, difficulty eating, and back pain (Mahmood et al. 2022). Although clinical trials are being conducted to identify effective strategies to screen for ovarian cancer, there are no recommended screening tests at present. Several FDA-approved companion diagnostic tests are available to diagnose the ovarian cancer, including BRACAnalysis CDx, FoundationFocus CDxBRCA Assay, FoundationOne CDx, FoundationOne Liquid CDx, and Myriad myChoice CDx. By analyzing genomic DNA obtained from whole blood specimens, BRACAnalysis CDx™ is an *in vitro* diagnostic test detects SNPs, insertions, deletions, and substitutions in the protein coding regions and intron/exon boundaries of the *BRCA1* and *BRCA2* genes. (LaFargue et al. 2019). FFPE ovarian tumor tissue has been shown to contain mutations in the *BRCA1* and *BRCA2* genes, which are highly relevant for treatment of ovarian cancer with rucaparib when the FoundationFocus™ CDxBRCA test is used *in vitro* for the qualitative detection of the mutations. (Ford et al. 2020). Myriad myChoice CDx, a diagnostic test based on NGS technology, operates similarly to the latter, as this test is suitable for ovarian cancer treatment with olaparib or niraparib (Arora et al. 2021). Treatment often includes surgery and chemotherapy for early-stage ovarian cancer and targeted therapy accounts for the late-stage ovarian cancer. The targeted drug therapies used for ovarian cancer including recombinant humanized mAb (bevacizumab) and PARP inhibitors (niraparib tosylate monohydrate, olaparib, and rucaparib camsylate) (Table 1).

5 Experiences from Clinical Trials

The progress of clinical trial designs for precision medicine revolves largely around advances in tailored targeted therapies, made possible through the simultaneous advancement and emergence of high-throughput technologies (i.e., NGS) enabling detailed knowledge about tumors, along with a global increase in cancer screening initiatives (Solis et al. 2022). It is important to molecularly characterize and rationally match antitumor agents with known mechanisms of action to molecularly targeted therapies in order to ultimately benefit patients (Lopez et al. 2015; Joshi et al. 2019a,b; Zhong et al. 2021; Lin et al. 2019). However, most cancers remain poorly managed despite the promise of individualizing antitumor treatments. It may be due to the need to recruit biomarker-specific populations, which makes traditional clinical trials of molecular-guided treatment more expensive and lengthy (Janiaud et al. 2019), making it more difficult to translate the findings into real-life benefits. It is imperative to screen large populations in clinical trials to achieve adequate power. An array of organizations from industry to academia to regulatory agencies to patient associations must be involved and collaborated with precision medicine to achieve high-quality evidence. In order to meet this challenge, clinical trials must be designed in a novel way that produces the necessary evidence as efficiently as possible, resulting in more lenient standards of drug approval. The selected precision based medicine trials include basket trials (for example, BRAFV600 trials), enrichment trials (e.g., ToGA trials), umbrella (e.g., Lung-MAP, ALCHEMIST, FOCUS-4, and ADAPT trials), and adaptive trials (e.g., BATTLE-1, I-SPY-2, and BATTLE-2 trials), as well as combinations of these trials are hybrid trials (e.g., SHIVA, MPACT, and TAPUR trials) (Janiaud et al. 2019; Ferrarotto et al. 2015; Alden et al. 2015; Liu and Lee 2015; Hofmann et al. 2013; Renfro et al. 2017; Hollingsworth 2015). The detailed version of planned, ongoing, and completed basket and umbrella trials for cancer treatment are mentioned in the Tables 3 and 4, respectively. Ultimately precision medicine trials will be beneficial to future cancer patients not that of the individuals participating, and trials cannot be regarded as a means of avoiding intensive clinical trials to accelerate access to treatment options that have not yet been proven. As trials become more efficient, innovative therapies will be available to cancer patients more rapidly.

6 Conclusion

With the increasingly sophisticated use of genomic and molecular biomarkers, precision oncology is rapidly expanding as a research field. In spite of its impact on treatment guidelines already being realized, the full implications of its findings for clinical medicine rely heavily on ensuring that diagnostics and therapeutics are accessible to patients in countries with limited access to academic centers. As a first step toward achieving this goal, we need to accelerate the implementation of clinical

Table 3 Selected precision medicine based basket clinical trials in oncology

Clinical trial	Cancer	Phase	Status	Sample size	Target	Treatment
Basket trials are conducted by targeting single nucleotide mutation using an individual drug in multiple cancers						
NCT05483400 (TIRACAN trial)	HNC, MSI-H cancer, melanoma	Phase II	Planned	97	PD-L1, TIGIT	Tiragolumab and atezolizumab
NCT04174716 (VASTUS trial)	Solid tumors	Phase I/II	Ongoing	310	PARP enzyme	IDX-1197
NCT05103358 (PRECISION 1 trial)	Metastatic solid tumors	Phase II	Ongoing	120	*TSC1* and *TSC2* genes mutations	Nab-sirolimus
NCT04690699	Muscle invasive bladder cancer, non-muscle-invasive bladder cancer, and solid tumors	Phase I/II	Ongoing	15	PD-L1	Lerapolturev
NCT05221320	Solid tumors and gastrointestinal cancer	Phase II	Ongoing	215	MAPK mutations	Ulixertinib and hydroxychloroquine
NCT03982173 (MATILDA trial)	Esophageal adenocarcinoma, metastatic CRC, prostate adeno-carcinoma, triple negative breast cancer (TNBC), and stomach adenocarcinoma	Phase II	Planned	88	PD-L1, CTLA-4	Tremelimumab and durvalumab
NCT02509507 (MASTERKEY-318 trial)	Hepatocellular carcinoma, liver metastases, and cutaneous or subcutaneous lymph node	Phase I/II	Planned	127	–	Talimogene laherparepvec and pembrolizumab
NCT03491345 (K-BASKET trial)	Solid tumors	Phase II	Planned	60	*PD-L1*, *EBV*, *MSI-H* or *POLE/POLD1* mutations	Avelumab
NCT02428270	Pancreatic cancer and adenocarcinoma	Phase II	Planned	16	MEK1 & 2 and FAK-1	GSK2256098 and trametinib

NCT03767075 (BoB trial)	Advanced solid tumors	Phase II	Ongoing	1000	PD-L1	Atezolizumab
NCT04276376 (ARIANES trial)	Solid tumors	Phase II	Ongoing	1000	PD-L1 and PARP	Atezolizumab and rucaparib
NCT03017521	Solid tumors	Phase II	Completed	13	*P13K/AKT* gene mutations	TAS-117
NCT02399943	CRC	Phase II	Completed	14	EGFR and MEK1 & 2	Trametinib and panitumumab
NCT03518606 (MOVIE trial)	BCa, advanced solid tumours, PCa, HNC, and cervix cancer	Phase I/II	Planned	150	PD-L1, CTLA-4	Durvalumab + tremelimumab + metronomic vinorelbine
NCT03872427	Advanced solid tumors, and unresectable malignant solid tumors	Phase II	Planned	108	*NF1*, *KEAP1/NRF2*, or *STK11/LKB1* mutation	Telaglenastat hydrochloride
NCT02478320	Advanced cancers	Phase II	Completed	12	*CDKN2A* Deletion or Mutation	Ilorasertib
NCT02506517	Solid tumors	Phase II	Planned	30	HER	Afatinib
NCT04779151 (NIRADO trial)	Urothelial bladder cancer, biliary tract cancer, gastric adenocarcinoma, clear cell renal cell carcinoma, and HNC	Phase II	Ongoing	112	PD-L1 and PARP	Dostarlimab and niraparib
NCT04534283	Metastatic cancers	Phase II	Ongoing	35	*BRAF, RAF1, MEK1/2, ERK1/2*, and *NF1* mutations	Abemaciclib and LY3214996
NCT03428802	Metastatic malignant solid neoplasm, recurrent ovarian carcinoma, and recurrent or advanced BCa	Phase II	Ongoing	40	*POLE/POLD1* and *BRCA1/2* mutations	Pembrolizumab
NCT04400474 (CABATEN trial)	Paraganglioma, pheochromocytoma, adenocarcinoma,	Phase II	Ongoing	144	VEGF	Cabozantinib

(continued)

Table 3 (continued)

Clinical trial	Cancer	Phase	Status	Sample size	Target	Treatment
	anaplastic thyroid cancer, and neuroendocrine tumor					
NCT03915678 (AGADIR trial)	Solid tumors, pancreatic cancer, NSCLC, melanoma, bladder cancer, and TNBC	Phase II	Ongoing	247	PD-1/PD-L1	Atezolizumab + BDB001 + radiotherapy
NCT02955290	Advanced and recurrent HNC and metastatic NSCLC	Phase I/II	Ongoing	193	PD-1	Nivolumab, pembrolizumab, and recombinant human EGF-rP64K/montanide vaccine
NCT04722133	Biliary tract cancer	Phase II	Ongoing	34	HER2	HERZUMA+mFOLFOX
NCT03386721	Cervical cancers, esophageal cancer, and advanced HNC	Phase II	Completed	256	Fibroblast Activation Protein-A (FAP)	Simlukafusp alfa, atezolizumab, gemcitabine, and vinorelbine
NCT04357873 (PEVOsq trial)	Anal cancer, HNC, penile cancer, vulvar cancer, cervix cancer, and squamous cell lung cancer	Phase II	Planned	112	-	Pembrolizumab and vorinostat
NCT02372006	Neuroectodermal tumors and rhabdomyosarcoma	Phase I/II	Completed	56	HER	Afatinib
NCT05152797	Tumor infiltrating lymphocytes	Phase I/II	Planned	91	–	Autologous tumor-infiltrating lymphocytes
NCT02813135 (ESMART trial)	Refractory or recurrent tumors	Phase I/II	Ongoing	397	–	Ribociclib, topotecan, temozolomide, AZD1775, carboplatin, olaparib, irinotecan, nivolumab, enasidenib, and lirilumab
NCT05329792 (IntriNSiC trial)	Skin keratoacanthoma, Basal and squamous cell carcinoma, malignant adnexal skin tumors,	Phase II	Planned	70	–	L19IL2/L19TNF

	kaposi sarcoma, cutaneous T-cell lymphoma, and merkel cell carcinoma					
NCT03810872	Cancers harbouring an *EGFR* mutation	Phase II	Ongoing	87	*EGFR*, *HER2* or *HER3* Mutation	Afatinib and paclitaxel
NCT04122625 (SMARTPLUS-106 trial)	Solid tumor	Phase I/II	Completed	46	PD-1/PD-L1	Debio 1143 and nivolumab
NCT03339843 (MiMe-A trial)	Esophageal adenocarcinoma, cholangiocarcinoma, urothelial/bladder cancer, endometrial cancer	Phase II	Planned	85	–	Abemaciclib
NCT03297606 (CAPTUR trial)	Advanced solid tumors, multiple myeloma, and NHL	Phase II	Ongoing	720	–	Ipilimumab, dasatinib, olaparib, and nivolumab; Sunitinib, axitinib, bosutinib, palbociclib and crizotinib; Pertuzumab, temsirolimus, trastuzumab, and erlotinib; Cobimetinib and vemurafenib; Tucatinib and vismodegib
NCT04185831 (MEGALiT trial)	Solid tumors	Phase II	Ongoing	154	MEK, MTOR, and PD-L1	Atezolizumab, everolimus, and cobimetinib
NCT04617054	Solid tumors	Phase II	Ongoing	40	–	AB-106
NCT04579380	BCa, biliary tract cancer, NSCLC, urologic neoplasms, and uterine cervical cancer	Phase II	Ongoing	270	*HER2* mutations	Tucatinib, trastuzumab, and fulvestrant
NCT04266912	Metastatic, refractory, and unresectable malignant solid neoplasm	Phase I/II	Ongoing	36	–	Avelumab and berzosertib

(continued)

Table 3 (continued)

Clinical trial	Cancer	Phase	Status	Sample size	Target	Treatment
NCT04551521	Metastatic or locally advanced malignancies	Phase I/II	Ongoing	175	*BRAF* V600E/K, *ERBB2* amplification and overexpression, *ALK* gene rearrangement, and *PIK3CA* or *AKT* mutations	Vemurafenib, cobimetinib, atezolizumab, trastuzumab, pertuzumab, alectinib, and ipatasertib
NCT02568267 (STARTRK-2 trial)	Adult solid tumors, cholangiocarcinoma, HNC, CRC, NSCLC, sarcomas, lymphoma, ovarian cancer, anaplastic melanoma, primary brain tumors, neuroendocrine tumors, renal cell carcinoma, pancreatic cancer, salivary gland cancers, papillary thyroid cancer, and BCa	Phase II	Ongoing	700	*NTRK1/2/3*, *ALK* gene rearrangements, or *ROS1*	Entrectinib
NCT05077709	NSCLC, HNC, and urothelial bladder carcinoma	Phase II	Ongoing	90	PD-L1	IO102-IO103 and pembrolizumab
NCT03851614 (DAPPER trial)	Leiomyosarcoma, pancreatic adenocarcinoma, and mismatch repair proficient CRC	Phase II	Planned	90	PD-L1, VEGFR, and PARP	Durvalumab, olaparib, and cediranib
NCT04993677	Melanoma and non-small- cell lung carcinoma	Phase II	Ongoing	200	–	SEA-CD40, pembrolizumab, pemetrexed, and carboplatin
NCT04374305 (INTUITT-NF2 trial)	Ependymoma, vestibular and non-vestibular schwannoma, neurofibromatosis type 2, and meningioma	Phase II	Ongoing	80	*NF2* gene	Brigatinib

Table 4 Various precision medicine based umbrella clinical trials for the treatment of cancers

Clinical trial	Cancer	Phase	Status	Sample size	Target	Treatment
Umbrella trials are conducted by targeting multiple genetic mutations using multiple therapeutic agents in a single cancer type						
NCT04395989 (FUTURE SUPER trial)	TNBC	Phase II	Ongoing	138	*HER2* mutation, PD-1, VEGFR, EGFR	Pyrotinib with capecitabine and nab-paclitaxel; everolimus with nab-paclitaxel; PD-1 with nab-paclitaxel and famitinib; VEGFR and nab-paclitaxel; everolimus with nab-paclitaxel
NCT05270889 (K-Umbrella trial)	*HER2*-positive gastric cancer	Phase II	Planned	50	*HER2* mutation	Tislelizumab + zanidatamab
NCT03292250 (SUKSES trial)	HNC	Phase II	Completed	180	PI3K, EGFR/HER2, FGFR, CDK4/6, and PD1/PD-L1	BYL719, poziotinib, nintedanib, abemaciclib, durvalumab, and tremelimumab
NCT05101564	BCa	Phase II	Planned	264	ER	Alpelisib, tamoxifen, infigratinib, amcenestrant, and letrozole
NCT02951091	Gastric cancer	Phase II	Planned	400	HER2, EGFR, c-MET, and PTEN	Biomarker screening
NCT03267589	Ovarian cancer	Phase II	Completed	25	–	Durvalumab, tremelilumab, MEDI 9447, and MEDI 0562
NCT03805399 (FUTURE trial)	TNBC	Phase I/II	Ongoing	140	AR, CDK4/6, EZH2, PD-1, VEGFR, mTOR	mTOR inhibitor plus nab-paclitaxel; Pyrotinib plus capecitabine; anti PD-1 plus nab-paclitaxel; AR inhibitor combined with CDK4/6 inhibitor or everolimus or EZH2 inhibitor; mesenchymal-like TNBC with anti-VEGFR included therapy; PARP

(continued)

Table 4 (continued)

Clinical trial	Cancer	Phase	Status	Sample size	Target	Treatment
						inhibitor included therapy; basal-like immune suppressed TNBC with anti-VEGFR included therapy
NCT03182634 (plasmaMATCH trial	Advanced BCa	Phase II	Ongoing	1150	–	Fulvestrant, neratinib, AZD5363, olaparib, and AZD6738
NCT04754191 (ENCORE trial)	Metastatic castration-resistant PCa	Phase II	Ongoing	34	–	Enfortumab vedotin
NCT04185831 (MEGALiT trial)	Solid tumors	Phase II	Ongoing	154	MEK, mTOR, PD-L1	Atezolizumab, everolimus, and cobimetinib
NCT04215003	BCa	Phase I/II	Ongoing	20	HER2, AR, PI3K-AKT, CDK4/6	Paclitaxel, carboplatin, herceptin, and pertuzumab; paclitaxel and carboplatin; SHR6390, fulvestrant, and goserelin; SHR6390, alpelisib, and goserelin; SHR6390, fulvestrant, herceptin, pertuzumab, and goserelin; SHR6390, alpelisib, herceptin, pertuzumab, and goserelin; pyrotinib, herceptin, and capecitabine; paclitaxel, carboplatin, herceptin, alpelisib, pertuzumab, and goserelin; SHR6390 and SHR3680
NCT04643002	Plasma cell myeloma refractory	Phase I/II	Ongoing	72	–	Isatuximab SAR650984, dexamethasone, pomalidomide, and belantamab mafodotin

NCT02664935	NSCLC, squamous cell carcinoma, adenocarcinoma	Phase II	Planned	423	FGFR, mTORC1/2, CDK4/6, ALK/MET/ROS1, MEK, AKT, PDL1, VEGFR, ATR	AZD4547, vistusertib, palbociclib, crizotinib, selumetinib, docetaxel, AZD5363, osimertinib, durvalumab, sitravatinib, and AZD6738
NCT04605562	Nasopharyngeal carcinoma	Phase II	Planned	206	PD-1	GP+DDP, palbociclib, intensity-modulated radiotherapy, PD-1 blocking antibody, and galunisertib
NCT04072107 (EP-STAR trial)	Nasopharyngeal carcinoma	Phase II	Planned	110	PD-1	Sintilimab and Capecitabine
NCT05563220 (ELEVATE trial)	Metastatic BCa	Phase I/II	Planned	322	ER	Elacestrant, alpelisib, everolimus, ribociclib, and palbociclib
NCT03574402 (TRUMP trial)	NSCLC	Phase II	Ongoing	400	*EGFR* mutation, *MET* 14 exon mutation, *MET* amplification, *ROS1* fusion, *Ntrk1/2/3* fusion, *HER2* mutation, *EGFR* amplification, *ALK* rearrangement	Avitinib maleate, afatinib, crizotinib, X-396, chidamide, pyrotinib maleate, AZD3759, pirotinib, nimotuzumab, pemetrexed, cisplatin, sintilimab, gemcitabine, and carboplatin
NCT04193722 (HONEY trial)	BCa and radiation toxicity	Phase III	Ongoing	120	–	Hyperbaric oxygen therapy
NCT05255653 (RAINBO trial)	Endometrial cancer	Phase II/III	Ongoing	1611	–	Megestrol acetate, olaparib, durvalumab, chemotherapy, medroxyprogesterone acetate, and pelvic external beam radiotherapy
NCT03574779 (OPAL trial)	Ovarian neoplasms	Phase I/II	Ongoing	125	PD-1, PARP	Niraparib, TSR-042, bevacizumab, carboplatin, and paclitaxel

(continued)

Table 4 (continued)

Clinical trial	Cancer	Phase	Status	Sample size	Target	Treatment
NCT03851445 (Lung-MAP trial)	NSCLC	Phase II/III	Ongoing	10000	–	Screening platform
NCT04005820	Kidney cancer	-	Planned	510	–	Radiologic centralized review and tissue sample
NCT05332561 (COGNITION-GUIDE trial)	Early-stage BCa	Phase II	Planned	240	PI3K, AKT, PARP, TROP-2	Atezolizumab, inavolisib, ipatasertib, olaparib, sacituzumab govitecan, and trastuzumab/pertuzumab
NCT04524871 (Morpheus-Liver trial)	Advanced liver cancers	Phase I/II	Ongoing	320	–	Atezolizumab, bevacizumab, tiragolumab, tocilizumab, TPST-1120, RO7247669, RO7247669, and RO7247669
NCT03424005 (Morpheus-TNBC trial)	TNBC	Phase I/II	Ongoing	280	PD-L1	Sacituzumab govitecan, atezolizumab, bevacizumab, chemotherapy (eribulin + carboplatin or gemcitabine), ipatasertib, tocilizumab, SGN-LIV1A, nab-Paclitaxel, selicrelumab, and capecitabine
NCT04812366 (GUNS trial)	PCa	Phase II	Ongoing	315	–	Apalutamide, abiraterone acetate, prednisone, docetaxel, niraparib, and atezolizumab
NCT04802759 (Morpheus-Breast cancer trial)	Unresectable metastatic BCa	Phase I/II	Ongoing	510	CDK4/6, ER, HER2	Atezolizumab, inavolisib, abemaciclib, ribociclib, ipatasertib, everolimus, samuraciclib, palbociclib, PH FDC SC, and giredestrant

NCT03983226 (SOC-3 trial)	Primary peritoneal cancer, fallopian tube cancer, and recurrent ovarian cancer	Phase II	Ongoing	167	PARP	Surgery, niraparib, liposome doxorubicin or carboplatin, carboplatin/gemcitabine, carboplatin or taxane, and cis-platin or gemcitabine
NCT04739202 (IMMUNOGAST trial)	Gastric adenocarcinoma, metastatic gastric cancer, and metastatic adenocarcinoma	Phase II	Ongoing	60	–	Atezolizumab + ipatasertib and atezolizumab + bevacizumab
NCT04161417	Pancreatic neoplasms	-	Ongoing	2500	–	Molecular profiling of biopsy material
NCT05459129 (Morpheus-Head and Neck Cancer trial)	HNC	Phase I/II	Ongoing	180	–	Atezolizumab, tiragolumab, immune-modulating stereotactic radiotherapy (iSBRT), carboplatin, and paclitaxel
NCT05251948	Gastric and gastroesophageal junction carcinoma	Phase I/II	Planned	90	–	Atezolizumab, capecitabine, oxaliplatin, and tiragolumab
NCT03158389 (N^2M^2 trial)	Adult glioblastoma	Phase I/II	Ongoing	350	–	APG101, alectinib, idasanutlin, atezolizumab, vismodegib, temsirolimus, palbociclib

advances based on research, to invest in transforming healthcare infrastructures to accommodate new resources free of health inequalities, and to provide information to patients to make informed decisions. It will be critical to put resources into advancing assets for healthcare systems, for instance clinical genomic data repositories, and to enhance the emergence of standardized decision support tools and multidisciplinary team training. There are inevitably many challenges to address, including how different clinical trial studies can contribute to the approval of new precision medicine products on the market, so collaboration and crossing boundaries will be essential to overcome these challenges. Nevertheless, the evolution of precision medicine has led to significant advances in creating truly patient-centered clinical trial designs in the drug development process, which allows us to tailor the trial that best fits the patient as opposed to the patient being selected for the trial.

References

Alden RS, Mandrekar SJ, Oxnard GR (2015) Designing a definitive trial for adjuvant targeted therapy in genotype defined lung cancer: the ALCHEMIST trials. Chin Clin Oncol 4:3

Alix-Panabières C, Pantel K (2013) Circulating tumor cells: liquid biopsy of cancer. Clin Chem 59(1):110–118

Al-Nedawi K, Meehan B, Micallef J, Lhotak V, May L, Guha A, Rak J (2008) Intercellular transfer of the oncogenic receptor EGFRvIII by microvesicles derived from tumour cells. Nat Cell Biol 10(5):619–624

American Cancer Society (2022) Cancer facts & figures. American Cancer Society, Atlanta

Antonarakis ES, Lu C, Luber B, Wang H, Chen Y, Nakazawa M, Nadal R, Paller CJ, Denmeade SR, Carducci MA, Eisenberger MA (2015) Androgen receptor splice variant 7 and efficacy of taxane chemotherapy in patients with metastatic castration-resistant prostate cancer. JAMA Oncol 1(5):582–591

Arora S, Balasubramaniam S, Zhang H, Berman T, Narayan P, Suzman D, Bloomquist E, Tang S, Gong Y, Sridhara R, Turcu FR (2021) FDA approval summary: olaparib monotherapy or in combination with bevacizumab for the maintenance treatment of patients with advanced ovarian cancer. Oncologist 26(1):164–172

Bauml JM, Li BT, Velcheti V, Govindan R, Curioni-Fontecedro A, Dooms C, Takahashi T, Duda AW, Odegaard JI, Cruz-Guilloty F, Jin L (2022) Clinical validation of Guardant360 CDx as a blood-based companion diagnostic for sotorasib. Lung Cancer 166:270–278

Berrondo C, Flax J, Kucherov V, Siebert A, Osinski T, Rosenberg A, Fucile C, Richheimer S, Beckham CJ (2016) Expression of the long non-coding RNA HOTAIR correlates with disease progression in bladder cancer and is contained in bladder cancer patient urinary exosomes. PLoS One 11(1):e0147236

Bettegowda C, Sausen M, Leary RJ, Kinde I, Wang Y, Agrawal N, Bartlett BR, Wang H, Luber B, Alani RM, Antonarakis ES (2014) Detection of circulating tumor DNA in early-and late-stage human malignancies. Sci Transl Med 6(224):224

Bohorquez M, Sahasrabudhe R, Criollo A, Sanabria-Salas MC, Velez A, Castro JM, Marquez JR, Mateus G, Bolanos F, Panqueva C, Restrepo JI (2016) Clinical manifestations of colorectal cancer patients from a large multicenter study in Colombia. Medicine 95(40):e4883

Brandão M, Pondé N, Piccart-Gebhart M (2019) Mammaprint™: a comprehensive review. Future Oncol 15(2):207–224

Brown JR, Skarin AT (2004) Clinical mimics of lymphoma. Oncologist 9(4):406–416

Bryan RT, Wei W, Shimwell NJ, Collins SI, Hussain SA, Billingham LJ, Murray PG, Deshmukh N, James ND, Wallace DM, Johnson PJ (2011) Assessment of high-throughput high-resolution MALDI-TOF-MS of urinary peptides for the detection of muscle-invasive bladder cancer. Proteomics Clin Appl 5(9-10):493–503

Buckley AF, Kakar S (2007) Comparison of the Dako EGFR pharmDx kit and Zymed EGFR antibody for assessment of EGFR status in colorectal adenocarcinoma. Appl Immunohistochem Mol Morphol 15(3):305–309

Butz H, Nofech-Mozes R, Ding Q, Khella HW, Szabó PM, Jewett M, Finelli A, Lee J, Ordon M, Stewart R, Krylov S (2016) Exosomal microRNAs are diagnostic biomarkers and can mediate cell–cell communication in renal cell carcinoma. Eur Urol Focus 2(2):210–218

Carlson JJ, Roth JA (2013) The impact of the Oncotype Dx breast cancer assay in clinical practice: a systematic review and meta-analysis. Breast Cancer Res Treat 141(1):13–22

Casadio V, Calistri D, Tebaldi M, Bravaccini S, Gunelli R, Martorana G, Bertaccini A, Serra L, Scarpi E, Amadori D, Silvestrini R (2013) Urine cell-free DNA integrity as a marker for early bladder cancer diagnosis: preliminary data. In: Urologic oncology: seminars and original investigations, vol 31. Elsevier, Amsterdam, pp 1744–1750

Chen W, Cai F, Zhang B, Barekati Z, Zhong XY (2013) The level of circulating miRNA-10b and miRNA-373 in detecting lymph node metastasis of breast cancer: potential biomarkers. Tumor Biol 34(1):455–462

Chen Y, Ma X, Lou C, Zhou C, Zhao X, Li N, Tian H, Meng X (2022) PLA2G10 incorporated in exosomes could be diagnostic and prognostic biomarker for non-small cell lung cancer. Clin Chim Acta 530:55–65

Cheng F, Su L, Qian C (2016) Circulating tumor DNA: a promising biomarker in the liquid biopsy of cancer. Oncotarget 7(30):48832

Cheng L, Lopez-Beltran A, Massari F, MacLennan GT, Montironi R (2018) Molecular testing for BRAF mutations to inform melanoma treatment decisions: a move toward precision medicine. Mod Pathol 31(1):24–38

Chiong E, Wong AF, Chan YH, Chin CM (2005) Review of clinical manifestations of biochemically-advanced prostate cancer cases. Asian J Surg 28(3):202–206

Cohen SJ, Punt CJ, Iannotti N, Saidman BH, Sabbath KD, Gabrail NY, Picus J, Morse MA, Mitchell E, Miller MC, Doyle GV (2009) Prognostic significance of circulating tumor cells in patients with metastatic colorectal cancer. Ann Oncol 20(7):1223–1229

Collins LG, Haines C, Perkel R, Enck RE (2007) Lung cancer: diagnosis and management. Am Fam Physician 75(1):56–63

Cooper L, Chen J (2022) Changes in companion diagnostic labelling: implementation of FDA's April 2020 guidance for industry for in vitro CDx labeling for specific oncology therapeutic groups. Ther Innov Regul Sci 10:1–9

Corcoran C, Rani S, O'Driscoll L (2014) miR-34a is an intracellular and exosomal predictive biomarker for response to docetaxel with clinical relevance to prostate cancer progression. Prostate 74(13):1320–1334

Cristofanilli M, Budd GT, Ellis MJ, Stopeck A, Matera J, Miller MC, Reuben JM, Doyle GV, Allard WJ, Terstappen LW, Hayes DF (2004) Circulating tumor cells, disease progression, and survival in metastatic breast cancer. N Engl J Med 351(8):781–791

De Bono JS, Scher HI, Montgomery RB, Parker C, Miller MC, Tissing H, Doyle GV, Terstappen LW, Pienta KJ, Raghavan D (2008) Circulating tumor cells predict survival benefit from treatment in metastatic castration-resistant prostate cancer. Clin Cancer Res 14(19):6302–6309

De T, Goyal S, Balachander G, Chatterjee K, Kumar P, Babu KG, Rangarajan A (2019) A novel ex vivo system using 3D polymer scaffold to culture circulating tumor cells from breast cancer patients exhibits dynamic EM phenotypes. J Clin Med 8(9):1473

Deligezer U, Erten N, Akisik EE, Dalay N (2006) Circulating fragmented nucleosomal DNA and caspase-3 mRNA in patients with lymphoma and myeloma. Exp Mol Pathol 80(1):72–76

Di Meo A, Pasic MD, Yousef GM (2016) Proteomics and peptidomics: moving toward precision medicine in urological malignancies. Oncotarget 7(32):52460

Di Meo A, Bartlett J, Cheng Y, Pasic MD, Yousef GM (2017) Liquid biopsy: a step forward towards precision medicine in urologic malignancies. Mol Cancer 16(1):1–4

Diamandis M, White N, Yousef GM (2010) Personalized medicine: marking a new epoch in cancer patient management personalized medicine in cancer management. Mol Cancer Res 8(9): 1175–1187

Durand X, Moutereau S, Xylinas E, De La Taille A (2011) Progensa™ PCA3 test for prostate cancer. Expert Rev Mol Diagn 11(2):137–144

Elazezy M, Joosse SA (2018) Techniques of using circulating tumor DNA as a liquid biopsy component in cancer management. Comput Struct Biotechnol J 16:370–378

El-Kenawi A, Hänggi K, Ruffell B (2020) The immune microenvironment and cancer metastasis. Cold Spring Harb Perspect Med 10(4):a037424

Elshimali YI, Khaddour H, Sarkissyan M, Wu Y, Vadgama JV (2013) The clinical utilization of circulating cell free DNA (CCFDNA) in blood of cancer patients. Int J Mol Sci 14(9): 18925–18958

Fernandez-del Castillo C, Howell DA, Robson KM (2020) Clinical manifestations, diagnosis, and staging of exocrine pancreatic cancer. Up to Date

Fernández-Lázaro D, García Hernández JL, García AC, Córdova Martínez A, Mielgo-Ayuso J, Cruz-Hernández JJ (2020) Liquid biopsy as novel tool in precision medicine: Origins, properties, identification and clinical perspective of cancer's biomarkers. Diagnostics 10(4):215

Fernandez-Mercado M, Manterola L, Larrea E, Goicoechea I, Arestin M, Armesto M, Otaegui D, Lawrie CH (2015) The circulating transcriptome as a source of non-invasive cancer biomarkers: concepts and controversies of non-coding and coding RNA in body fluids. J Cell Mol Med 19(10):2307–2323

Ferrarotto R, Redman MW, Gandara DR, Herbst RS, Papadimitrakopoulou VA (2015) Lung-MAP—framework, overview, and design principles. Chin Clin Oncol 4:3

Fidler IJ (2003) The pathogenesis of cancer metastasis: the 'seed and soil' hypothesis revisited. Nat Rev Cancer 3(6):453–458

Ford L, Wolford JE, Brown SM, Randall LM (2020) A profile on the FoundationFocus CDxBRCA tests. Expert Rev Mol Diagn 20(3):285–292

Friedlander TW, Fong L (2014) The end of the beginning: circulating tumor cells as a biomarker in castration-resistant prostate cancer. J Clin Oncol 32(11):1104–1106

Gang F, Guorong L, An Z, Anne GP, Christian G, Jacques T (2010) Prediction of clear cell renal cell carcinoma by integrity of cell-free DNA in serum. Urology 75(2):262–265

Goessl C, Heicappell R, Münker R, Anker P, Stroun M, Krause H, Müller M, Miller K (1998) Microsatellite analysis of plasma DNA from patients with clear cell renal carcinoma. Cancer Res 58(20):4728–4732

Gordon R (2013) Skin cancer: an overview of epidemiology and risk factors. In: Seminars in oncology nursing, vol 29. WB Saunders, Philadelphia, pp 160–169

Guo B, Luo C, Xun C, Xie J, Wu X, Pu J (2009) Quantitative detection of cytokeratin 20 mRNA in urine samples as diagnostic tools for bladder cancer by real-time PCR. Exp Oncol 31(1):43–47

Hanke M, Hoefig K, Merz H, Feller AC, Kausch I, Jocham D, Warnecke JM, Sczakiel G (2010) A robust methodology to study urine microRNA as tumor marker: microRNA-126 and microRNA-182 are related to urinary bladder cancer. In: Urologic oncology: seminars and original investigations, vol 28. Elsevier, Amsterdam, pp 655–661

He H, Han C, Hao L, Zang G (2016) ImmunoCyt test compared to cytology in the diagnosis of bladder cancer: A meta-analysis. Oncol Lett 12(1):83–88

Heitzer E, Haque IS, Roberts CE, Speicher MR (2019) Current and future perspectives of liquid biopsies in genomics-driven oncology. Nat Rev Genet 20(2):71–88

Hofmann D, Nitz U, Gluz O, Kates RE, Schinkoethe T, Staib P, Harbeck N (2013) WSG ADAPT–adjuvant dynamic marker-adjusted personalized therapy trial optimizing risk assessment and therapy response prediction in early breast cancer: study protocol for a prospective, multi-center, controlled, non-blinded, randomized, investigator initiated phase II/III trial. Trials 14(1):1–6

Hollingsworth SJ (2015) Precision medicine in oncology drug development: a pharma perspective. Drug Discov Today 20(12):1455–1463
Imamura T, Komatsu S, Ichikawa D, Kawaguchi T, Miyamae M, Okajima W, Ohashi T, Arita T, Konishi H, Shiozaki A, Morimura R (2016) Liquid biopsy in patients with pancreatic cancer: Circulating tumor cells and cell-free nucleic acids. World J Gastroenterol 22(25):5627
Janiaud P, Serghiou S, Ioannidis JP (2019) New clinical trial designs in the era of precision medicine: an overview of definitions, strengths, weaknesses, and current use in oncology. Cancer Treat Rev 73:20–30
Jia Y, Chen Y, Wang Q, Jayasinghe U, Luo X, Wei Q, Wang J, Xiong H, Chen C, Xu B, Hu W (2017) Exosome: emerging biomarker in breast cancer. Oncotarget 8(25):41717
Joshi H, Malik A, Aggarwal S, Munde M, Maitra SS, Adlakha N, Bhatnagar R (2019a) In-vitro detection of phytopathogenic fungal cell wall by polyclonal sera raised against trimethyl chitosan nanoparticles. Int J Nanomed 14:10023
Joshi H, Verma A, Soni DK (2019b) Impact of microbial genomics approaches for novel antibiotic target. In: Microbial genomics in sustainable agroecosystems. Springer, Singapore, pp 75–88
Joshi H, Kumar G, Tuli HS, Mittal S (2023) Inhibition of cancer cell metastasis by nanotherapeutics: current achievements and future trends. In: Nanotherapeutics in cancer. Jenny Stanford Publishing, Singapore, pp 161–209
Junqueira-Neto S, Batista IA, Costa JL, Melo SA (2019) Liquid Biopsy beyond Circulating Tumor Cells and Cell-Free DNA. Acta Cytol 63:479–488
Kashyap D, Mondal R, Tuli HS, Kumar G, Sharma AK (2016) Molecular targets of gambogic acid in cancer: recent trends and advancements. Tumor Biol 37(10):12915–12925
Kashyap D, Kumar G, Sharma A, Sak K, Tuli HS, Mukherjee TK (2017) Mechanistic insight into carnosol-mediated pharmacological effects: Recent trends and advancements. Life Sci 169:27–36
Kim WT, Jeong P, Yan C, Kim YH, Lee IS, Kang HW, Kim YJ, Lee SC, Kim SJ, Kim YT, Moon SK (2016) UBE2C cell-free RNA in urine can discriminate between bladder cancer and hematuria. Oncotarget 7(36):58193
Kinde I, Munari E, Faraj SF, Hruban RH, Schoenberg M, Bivalacqua T, Allaf M, Springer S, Wang Y, Diaz LA, Kinzler KW (2013) TERT promoter mutations occur early in urothelial neoplasia and are biomarkers of early disease and disease recurrence in urine TERT promoter mutations in early bladder cancers. Cancer Res 73(24):7162–7167
Kiriyama S, Gabata T, Takada T, Hirata K, Yoshida M, Mayumi T, Hirota M, Kadoya M, Yamanouchi E, Hattori T, Takeda K (2010) New diagnostic criteria of acute pancreatitis. J Hepatobiliary Pancreat Sci 17(1):24–36
Kumar G, Tuli HS, Mittal S, Shandilya JK, Tiwari A, Sandhu SS (2015) Isothiocyanates: a class of bioactive metabolites with chemopreventive potential. Tumor Biol 36(6):4005–4016
Kumar G, Mittal S, Sak K, Tuli HS (2016) Molecular mechanisms underlying chemopreventive potential of curcumin: Current challenges and future perspectives. Life Sci 148:313–328
LaFargue CJ, Dal Molin GZ, Sood AK, Coleman RL (2019) Exploring and comparing adverse events between PARP inhibitors. Lancet Oncol 20(1):e15–e28
Le Tourneau C, Borcoman E, Kamal M (2019) Molecular profiling in precision medicine oncology. Nat Med 25(5):711–712
Lee YT, Tan YJ, Oon CE (2018) Molecular targeted therapy: Treating cancer with specificity. Eur J Pharmacol 834:188–196
Leong SP, Ballesteros-Merino C, Jensen SM, Marwitz S, Bifulco C, Fox BA, Skoberne M (2018) Novel frontiers in detecting cancer metastasis. Clin Exp Metastasis 35(5):403–412
Lin A, Giuliano CJ, Palladino A, John KM, Abramowicz C, Yuan ML, Sausville EL, Lukow DA, Liu L, Chait AR, Galluzzo ZC (2019) Off-target toxicity is a common mechanism of action of cancer drugs undergoing clinical trials. Sci Transl Med 11(509):8412
Liu S, Lee JJ (2015) An overview of the design and conduct of the BATTLE trials. Chin Clin Oncol 4:3

Loeb S, Catalona WJ (2014) The Prostate Health Index: a new test for the detection of prostate cancer. Ther Adv Urol 6(2):74–77

Lopez J, Harris S, Roda D, Yap TA (2015) Precision medicine for molecularly targeted agents and immunotherapies in early-phase clinical trials. Transl Oncogenomics 7(1):1

Lötvall J, Hill AF, Hochberg F, Buzás EI, Di Vizio D, Gardiner C, Gho YS, Kurochkin IV, Mathivanan S, Quesenberry P, Sahoo S (2014) Minimal experimental requirements for definition of extracellular vesicles and their functions: a position statement from the International Society for Extracellular Vesicles. J Extracellular Ves 3(1):26913

Mahmood T, Ved A, Siddiqui MH, Ahsan F, Shamim A, Ansari VA, Ahmad A, Kashyap MK (2022) An in-depth analysis of ovarian cancer: pathogenesis and clinical manifestation. Drug Res 72(8):424–434

Majumder S, Ranjan Dahiya U, Yadav S, Sharma P, Ghosh D, Rao GK, Rawat V, Kumar G, Kumar A, Srivastava CM (2020) Zinc oxide nanoparticles functionalized on hydrogel grafted silk fibroin fabrics as efficient composite dressing. Biomol Ther 10(5):710

Malentacchi F, Vinci S, Della Melina A, Kuncova J, Villari D, Nesi G, Selli C, Orlando C, Pazzagli M, Pinzani P (2016) Urinary carbonic anhydrase IX splicing messenger RNA variants in urogenital cancers. In: Urologic oncology: seminars and original investigations, vol 34. Elsevier, Amsterdam, p 292

Maltoni R, Gallerani G, Fici P, Rocca A, Fabbri F (2016) CTCs in early breast cancer: a path worth taking. Cancer Lett 376(2):205–210

McKiernan J, Donovan MJ, O'Neill V, Bentink S, Noerholm M, Belzer S, Skog J, Kattan MW, Partin A, Andriole G, Brown G (2016) A novel urine exosome gene expression assay to predict high-grade prostate cancer at initial biopsy. JAMA Oncol 2(7):882–889

Miller MC, Doyle GV, Terstappen LW (2010) Significance of circulating tumor cells detected by the CellSearch system in patients with metastatic breast colorectal and prostate cancer. J Oncol 2010:617421

Millner LM, Strotman LN (2016) The future of precision medicine in oncology. Clin Lab Med 36(3):557–573

Morrissey JJ, Mobley J, Figenshau RS, Vetter J, Bhayani S, Kharasch ED (2015) Urine aquaporin 1 and perilipin 2 differentiate renal carcinomas from other imaged renal masses and bladder and prostate cancer. In: Mayo clinic proceedings, vol 90. Elsevier, Amsterdam, pp 35–42

Mourah S, Denis MG, Narducci FE, Solassol J, Merlin JL, Sabourin JC, Scoazec JY, Ouafik LH, Emile JF, Heller R, Souvignet C (2015) Detection of BRAF V600 mutations in melanoma: evaluation of concordance between the Cobas® 4800 BRAF V600 mutation test and the methods used in French National Cancer Institute (INCa) platforms in a real-life setting. PLoS One 10(3):e0120232

Nagrath S, Sequist LV, Maheswaran S, Bell DW, Irimia D, Ulkus L, Smith MR, Kwak EL, Digumarthy S, Muzikansky A, Ryan P (2007) Isolation of rare circulating tumour cells in cancer patients by microchip technology. Nature 450(7173):1235–1239

Nagy Á, Bátai B, Balogh A, Illés S, Mikala G, Nagy N, Kiss L, Kotmayer L, Matolcsy A, Alpár D, Masszi T (2020) Quantitative analysis and monitoring of EZH2 mutations using liquid biopsy in follicular lymphoma. Gene 11(7):785

Nelson NJ (2010) Circulating tumor cells: will they be clinically useful? J Natl Cancer Inst 102: 146–148

Otsuji K, Sasaki T, Tanaka A, Kunita A, Ikemura M, Matsusaka K, Tada K, Fukayama M, Seto Y (2017) Use of droplet digital PCR for quantitative and automatic analysis of the HER2 status in breast cancer patients. Breast Cancer Res Treat 162(1):11–18

Oyan B, Sonmez O, Yazar A, Teomete M (2018) Comparison of SP142 (Ventana) and SP263 (Ventana) assays to test PD-L1 expression in metastatic cancer patients to be treated with immune checkpoint inhibitors. ASCO J 2018:e24306

Parekh DJ, Punnen S, Sjoberg DD, Asroff SW, Bailen JL, Cochran JS, Concepcion R, David RD, Deck KB, Dumbadze I, Gambla M (2015) A multi-institutional prospective trial in the USA

confirms that the 4Kscore accurately identifies men with high-grade prostate cancer. Eur Urol 68(3):464–470
Pasic MD, Samaan S, Yousef GM (2013a) Genomic medicine: new frontiers and new challenges. Clin Chem 59(1):158–167
Pasic MD, Yousef GM, Diamandis EP (2013b) The proteomic revolution in laboratory medicine. Clin Biochem 46(6):397–398
Qin X, Zhang H, Ye D, Dai B, Zhu Y, Shi G (2013) B7-H3 is a new cancer-specific endothelial marker in clear cell renal cell carcinoma. Onco Targets Ther 6:1667
Qin Z, Ljubimov VA, Zhou C, Tong Y, Liang J (2016) Cell-free circulating tumor DNA in cancer. Chin J Cancer 35(1):1–9
Reddi KK, Holland JF (1976) Elevated serum ribonuclease in patients with pancreatic cancer. Proc Natl Acad Sci 73(7):2308–2310
Renard I, Joniau S, Van Cleynenbreugel B, Collette C, Naômé C, Vlassenbroeck I, Nicolas H, de Leval J, Straub J, Van Criekinge W, Hamida W (2010) Identification and validation of the methylated TWIST1 and NID2 genes through real-time methylation-specific polymerase chain reaction assays for the noninvasive detection of primary bladder cancer in urine samples. Eur Urol 58(1):96–104
Renfro LA, An MW, Mandrekar SJ (2017) Precision oncology: a new era of cancer clinical trials. Cancer Lett 387:121–126
Rosell R, Wei J, Taron M (2009) Circulating MicroRNA signatures of tumor-derived exosomes for early diagnosis of non-small-cell lung cancer. Clin Lung Cancer 10(1):8–9
Sasaki H, Yoshiike M, Nozawa S, Usuba W, Katsuoka Y, Aida K, Kitajima K, Kudo H, Hoshikawa M, Yoshioka Y, Kosaka N (2016) Expression level of urinary microRNA-146a-5p is increased in patients with bladder cancer and decreased in those after transurethral resection. Clin Genitourin Cancer 14(5):e493–e499
Schiffer E, Vlahou A, Petrolekas A, Stravodimos K, Tauber R, Geschwend JE, Neuhaus J, Stolzenburg JU, Conaway MR, Mischak H, Theodorescu D (2009) Prediction of muscle-invasive bladder cancer using urinary proteomics. Clin Cancer Res 15(15):4935–4943
Sharkey RM, Goldenberg DM (2006) Targeted therapy of cancer: new prospects for antibodies and immunoconjugates. CA Cancer J Clin 56(4):226–243
Siemens DR, Morales A, Johnston B, Emerson L (2003) A comparative analysis of rapid urine tests for the diagnosis of upper urinary tract malignancy. Can J Urol 10(1):1754–1758
Singh K, Thakur R, Kumar G (2016) Synthesis, Characterization and in vitro anticancer activity of Co (II), Ni (II), Cu (II), and Zn (II) complexes with 4-[{3-(4-bromophenyl)-1-phenyl-1H-pyrazol-4-ylmethylene}-amino]-3-mercapto-1, 2, 4-triazin-5-one. Eur Chem Bull 5(2):46–53
Solis RN, Silverman DA, Birkeland AC (2022) Current Trends in Precision Medicine and Next-Generation Sequencing in Head and Neck Cancer. Curr Treat Options in Oncol 2022:1–4
Stracci F, Zorzi M, Grazzini G (2014) Colorectal cancer screening: tests, strategies, and perspectives. Front Public Health 2:210
Sugie S, Mukai S, Yamasaki K, Kamibeppu T, Tsukino H, Kamoto T (2015) Significant association of caveolin-1 and caveolin-2 with prostate cancer progression. Cancer Genomics Proteomics 12(6):391–396
Sunami E, Shinozaki M, Higano CS, Wollman R, Dorff TB, Tucker SJ, Martinez SR, Singer FR, Hoon DS (2009) Multimarker circulating DNA assay for assessing blood of prostate cancer patients. Clin Chem 55(3):559–567
Swensen SJ, Jett JR, Sloan JA, Midthun DE, Hartman TE, Sykes AM, Aughenbaugh GL, Zink FE, Hillman SL, Noetzel GR, Marks RS (2002) Screening for lung cancer with low-dose spiral computed tomography. Am J Respir Crit Care Med 165(4):508–513
Syed YY (2016) Therascreen® EGFR RGQ PCR Kit: a companion diagnostic for afatinib and gefitinib in non-small cell lung cancer. Mol Diagn Ther 20(2):191–198
Tan AR (2016) Cutaneous manifestations of breast cancer. In: Seminars in oncology, vol 43. WB Saunders, Philadelphia, pp 331–334

Teixeira AL, Ferreira M, Silva J, Gomes M, Dias F, Santos JI, Mauricio J, Lobo F, Medeiros R (2014) Higher circulating expression levels of miR-221 associated with poor overall survival in renal cell carcinoma patients. Tumor Biol 35(5):4057–4066

Thakur BK, Zhang H, Becker A, Matei I, Huang Y, Costa-Silva B, Zheng Y, Hoshino A, Brazier H, Xiang J, Williams C (2014) Double-stranded DNA in exosomes: a novel biomarker in cancer detection. Cell Res 24(6):766–769

Tuli HS, Kumar G, Sandhu SS, Sharma AK, Kashyap D (2015a) Apoptotic effect of cordycepin on A549 human lung cancer cell line. Turk J Biol 39(2):306–311

Tuli HS, Kashyap D, Bedi SK, Kumar P, Kumar G, Sandhu SS (2015b) Molecular aspects of metal oxide nanoparticle (MO-NPs) mediated pharmacological effects. Life Sci 143:71–79

Udar N, Iyer A, Porter M, Haigis R, Smith S, Dhillon S, Meier K, Ward D, Lu J, Wenz P, Buchner L (2020) Development and analytical validation of a DNA dual-strand approach for the US Food and Drug Administration–approved next-generation sequencing–based praxis extended RAS panel for metastatic colorectal cancer samples. J Mol Diagn 22(2):159–178

Ulz P, Belic J, Graf R, Auer M, Lafer I, Fischereder K, Webersinke G, Pummer K, Augustin H, Pichler M, Hoefler G (2016) Whole-genome plasma sequencing reveals focal amplifications as a driving force in metastatic prostate cancer. Nat Commun 7(1):1–2

Uruga H, Mino-Kenudson M (2018) ALK (D5F3) CDx: an immunohistochemistry assay to identify ALK-positive NSCLC patients. Pharmacogenomics Personal Med 11:147

von Brandenstein M, Pandarakalam JJ, Kroon L, Loeser H, Herden J, Braun G, Wendland K, Dienes HP, Engelmann U, Fries JW (2012) MicroRNA 15a, inversely correlated to PKCα, is a potential marker to differentiate between benign and malignant renal tumors in biopsy and urine samples. Am J Pathol 180(5):1787–1797

Wallden B, Storhoff J, Nielsen T, Dowidar N, Schaper C, Ferree S, Liu S, Leung S, Geiss G, Snider J, Vickery T (2015) Development and verification of the PAM50-based Prosigna breast cancer gene signature assay. BMC Med Genet 8(1):1–4

Wang Y, Yu Y, Ye R, Zhang D, Li Q, An D, Fang L, Lin Y, Hou Y, Xu A, Fu Y (2016) An epigenetic biomarker combination of PCDH17 and POU4F2 detects bladder cancer accurately by methylation analyses of urine sediment DNA in Han Chinese. Oncotarget 7(3):2754

White NM, Masui O, DeSouza LV, Krakovska-Yutz O, Metias S, Romaschin AD, Honey RJ, Stewart R, Pace K, Lee J, Jewett MA (2014) Quantitative proteomic analysis reveals potential diagnostic markers and pathways involved in pathogenesis of renal cell carcinoma. Oncotarget 5(2):506

Woodhouse R, Li M, Hughes J, Delfosse D, Skoletsky J, Ma P, Meng W, Dewal N, Milbury C, Clark T, Donahue A (2020) Clinical and analytical validation of FoundationOne Liquid CDx, a novel 324-Gene cfDNA-based comprehensive genomic profiling assay for cancers of solid tumor origin. PLoS One 15(9):e0237802

Xie XY, Yang X, Zhang JH, Liu ZJ (2007) Analysis of hTERT expression in exfoliated cells from patients with bladder transitional cell carcinomas using SYBR green real-time fluorescence quantitative PCR. Ann Clin Biochem 44(6):523–528

Yang L, Jiang Q, Li DZ, Zhou X, Yu DS, Zhong J (2019) TIMP1 mRNA in tumor-educated platelets is diagnostic biomarker for colorectal cancer. Aging 11(20):8998

Yoshioka Y, Kosaka N, Konishi Y, Ohta H, Okamoto H, Sonoda H, Nonaka R, Yamamoto H, Ishii H, Mori M, Furuta K (2014) Ultra-sensitive liquid biopsy of circulating extracellular vesicles using ExoScreen. Nat Commun 5(1):1–8

Yu S, Cao H, Shen B, Feng J (2015) Tumor-derived exosomes in cancer progression and treatment failure. Oncotarget 6:37151–37168

Zhang HL, Yang LF, Zhu Y, Yao XD, Zhang SL, Dai B, Zhu YP, Shen YJ, Shi GH, Ye DW (2011) Serum miRNA-21: Elevated levels in patients with metastatic hormone-refractory prostate cancer and potential predictive factor for the efficacy of docetaxel-based chemotherapy. Prostate 71(3):326–331

Zhang Y, Yang P, Sun T, Li D, Xu X, Rui Y, Li C, Chong M, Ibrahim T, Mercatali L, Amadori D (2013) miR-126 and miR-126* repress recruitment of mesenchymal stem cells and inflammatory monocytes to inhibit breast cancer metastasis. Nat Cell Biol 15(3):284–294

Zhang H, Mao F, Shen T, Luo Q, Ding Z, Qian L, Huang J (2017) Plasma miR-145, miR-20a, miR-21 and miR-223 as novel biomarkers for screening early-stage non-small cell lung cancer. Oncol Lett 13(2):669–676

Zhang H, Lin X, Huang Y, Wang M, Cen C, Tang S, Dique MR, Cai L, Luis MA, Smollar J, Wan Y (2021) Detection methods and clinical applications of circulating tumor cells in breast cancer. Front Oncol 11:1816

Zhong L, Li Y, Xiong L, Wang W, Wu M, Yuan T, Yang W, Tian C, Miao Z, Wang T, Yang S (2021) Small molecules in targeted cancer therapy: advances, challenges, and future perspectives. Signal Transduct Target Ther 6(1):1–48

Pharmacogenetics and Pharmacogenomics in Precision Medicine

Eren Demirpolat

1 Introduction

Precision medicine is a definition that arises from the use of all individual parameters in choosing the right dose and the right drug for the patients. Today, clinicians use many different biochemical parameters, weight and body surface data to individualize the treatment and choosing an appropriate medicine at the right dose which has the best benefit/harm balance. In other words, the effectiveness and safety of the drug must be at an optimum level. It can be thought that all the data required to prescribe drugs have been collected so far and the appropriate drug has been prescribed. However, not all individuals in the population give the same response in terms of pharmcodynamics and pharmakokinetic parameters after taking the medication. Some patients experience the therapeutic and unwanted effects of the drug differently in terms of severity, frequency and its nature when the medication is prescribed using those parameters at standart doses that mentioned in previous sentences. This is where pharmacogenomics comes into play and a question comes to mind about "Should we consider pharmacogenomics data and genetic polymorphisms before prescription?" The answe is yes for some of pharmacogenomic parameters which were implemented to clinic. These parameters are presented in The Pharmacogenomics Knowledge Base (pharmgkb.org) and Clinical Pharmacogenetics Implementation Consortium (cpicpgx.org). On the other side if physicians are not provided with any pharmacogenomic data, one pillar of individual therapy will be missing. It means that no data no appropriate dose adjustment or choosing the proper medication for the patient.

E. Demirpolat (✉)
Faculty of Pharmacy, Department of Pharmacology, Erciyes University, Kayseri, Turkey
e-mail: edemirpolat@erciyes.edu.tr

H. S. Tuli, M. B. Yerer Aycan (eds.), *Oncology: Genomics, Precision Medicine and Therapeutic Targets*, https://doi.org/10.1007/978-981-99-1529-3_5

2 Association of Pharmacogenomics with Pharmacodynamics

Presicion medicine has two main parts which are choosing the right medicine and choosing the right dose. In addition, checking the blood concentration levels of medication can verify that everything is allright. The best examples of choosing the right drug can be observed in cancer treatment. Nearly all clinics choose tumor profiling methods such as molecular, histologic, radiographic and physiologic characteristis. For example in breast cancer patients, tumor is analysed in terms of receptor expression levels and types of receptors. If the tumor is expressing human epidermal growth factor receptor-2 (HER-2) then trastuzumab which targets this receptor are included in chemotherapy regimen. In other case, when the estrogen receptor is highly expressed, estrogen supression hormonal therapy are the best choice. At the end of the day, prescriber makes sure that patient is taking the right medication in terms of efficacy. In terms of safety, carbamazepine and abacavir are good examples because patient's who carries Human Leucocyte Antigen (HLA) genetic polymorphisms are more susceptible to Steven Johnson Syndrome (SJS). According to cpic databese, HLA genetic testing is required before prescribing the medication (Saito et al. 2016).

3 Association of Pharmacogenomics with Pharmakokinetics

Genetic polymorphisms may alter the pharmacokinetics or pharmacodynamics of the drug. Genetic polymorphism of proteins which are responsible for metabolism and elimination events may alter pharmacokinetic profile of the drug. For example CYP2C19 enzyme is responsible for inactivation of voriconazole. Voriconazole is an antifungal drug prescribed for systemic fungal infections and dose related side effects are hallucination, photophobia and liver toxicity. When patients have slow metaboliser allele (CYP2C19*2, CYP2C19*3), voriconazole inactivation slows down and blood concentration might be at supratherapeutic level (Moriyama et al. 2017). This causes hallucination in most of the patients (Theuretzbacher et al. 2006). In this example genetic polymorphism have an impact on drug's metabolism and high blood conentration related side effect might occur. Genetic polymorphis effected the blood concentration level directly (Hamadeh et al. 2017). In previos examples HLA genetic polymorphism have changed the safety profile of carbamazepine. When this allele is carried incidence of SJS increases after taking the drug without the impact of blood concentration change because HLA's physiological role is recognizing and creating a defense mechanism against the nonself from self. People who carries HLA-B/*57:01 are at great risk for SJS.

In general patient's genotype may alter both pharmacokinetic and pharmacodynamic properties of a drug. Genotype is expressed as phenotype and it should be considerd that genotype data might be a good guide to some extent because epigenetics play an important role in patient's real phenotype. After a genomic data is analyzed, we expect a potential phenotype which is affected by regarding gene but in real life phenotype or patient's real response to the medication in terms of efficacy and safety might be affected by drug-drug interactions, life style, supplement use so these parameters should be considered when therapy is being personalized. Analyzing the patient's phenotype is not very difficult bu it is not always practical. There must be a time point when the patient is not taking any medication and after patients take probe drugs for CYP450 enzymes, parent molecule and its metabolite is determined in blood. The ratio of parent and metabolite shows the enzyme real activity named as phenotype. If the ratio is high, activity of enzyme is low and the patient is poor metaboliser. While the ratio is low, the patient is rapid metabolizer.

4 Pharmacogenomics and Therapeutic Drug Monitoring

If genotype is affecting the pharmacokinetic properties, therapeutic drug monitoring (TDM) comes into play and might be beneficial for personalized medicine. Whenever drug dose is adjusted, drug interactions occur or pharmacogenomic background effects the pharmacokinetics, blood level of medications should be checked using TDM. If validated methods are available for regarding drugs then adding TDM to clinical routine is so logical. TDM is a great tool to see whether the medication is in the area of sub, supra therapeutic range. Whatever the cause, monitoring the blood concentration of the drug is a good approach to monitor and predict the patient's clinical response. Especially if there is risk of blood concentration of the drug being out of therapeutic level (supra or sub). When the blood concentration is at supratherapeutic level, dose related side effects incrases and patient might not be able to tolerate the treatment. On the contrary when blood concentration is at subtherapeutic level, effectiveness of the drug is reduced and treatment might fail. Clinical protocols should be very clear to carry out a proper TDM analysis. Drug's purpose of use (prophylactic or treatment), blood sampling time, validated methods, device's sensitivity should be considered. Time to reach steady state concentration is effected by dosing scheme. When drug is dosed as standart, it reaches steady state concentration approximately 3–5 half lives. When loading doses are used, steady state might be achieved under 24 h time period. Because of these reasons sampling time is important. On the other hand analysis laboratories should work all day and send the TDM results as fast as possible to give the prescriber an opportunity for adjusting the treatment.

In some cases patients benefit mostly by biochemical tests which reflect drug's pharmacodynamic properties. For example warfarin is a protoype drug which is personalized by using international normalized ratio (INR) score. Warfarin is an antithrombotic drug which inhibits VKORC1 enzyme in liver which is responsible for coagulation factor production. Warfarin decreases the amount of coagulation factors including factor 2, 7, 9 etc. The drug is inactivated by CYP2C9 and patients who have genetic polymorphisms experince different kind of drug safety and efficacy outcomes when compared with normal metabolizers. Vitamin K is inactivated by CY4F2 (Johnson et al. 2017). As it is known VKORC1 have genetic variants among population. According to these data it is not surprising that patients have different response to warfarin. There are a lot of variables for warfarin and vitamin K metabolism. Interindividual variables causes different warfarin responses among patients but at the end drug's efficacy is meausured using INR because for antithrombotic efficacy INR should be incresed by twofold at least and do not exceed threefold. Warfarin is a drug that can not personalized using pharmacogenomics or TDM. The gold standart for monitoring warfarin efficacy and safety is INR. According to guidelines INR level should be between 2–3. If it is below 2, warfarin's efficacy is not enough and patients might experience thromboembolic event. While INR is above 3, safety concerns arise and patients might have internal bleeding ranging from minor intradermal bleeding to life threatining organ bleeding (Agrawal et al. 2020).

5 Pharmacogenomics and Drug Interactions

Pharmacogenomic approach is so beneficial in terms of drug efficacy and safety. When a prescriber has the patient' s pharmacogenomic data like other standart biochemical parameters, it will be used actively for personalizing the medicine. If there is no genotype data, it can not used. Because of this Laboratories that will provide hospital-level service are required and they need to be qualified for his service. Due to the nature of genetic polymorphisms, there is a frequency in every population. If there is a genetic screening for whole population and genotype data are readily available to prescrbers, personalizing the therapy according to this data will be so much easier. The most important thing about this screening is public reimbursement agencies, private insurance companies and individual interests of patients. When these agencies and companies support pharmacogenomic testing, genotyping culture will become widespread. Therefore readily available genotyping data, individual genecards and genetic mapping are important for personalized medicine. If there is no genotype data, prescribers can not use it. On the other hand, who will evaluate this data and integrate it into the clinic is another matter. Clinical pharmacists and precribers should work together, provide a better pharmaceutical care for the patient. In fact clinical pharmacists are helping all other healthcare professionals for detecting and solving drug related problems such as drug-drug interactions. As we know and according to the growing number of publications, genetic polymorphisms are effecting the drug-drug interaction characteristic. Although some of drug-drug interactions are not observed in patients who have wild type genotype, people who have genetic polymorphisms might experience these drug interactions. For example omeprazole and tacrolimus interaction is observed in patients who have CYP2C19 slow metaboliser genoytpe because of competitive intihibtion on CYP3A4 and alternative metabolism of omeprazole with CYP3A4. Primarily omeprazole is metabolised by CYP2C19 but in slow metaboliser patients omeprazole are metabolized by CYP3A4 and this cause competitive inhition of tacrolimus metabolism because tacrolimus is mainly metabolised by CYP3A4 too (Pascual et al. 2005). As a result of this drug interaction tacrolimus blood concentration level increases and may be reach supratherapeutic level. This might cause increased efficacy but at the same time increases the dose induced side effects and patients might not tolerate tacrolimus although prescriber personalize the dose according to standart prescription and biochemical parameters such as renal-liver function, albumin level, weight, age. It is really interesting and logical to investigate that this durg interaction only observed in patients who have CY2C19 slow metaboliser genotype (Itagaki et al. 2002). On the other hand wild type individuals are treated as they should and benefit/harm ratio is as expected. Interactions like we saw in this example, might be prevented if patient genotype is readily available to prescriber, if pharmacogenomic testing become a routine biochemical parameter. Beyond that drug-drug interaction clinical trials should continue and the number of trials should increase to gather information on how genetic polymorphism affect these patients. When these trials and guidelines are increased, alternative prescription medication can be presented and advised to doctors. In this example pantoprazol is much more suitable then omeprazole while prescribing to a patient who takes tacrolimus at the same time. As it is known pantoprazole does not effect CYP450 enzyme sytem and is not expected to cause such an interaction (Lorf et al. 2000).

As a result we can say that personalized medicine consists of three main components: pharmacogenomic data, theraputic drug monitoring, standart prescription methods (using weight, body surface, standard biochemical parameters including functions of elimination organs). Before choosing the appropriate medication pharmacogenomic data is a great tool because all of the potential side effects might be prevented and potential benefit migh be increased by this way While using the medication, TDM is a good guide to see what really happens in clinic routine and detecting phenotype related situations directly affected by blood concentration. TDM helps to take the decision of adjusting the dose. Standard biochemical parameters have already been in clinic routine for so many years and helps doctors to personalize the medication to some extent. In addition these routine parameters are not complete without pharmacogenomics. This is why pharmacogenomics emerged and another component of personalized medicine is completed. General approach for personalized medicine for all diseases are similar. What has been written to this section also explains this main idea. Next sections are focued on pharmacogenomic's role in cancer.

6 Pharmacogenomics of Fluoropyrimidine Drugs

Fluoropyrimidine drugs are 5-fluorouracil and capecitabin. They are prescribed for solid tumors for breast and gastrointestinal cancers such as colorectal cancer. Patients migt experience serious and dose limiting side effects. Some of these side effects are hand-foot syndrome, mucositis, vomiting, nasuea, neutropenia, diarrhea, stomatitis. These drugs are inactivated by dihydropyrimidine dehydrogenase (DPD) after they enter to cell. For example 5-FU is metabolized to dihydrofluororacil (DHFU) and capesitabine is converted to 5-FU in the cell (Heggie et al. 1987). For this reason after a certain metabolism steps, capesitabin and 5-FU have common metabolic pathways. DPD protein is expressed by DPYD gene and 80–90% of 5-FU is degraded by this enzyme. Genetic variants in DPYD changes the protein sequence and causes mRNA splicing. Patients who have decreased or no function DPD activity, can not inactive 5-FU and might experience regarding toxic effects (Harris et al. 1991).

5-florouracil (5-FU) inhibits thymidylate synthase enzyme which catalyzes the conversion of deoxyuridine monophosphate (dUMP) to deoxythymidine monophosphate (dTMP). 5-FU is metabolised to 5-FU deoxythymidylic acid (5-FdUMP) and it decreases the activity of thymidylate synthase. As a result purine synthesis is suppressed, repair and replication of DNA decreases. If Methylene tetrahydrofolate reductase (MTHFR) is available and 5-dUMP forms a complex with thymidylate synthase (Derwinger et al. 2009). 5-FU is absorbed into cells via nucleoside transporters. A detailed metabolism of 5-FU and Capesitabin are presented in this study (Loganayagam et al. 2013). Therefore there are three enzymes that changes the anticancer activity of 5-FU. These are DPD, MTHFR and TS. The genetic polymorphisms most associated with 5-FU toxicity belongs to DPYD (Dolnick et al. 2003).

When the expression is increased DPD protein level increases. This causes enhanced inactivation of 5-FU and resistance to the drug. Patients who have high expression of DPYD gene have shorter survival rates compared to the patients who have low expression of DPYD gene (Rosmarin et al. 2015). Genetic polymorphism in DPYD (IVS14+1G>A, 496A>G, and 2194G>A) gene might cause lack of activity and this leads less inactivation of 5-FU and as a result 5-FU toxicity might be increased (Ma et al. 2004).

c.190511G>A (rs3918290, also known as DPYD*2A, DPYD:IVS14 1 1G>A), c.1679T>G (rs55886062, DPYD *13, p.I560S), c.2846A>T (rs67376798, p. D949V), and c.1129–5923C>G (rs75017182, HapB3). Of these variants, c.190511G>A and c.1679T>G effects the enzyme activity and has some deleterious impact on it (Ma et al. 2004). While, c.2846A>T and c.1129–5923C>G causes DPYD to have moderate activity. In caucasians, the most common polymorphism is c.1129–5923C>G with carrier frequency of 4.7% (Rosmarin et al. 2015). DPYD genetic polymorphisms and outcomes are presented in Table 1.

Table 1 DPYD genetic polymorphisms and outcomes

DPYD				
Expected phenotype	Activity	Genotype and phenotype chacacteristics	Allele examples	Dosing recommendations
Normal metabolizer	2	Carrying two wild type alleles, expected DPD activity and toxicity	85T>C, 1627A>G	Use label recommended dose, because these patients benefit the drug most. Benefit/harm ratio is very high
Intermediate metabolizer	1 or 1.5	Individual who carries one decreased function allele, 30–70% decreased DPD activity and increased toxicity	190511G>A, 1679T>G, 2846A>T, 1129–5923C>G, 2846A>T	Dose should be reduced if the patients experince untolerable side effects 50% dose reduction for activity score 1, 25–50% dose reduction for activity score 1.5
Poor metabolizer	0 or 0.5	Individual who carries two decreased alleles No DPD activity and more toxicity compared to intermediate metabolizers	190511G>A, 1679T>G, 2846A>T	Avoid or reduce the 5-FU dose seriously according to clinical outcomes if the activity score is 0.5 5-FU based drugs should not be used if the activity score is 0

7 Pharmacogenomics of Abiraterone Acetate

Abiraterone acetae is an anticancer drug prescribed for prostate cancer. In combination with prednisone it is used for metastatic prostate cancer patients. According to clinical trials it increases the survival time compared to placebo group. Abiraterone acetate is a prodrug and converted in the body to active metabolite abiraterone. Abiraterone is a testosterone receptor blocker and an androgen synthesis intihibitor. Because of this it is a key drug for androgen decreasing therapy (ADT). The drug inhibits testosterone production in peripheral tissues, adrenal glands, testis and tumor cells via CYP450 17α-hydroxy/17,20-lyase (CYP17A1) inhibition and it shares similar steroid structures as dehydroepiandrosterone (DHEA) and ohter 3-β-hydroxysteroid dehydrogenase (3β-HSD) catalytic substrates. CYP17A1 are highly expressed in testosteroen producing tissues. In addition, it decreases the level of dehydroepiandrosterone and serum levels of testosterone. Abiraterone acetate is metabolised by CYP3A4 and SULT2A1 into two active metabolites abiraterone sulfate and N-oxide abiraterone sulfate. Metabolism of abiraterone is presented in this study (Liu et al. 2021).

Table 2 Polymorphisms associated with abiraterone and outcomes

Expected phenotype	Allele examples	Expected outcomes	Recommendations
HSD3B1			
Increased enzyme activity because of accumulation in the cell	1100C>A, 1245C	DHT production is increased in homozygotic patients	Worse prognostic outcomes after ADT therapy, decrease in adt treatment efficacy, docetaxel therapy might be considered
SLCO2B1			
Increased transportation of androgen precursors into the cell	75172532 G>T, G>A 75176120 T>A, T>C	Androgen precursors enters prostate tissue more, increased production	Worsening in the disease free survival rate
CYP17A1			
Increased transcription of CYP17A1	102837395A>G, A>T 102837723 T>C, T>G	Increased production of androgens	It is accepted as a prognostic factor and a predictor for survival rate. Resistance to abirateron therapy

3β-HSD1 enzyme is responsible for catalyzing androgen precursors into dihydrotestosterone (DHT) and it is encoded by HSD3B1 gene (Blanchet et al. 2018). Genetic polymorphism in this gene causes a resistance to degradation of the enzyme and 3β-HSD1 enzyme accumulates in the environment. As a result androgen precursors are converted to DHT so much more (van Nuland et al. 2020).

Solute carrier organic anion transporter family member 2B1 (SCLO2B1) are responsible for transportation of androgen precursors and abiraterone acetate into the cell. When this protein is highly expressed, androgen precursors are presented to cell so much more and this leads to shorten the progression free survival of patients (Emamekhoo et al. 2018).

CYP17A1 produces androgens such as testosteron, DHEA and glucocorticoids such as cortisol and it is highly expressed in prostate cancer cells. Adrenocorticotropic hormone controls the expression of CYP17A1. Genetic polymorphisms cause altered gene expression and this lead to increased androgen synhtesis, worsening survival rate in patients (Salvi et al. 2016) (Table 2).

8 Pharmacogenomics of Bevasizumab

Bevacizumab is an angiogenesis inhibitor which targets vascular endothelial growth factor (VEGF) and it is a monoclonal antibody. It is used in various cancers such as lung cancer, metastatic colorectal cancer, ovarian cancer etc. VEGF activates VEGFR receptor and when bevacizumab binds to VEGF, it prevents VEGF to activate its receptor and signalling pathway is inhibited. Proliferation, metastases and angiogenesis events are inhibited by this way. Bevacizumab might be used as monotherapy and combination with other anticancer drugs. There is a lack of parameters which is used for personalizing bevacizumab therapy. In this section, pharmacogenomic variables are going to be discussed. According to the studies, single nucleotide polymorphism causes interindividual differences to treatment because patients have variable angiogenesis capacity.

In metastatic cancer patients, rs2010963 C>T polymorphism changes the expression level VEGF and patients who have T allele have longer time to progression. Bevasizumab efficacy is increased. On the other hand patients who have rs2010963 G>C polymorphism experienced toxicity such as thromboembolism and hypertension (Etienne-Grimaldi et al. 2011). In a phase III trial, patients who are treated with neoadjuvant chemotherapy were included the study and 4 polymorphisms (rs833058, rs699947, rs3025039, rs3025030, rs79995976) were associated with better response to bevasizumab (Hein et al. 2015).

In metastatic colorectal cancer patients, patients who have rs699947 A/A are associated with polymorphism progression and free survival overall survival compared to C/C allelle. This polymorphism was accepted as a predictor factor in the regarding study (Papachristos et al. 2019). Patients who this polymorphism might benefit more and response better to treatment.

In lung cancer patients who have rs699947 C>A polymorphism, the response rate was %50 higher and patients who have rs2010963 C>T polysmorphism experience hypertension. In addition, rs9554316 G/T polymorphism caused 30% higher risk of progession. These patients were also under risk of death 40% higher (Pallaud et al. 2014).

In ovarian cancer patients, patients who have rs3025039 C/T polymorphism have longer progression free survival (Schneider et al. 2008).

9 Pharmacogenomics of Imatinib

Imatinib is a specifif tyrosine kinase inhibitor. It targets BCR-ABL tyrosine kinase. This type of kinase is expressed by BCR-ABL gene in patients who have Philedelphia chromosome. Patients who have this type of chromose are diagnosed with chronic myeloid leukemia. It is prescribed for both pediatric and adult patients. Imatinib is metabolized mainly by CYP3A4, and to a lesser extent by other CYP450 enzymes. Main circulating metabolite of imatinib is N-demethylated piperazine which is formed by CYP3A4. Most frequent adverse reactions are edema, nausea, fatigue, rash, musculoskeletal pain, diarrhea etc (Baccarani et al. 2009).

In terms of pharmacokinetic parameters, imatinib is a substrate for p-glycoprotein and it could be effected by multi drug resistance protein activities. C3435T polymorphism causes lower expression of p-glycoprotein and decreases substrate sensitivity. According to studies this increases the bioavailability of imatinib and these patients might experience toxicity. BCRP similar to p-glycoprotein, is responsible for removal of imatinib and expressed in liver, intestines and hematopoietic stem cells. C421A polymorphism is one of the most common one and causes greater accumulation of imatinib, decreases elimination. In addition to these, patients who have CYP3A5 polymorphism A6986G, have reduced enzyme activity. According to studies 6986GG genotype patients might have poor hematological response to imatinib (Takahashi et al. 2010).

In terms of pharmacodynamic parameters, tyrosine kinase enzyme is the main target for imatinib and any change in this enzyme might effect drug efficacy. Point mutations are responsible for imatinib resistance. Some patients have c-KIT activating mutations and these mutations enables tyrosine kinase enzyme activate itself without needing a substrate. As a result patient response to imatinib decreases.

As a result imatinib response is effected by pharmacokinetic parameters and therapeutic drug monitoring might help whether blood concentration of imatinib is in therapeutic window. Pharmacogenomic markers which effects pharmacodynamics of imatinib might help to predict the drug efficacy and toxicity (Pardanani 2010).

References

Agrawal S, Heiss MS, Fenter RB, Abramova TV, Perera MA, Pacheco JA, Smith ME, Rasmussen-Torvik LJ, George AL (2020) Impact of CYP2C9-interacting drugs on warfarin pharmacogenomics. Clin Transl Sci 13(5):941–949. https://doi.org/10.1111/cts.12781. Epub 2020 Apr 9. PMID: 32270628; PMCID: PMC7485961.

Baccarani M, Cortes J, Pane F, Niederwieser D, Saglio G, Apperley J, Cervantes F, Deininger M, Gratwohl A, Guilhot F, Hochhaus A, Horowitz M, Hughes T, Kantarjian H, Larson R, Radich J, Simonsson B, Silver RT, Goldman J, Hehlmann R (2009) European LeukemiaNet: chronic myeloid leukemia: an update of concepts and management recommendations of European LeukemiaNet. J Clin Oncol 27:6041–6051

Blanchet B, Carton E, Alyamani M, Golmard L, Huillard O, Thomas-Scheomann A, Vidal M, Goldwasser F, Sharifi N, Alexandre J (2018) A PK/PD study of Delta-4 abiraterone metabolite in metastatic castration-resistant prostate cancer patients. Pharmacol Res 136:5661

Derwinger K, Wettergren Y, Odin E, Carlsson G, Gustavsson B (2009) A study of the MTHFR gene polymorphism C677T in colorectal cancer. Clin Colorectal Cancer 8:43–48

Dolnick BJ, Angelino NJ, Dolnick R, Sufrin JR (2003) A novel function for the rTS gene. Cancer Biol Ther 2:364–369

Emamekhoo H, Alyamani M, Park S, Taylor J, Almassi N, Upadhyay S, Tyler AJ, Berk M, Hwang TH, Grivas P, Rini BI, Garcia JA, Auchus RJ, Sharifi N (2018) HSD3B1 genotype and abiraterone (Abi) metabolites in patients (pts) with prostate cancer (PCa). J Clin Oncol 36:325

Etienne-Grimaldi MC, Formento P, Degeorges A, Pierga JY, Delva R, Pivot X, Dalenc F, Espié M, Veyret C, Formento JL (2011) Prospective analysis of the impact of VEGF-A gene polymorphisms on the pharmacodynamics of bevacizumab-based therapy in metastatic breast cancer patients. Br J Clin Pharmacol 71:921–928

Hamadeh IS, Klinker KP, Borgert SJ, Richards AI, Li W, Mangal N, Hiemenz JW, Schmidt S, Langaee TY, Peloquin CA, Johnson JA, Cavallari LH (2017) Impact of the CYP2C19 genotype on voriconazole exposure in adults with invasive fungal infections. Pharmacogenet Genomics 27(5):190–196

Harris BE, Carpenter JT, Diasio RB (1991) Severe 5-fluorouraciltoxicity secondary to dihydropyrimidine dehydrogenase deficiency. A potentially more common pharmacogenetic syndrome. Cancer 68:499–501

Heggie GD, Sommadossi JP, Cross DS, Huster WJ, Diasio RB (1987) Clinical pharmacokinetics of 5-fluorouracil and its metabolites in plasma, urine, and bile. Cancer Res 47:2203–2206

Hein A, Lambrechts D, von Minckwitz G, Haberle L, Eidtmann H, Tesch H, Untch M, Hilfrich J, Schem C, Rezai M et al (2015) Genetic variants in VEGF pathway genes in neoadjuvant breast cancer patients receiving bevacizumab: results from the randomized phase III GeparQuinto study. Int J Cancer 137:2981–2988

Itagaki F, Homma M, Yuzawa K, Fukao K, Kohda Y (2002) Drug interaction of tacrolimus and proton pump inhibitors in renal transplant recipients with CYP2C19 gene mutation. Transplant Proc 34:2777–2778

Johnson JA, Caudle KE, Gong L, Whirl-Carrillo M, Stein CM, Scott SA, Lee MT, Gage BF, Kimmel SE, Perera MA, Anderson JL, Pirmohamed M, Klein TE, Limdi NA, Cavallari LH, Wadelius M (2017) Clinical pharmacogenetics implementation consortium (CPIC) guideline for pharmacogenetics-guided warfarin dosing: 2017 update. Clin Pharmacol Ther 102(3):397–404. https://doi.org/10.1002/cpt.668. Epub 2017 Apr 4. PMID: 28198005; PMCID: PMC5546947.

Liu M, Shi H, Yan J, Zhang Y, Ma Y, Le K, Li Z, Xing N, Li G (2021) Gene polymorphism-related differences in the outcomes of abiraterone for prostate cancer: a systematic overview. Am J Cancer Res 11(5):1873–1894. PMID: 34094659; PMCID: PMC8167691.

Loganayagam A, Arenas Hernandez M, Corrigan A, Fairbanks L, Lewis CM, Harper P, Maisey N, Ross P, Sanderson JD, Marinaki AM (2013) Pharmacogenetic variants in the DPYD, TYMS, CDA and MTHFR genes are clinically significant predictors of fluoropyrimidine toxicity. Br J Cancer 108(12):2505–2515. https://doi.org/10.1038/bjc.2013.262. Epub 2013 Jun 4. PMID: 23736036; PMCID: PMC3694243.

Lorf T, Ramadori G, Ringe B, Schworer H (2000) The effect of pantoprazole on tacrolimus and cyclosporin A blood concentration in transplant recipients. Eur J Clin Pharmacol 56:439–440

Ma T, Zhu ZG, Ji YB, Zhang Y, Yu YY, Liu BY, Yin HR, Lin YZ (2004) Correlation of thymidylate synthase, thymidine phosphorylase and dihydropyrimidine dehydrogenase with sensitivity of gastrointestinal cancer cells to 5-fluorouracil and 5-fluoro-2′-deoxyuridine. World J Gastroenterol 10:172–176

Moriyama B, Obeng AO, Barbarino J et al (2017) Clinical pharmacogenetics implementation consortium (CPIC) guidelines for CYP2C19 and voriconazole therapy. Clin Pharmacol Ther 102:45–51

Pallaud C, Reck M, Juhasz E, Szima B, Yu CJ, Burdaeva O, Orlov S, Hilton M, Archer V, Mok T (2014) Clinical genotyping and e_cacy outcomes: exploratory biomarker data from the phase II ABIGAIL study of first-line bevacizumab plus chemotherapy in non-squamous non-small-cell lung cancer. Lung Cancer 86:67–72

Papachristos A, Kemos P, Katsila T, Panoilia E, Patrinos GP, Kalofonos H, Sivolapenko GB (2019) VEGF-A and ICAM-1 gene polymorphisms as predictors of clinical outcome to first-line bevacizumab-based treatment in metastatic colorectal cancer. Int J Mol Sci 20:5791

Pardanani A (2010) Systemic mastocytosis in adults: a review on prognosis and treatment based on 342 Mayo Clinic patients and current literature. Curr Opin Hematol 17:125–132

Pascual J, Marcen R, Orea OE et al (2005) Interaction between omeprazole and tacrolimus in renal allograft recipients: a clinical-analytical study. Transplant Proc 37:3752–3753

Rosmarin D, Palles C, Pagnamenta A, Kaur K, Pita G, Martin M, Domingo E, Jones A, Howarth K, Freeman-Mills L, Johnstone E, Wang H, Love S, Scudder C, Julier P, Fernandez Rozadilla C, Ruiz-Ponte C, Carracedo A, CastellviBel S, Castells A, Gonzalez-Neira A, Taylor J, Kerr R, Kerr D, Tomlinson I (2015) A candidate gene study of capecitabine-related toxicity in colorectal cancer identifies new toxicity variants at DPYD and a putative role for ENOSF1 rather than TYMS. Gut 64:111–120

Saito Y, Stamp LK, Caudle KE et al (2016) Clinical pharmacogenetics implementation consortium (CPIC) guidelines for human leukocyte antigen B (HLA-B) genotype and allopurinol dosing: 2015 update. Clin Pharmacol Ther 99:36–37

Salvi S, Casadio V, Burgio SL, Conteduca V, Rossi L, Menna C, Carretta E, Costantini M, Zoli W, De Giorgi U (2016) CYP17A1 polymorphisms and clinical outcome of castration-resistant prostate cancer patients treated with abiraterone. Int J Biol Markers 31:e264269

Schneider BP, Wang M, Radovich M, Sledge GW, Badve S, Thor A, Flockhart DA, Hancock B, Davidson N, Gralow J (2008) Association of vascular endothelial growth factor and vascular endothelial growth factor receptor-2 genetic polymorphisms with outcome in a trial of paclitaxel compared with paclitaxel plus bevacizumab in advanced breast cancer. ECOG 2100. J Clin Oncol 26:4672

Takahashi N, Miura M, Scott SA, Kagaya H, Kameoka Y, Tagawa H, Saitoh H, Fujishima N, Yoshioka T, Hirokawa M, Sawada K (2010) Infl uence of CYP3A5 and drug transporter polymorphisms on imatinib trough concentration and clinical response among patients with chronic phase chronic myeloid leukemia. J Hum Genet. https://doi.org/10.1038/jhg.2010.98

Theuretzbacher U, Ihle F, Derendorf H (2006) Pharmacokinetic/pharmacodynamic profile of voriconazole. Clin Pharmacokinet 45(7):649–663. https://doi.org/10.2165/00003088-200645070-00002

van Nuland M, Groenland SL, Bergman AM, Steeghs N, Rosing H, Venekamp N, Huitema ADR, Beijnen JH (2020) Exposure-response analy-ses of abiraterone and its metabolites in realworld patients with metastatic castration-resistant prostate cancer. Prostate Cancer Prostatic Dis 23: 244–251

Pharmacogenetic Factors in Pharmacodynamic/Pharmacokinetic (PD/PK) Modulations of Medicines

Mükerrem Betül Yerer Aycan

Abbreviations

ABCB1	ATP-binding cassette sub-family B member 1
ADME	Absorption, distribution, metabolism, and excretion
CDS	Clinical decision support
CYP	Cytochrom P
DPD	Dihydropyrimidine dehydrogenase
EGFR	Epidermal growth factor receptor
FDA	Food and Drug Administration
GFR	Glomerular filtration rate
GPS	Genomic prescribing system
HER-2	Human epidermal growth factor-2
MDR	Multidrug resistance
miRNAs	MicroRNAs
MRP1	Multidrug resistance-associated protein 1
PD-L1	Programmed cell death ligand 1
P-gp	Permeability glycoprotein
PGx	Pharmacogenomics
RNAi	RNA interference
siRNA	Small interfering RNA
SNPs	Single nucleotide polymorphisms
TDM	Therapeutic drug monitoring
TMB	Tumor mutation burden
TPMT	Thiopurine methyltransferase
UGT	Uridine diphosphate glucuronosyltransferase
VEGF	Vascular endothelial growth factor

M. B. Yerer Aycan (✉)
Department of Pharmacology, Erciyes University, Faculty of Pharmacy, Kayseri, Turkey

H. S. Tuli, M. B. Yerer Aycan (eds.), *Oncology: Genomics, Precision Medicine and Therapeutic Targets*, https://doi.org/10.1007/978-981-99-1529-3_6

1 Clinical Pharmacokinetics

Pharmacokinetics is currently defined as the road journey of a drug in the body identified by the time course of ADME. Clinical pharmacokinetics is the treatment management using pharmacokinetic principles to enhance the safety and efficacy of the drugs in an individual patient. There is a strong correlation between drug concentrations and their pharmacologic responses has enabled clinicians to apply pharmacokinetic principles to therapeutic monitoring of an individual patient. A drug's effect is often related to its concentration at the site of action, where it binds to its receptors and the concentration of a drug in this action site is very important to reflect its efficacy over these mechanisms. However, these receptor sides are practically inaccessible or are widely distributed in the body, and therefore direct measurement of drug concentrations at these sites is not practically possible. We can only measure the drug concentrations at a time course statement in the blood or plasma, urine, saliva, and other easily sampled fluids. One of the most followed parameter among them is the plasma concentration and these concentrations may reflect us how much drug exists in the body and the action side. Although the distribution of a drug alters from drug to drug according to the solubility and lipophilicity of the drug the plasma unbounded concentrations of the drug in the plasma is a good reflector to follow up the total drug concentration within the body and it is proportional to the concentrations in the action site. The plasma concentration is affected form every step of ADME. Therefore all the steps included to pharmacokinetics can alter the plasma concentration and should be followed to maintain an effective plasma steady state concentration which is relevant to maintain a certain efficacy of a drug (Alfarouk et al. 2015). To maintain an effective treatment with less harmful side effects has substantial implications for individuals and the resources of health systems including especially the psychiatric diseases and other many diseases including cancer (Fava et al. 2017; Pardiñas et al. 2021).

2 Pharmacogenetic Factors in Relation to Pharmacokinetics

2.1 *Absorption*

Pharmacogenetic data is used to follow up the pharmacokinetics of a drug with the help of a genetic data related to ADME. To give some simple samples to all steps absorption related genetic data is mainly integrated with the efflux pump functions including the P-gp (permeability glycoprotein). Over the past few decades, intensive studies have been conducted to find novel therapeutic molecules to reverse the multidrug resistance (MDR), which is often associated with overexpression of multidrug resistant protein family and P-gp is the most extensively studied drug efflux transporter among this family (Mora Lagares and Novič 2022). Any genetic polymorphism or mutations cause to overexpression in multidrug resistant proteins results in the lack of absorption and therefore a decrease in plasma concentration (Callaghan et al. 2014; AlAzzeh and Roman 2023).

2.2 *Distribution*

In the distribution phase of a drug any changes in the transporter molecules such as albumin (Goldwasser and Feldman 1997) and other transporters leads to heterogeneity of the free drug concentration. Most important plasma protein in the drug distribution is albumin. Drugs in the bloodstream are available in both free and bounded forms to the transporter molecules within the plasma. The free drug is responsible for pharmacological activities and protein binding determines the amount of free concentration of the drug in the blood, measuring the free concentration of a drug is essential in therapeutic drug monitoring of an individual patient in the concept of personalized medicine (Seyfinejad et al. 2021). Many factors related to transporter molecules can affect the distribution of a drug carried by albumin since albumin shows individual differences with age (Liliemark et al. 1996) and sex (Grimm et al. 2009).

2.3 *Metabolism*

During the metabolism of the drugs, although there are many mechanisms and organs are related to metabolism of drugs the most important way of drugs to be detoxified is the microsomal enzymes in the liver. In a personalized medicine approach any genetic polymorphism in these enzymes leads to different plasma concentrations of the same drug in individual patients (Rodriguez-Antona and Ingelman-Sundberg 2006).

Cytochrom P (CYP) is a protein that plays a key role in the metabolism of drugs and other xenobiotic compounds and the important feature of these proteins for personalized medicine is the genetic polymorphism of these proteins among people (Estabrook 2003). Genetic polymorphism of CYP affects the metabolism of certain drugs and therefore varies the response of the therapy within the patients.

The most important enzymes which show genetic polymorphism in liver are CYP2D6, CYP2C19, CYP2C9, CYP1B1, CYP2C8 (Narendra et al. 2022). For instance, CYP2D6 contributes to the metabolism of approximately 20–25% of drugs and is highly polymorphic and different alleles can lead to impacts ranging from null to toxicity. Furthermore any pheno-conventors which can inhibit or induce this enzymes can cause a phenotype alteration (Nahid and Johnson 2022). Basically, a genetic polymorphism in this enzymes involves several types including small insertions, deletions, and single nucleotide polymorphisms (SNPs). SNPs are the most common form of polymorphisms on these enzymes' gene sequences (Dai et al.

2001). These genetic changes contribute to population heterogeneity in drug responses (Wilkinson 2005). Many evidence exist about the presence of a SNP in these genes affect the metabolism of different drugs. The alterations in the genotype and phenotype of these enzymes differs the proportion and the rate of the drug molecule. For instance, the presence of CYP2C19 LOF alleles, especially CYP2C19*2, correlates with impaired activation of clopidogrel and the ultra-fast metabolizing of these drug in the patients with the genetic polymorphism of these allele results in variable platelet inhibition, followed by minimal or no antiplatelet effect, and treatment failure (Djordjevic 2023).

The genetic polymorphisms in these enzymes are critically important in also chronic diseases such as cancer. The resistance to taxanes and platins are associated with CYP1B1*3 allele which is related to CYP1B1 overexpression, while the resistance to 5-fluorouracil is related to mutation in DPD (dihydropyrimidine dehydrogenase) *2A, 496T>G is associated with lower levels of dihydropyrimidine dehydrogenase (Narendra et al. 2022). The most important examples to demonstrate the influence of pharmacogenomics on anticancer therapy are thiopurine methyltransferase (TPMT), for 6-mercaptopurine and UGT (uridine diphosphate glucuronosyltransferase) 1A1*28, for irinotecan (Bosch 2008).

2.4 Excretion

Excretion from the body is the final step in drug journey in the body. There are many different routes of excretion but excretion of drugs mainly occur through two main routes: biliary and renal excretion. In the biliary or bile duct excretion the MDR genes are important and overexpression of these transporters increases the biliary excretion (Kusuhara and Sugiyama 2002). Therefore, monitoring of these gene expressions are important when a drug is prescribed which is excreted through the bile. For the renal excretion, the kidney is the primary organ by which drugs are excreted and the according to the glomerular filtrations rate (GFR), tubular reabsorption and tubular secretion of the drugs can differ from patient to patient (Vree et al. 1993). An furthermore this Inter-individual renal drug excretion variability might be affected from even the gender differences (Morris et al. 2003) and ethnic differences (Johnson 2000).

3 Pharmacogenetic Factors in Relation to Pharmacodynamics

3.1 *Efficacy*

The efficacy of the drugs are related to the disease specific proteins, genes and receptor binding sides. The tumor specific proteins such as HER-2, VEGF, EGFR were the most investigated targets of the drugs for cancer. There are many different list of proteins related to targeted therapies in different diseases and summarized in chapter "Omics in Precision Medicine", Table 1, 2 and chapter "Oncogenic Genomic Changes in Cancer", Table 1.

Furthermore, over the past decade, non-coding RNA-based therapeutics having a great potential for the development of targeted therapies for cancer and other diseases. The discovery of the critical function of microRNAs (miRNAs) in both the genetic and the epigenetic factors in many diseases has generated great excitement in developing miRNA-based therapies. RNA interference (RNAi) is the main route that miRNAs and small interfering RNA (siRNA) can inhibit the expression of any cancer-related genes/mRNAs with high specificity. Thus these novel gene therapy strategies are a remarkable therapeutic tool for targeted therapies and precision medicine (Kara et al. 2022).

Over the past decade the innovative gene therapy reagents over siRNA and miRNA-based therapies have entered clinical trials and recently three novel siRNA-based therapeutics were approved by the Food and Drug Administration (FDA). The approval of these drugs opened a new era of targeted drugs relying on safe and effective nano-delivery systems especially in tumor cells or tumor microenvironment (Daka and Peer 2012; Gandhi et al. 2014).

3.2 *Precision Immunotherapy in Cancer Progress*

Advances in the immunotherapy has especially opened a new era in cancer treatment. To maintain an active development on the vaccines, specific targeted therapies increased the excitement in the immune-therapeutics which includes immune cytokines, immune checkpoint inhibition, intra-tumoral therapies, and cellular therapies. Active development continues in these areas and in the development of vaccines, bispecific therapies, and more refined cellular therapies (Buchbinder and Hodi 2023).

Advances in immunotherapy are changing the standard of curing especially in cancer which target to enhance effective immune responses toward a tumor. Moreover, independent immune-related biomarkers that are currently used, such as programmed cell death ligand 1 (PD-L1) and tumor mutation burden (TMB), have achieved clinical relevance for a selection of patients to some extent. However, still some more scientific evidence is needed to be established to warrant the development of an integrated tumor microenvironment–based marker taking into account multiple parameters to maximize treatment efficacy and guide more suitable therapeutic strategies (Yu et al. 2019).

3.3 *Drug Resistance*

Pharmacodynamic resistance involves mutations in the target site that consequently influence the drug response at the site of action (Wang et al. 2019; Goldstein et al. 1989). Some of the most important identified multidrug resistant proteins are MDR1 and MRP1 in drug resistance.

P-glycoprotein (P-gp) which is also known as MDR1 or ATP-binding cassette sub-family B member 1 (ABCB1) is localized at the plasma membrane of many organs and tissues including colon, jejunum, bile canaliculi, renal tubular cells, placenta, the luminal surface of capillary endothelial cells, testes, pancreas and blood–brain barriers (Ueda et al. 1987; Zhou 2008).

Multidrug resistance-associated protein MRP1 (ABCC1) is a xenobiotic-transporting MRP-related protein to be cloned. The MRP family consists of the four isoforms MRP1-4 included in the resistance of many different cancer types including resistant lung cancer cell lines (Cole et al. 1992).

3.4 *Pharmacogenomis Data in Clinical Decision Support*

Pharmacogenomics (PGx) is the study of how a person's genes and DNA may impact their response to certain medications. These pharmacogenomics data can be analyzed and used in Clinical Decision Support (CDS) systems by using the Genomic Prescribing System (GPS) (Danahey et al. 2017)

There are many benefits of pharmacogenomics on CDS systems where the clinicians can predict the possible drug interactions during a common therapy and control the expected side effects of the combined medications that the patient uses (Wen et al. 2022). To maintain this approach in the field of routine therapy of the patients will inevitably increase the life quality of the patients during their treatments.

4 Concluding Remarks

Pharmacogenetic parameters and pharmacogenomics are started to be used to in the last decades to make a Clinical Decision and enables to prescribe the correct medicine at the correct dosage and dosage form with the right intervals makes the clinicians to decide in a more accurate way according to these parameters. These data are being used both to identify and select the write drug for the patients in an individualized decision using the pharmacodynamic evaluation that the patient have both to diagnose, and cure and to apply the cure in a most appropriate way for this patient with using the pharmacokinetic data from the patients.

References

AlAzzeh O, Roman Y. The frequency of rs2231142 in ABCG2 among Native Hawaiian and Pacific Islander subgroups: implications for personalized rosuvastatin dosing. Pharmacogenomics. 2023. https://doi.org/10.2217/pgs-2022-0160. Epub ahead of print. PMID: 36661065.

Alfarouk KO, Stock C-M, Taylor S, Walsh M, Muddathir AK, Verduzco D et al (2015) Resistance to cancer chemotherapy: failure in drug response from ADME to P-gp. Cancer Cell Int 15(1): 1–13

Bosch TM (2008) Pharmacogenomics of drug-metabolizing enzymes and drug transporters in chemotherapy. Methods Mol Biol 448:63–76. https://doi.org/10.1007/978-1-59745-205-2_5. PMID: 18370231.

Buchbinder EI, Hodi FS (2023) Impact of precision medicine in oncology: immuno-oncology. Cancer J 29(1):15–19. https://doi.org/10.1097/PPO.0000000000000641. PMID: 36693153.

Callaghan R, Luk F, Bebawy M (2014) Inhibition of the multidrug resistance P-glycoprotein: time for a change of strategy? Drug Metab Dispos 42:623–631

Cole SP, Bhardwaj G, Gerlach JH, Mackie JE, Grant CE, Almquist KC et al (1992) Overexpression of a transporter gene in a multidrug-resistant human lung cancer cell line. Science 258:1650–1654

Dai D, Zeldin DC, Blaisdell JA, Chanas B, Coulter SJ, Ghanayem BI et al (2001) Polymorphisms in human CYP2C8 decrease metabolism of the anticancer drug paclitaxel and arachidonic acid. Pharmacogenetics 11:597–607

Daka A, Peer D (2012) RNAi-based nanomedicines for targeted personalized therapy. Adv Drug Deliv Rev 64(13):1508–1521. https://doi.org/10.1016/j.addr.2012.08.014. Epub 2012 Sep 5. PMID: 22975009.

Danahey K, Borden BA, Furner B, Yukman P, Hussain S, Saner D, Volchenboum SL, Ratain MJ, O'Donnell PH (2017) Simplifying the use of pharmacogenomics in clinical practice: building the genomic prescribing system. J Biomed Inform 75:110–121. https://doi.org/10.1016/j.jbi.2017.09.012. Epub 2017 Sep 28. PMID: 28963061.

Djordjevic N (2023) Genotyping genetic variants of CYP2C19 for precision antiplatelet dosing: state of the art and future perspectives. Expert Opin Drug Metab Toxicol 2023:1–14. https://doi.org/10.1080/17425255.2022.2166486. Epub ahead of print. PMID: 36606363.

Estabrook RW (2003) A passion for P450s (remembrances of the early history of research on cytochrome P450). Drug Metab Dispos 31(12):1461–1473. https://doi.org/10.1124/dmd.31.12.1461

Fava GA, Tomba E, Bech P (2017) Clinical pharmacopsychology: Conceptual foundations and emerging tasks. Psychother Psychosom 86:134–140

Gandhi NS, Tekade RK, Chougule MB (2014) Nanocarrier mediated delivery of siRNA/miRNA in combination with chemotherapeutic agents for cancer therapy: current progress and advances. J Control Release 194:238–256. https://doi.org/10.1016/j.jconrel.2014.09.001. Epub 2014 Sep 7. PMID: 25204288; PMCID: PMC4254052.

Goldstein LJ, Galski H, Fojo A, Willingham M, Lai SL, Gazdar A et al (1989) Expression of a multidrug resistance gene in human cancers. J Natl Cancer Inst 81:116–124

Goldwasser P, Feldman J (1997) Association of serum albumin and mortality risk. J Clin Epidemiol 50:693–703.

Grimm G, Haslacher H, Kampitsch T, Endler G, Marsik C, Schickbauer T et al (2009) Sex differences in the association between albumin and all-cause and vascular mortality. Eur J Clin Investig 39:860–865

Johnson JA (2000) Predictability of the effects of race or ethnicity on pharmacokinetics of drugs. Int J Clin Pharmacol Ther 38:53–60

Kara G, Calin GA, Ozpolat B (2022) RNAi-based therapeutics and tumor targeted delivery in cancer. Adv Drug Deliv Rev 182:114113. https://doi.org/10.1016/j.addr.2022.114113. Epub 2022 Jan 19. PMID: 35063535.

Kusuhara H, Sugiyama Y (2002) Role of transporters in the tissue-selective distribution and elimination of drugs: transporters in the liver, small intestine, brain and kidney. J Control Release 78:43–54

Liliemark E, Söderhäll S, Sirzea F, Gruber A, Osby E, Björkholm M et al (1996) Higher in vivo protein binding of etoposide in children compared with adult cancer patients. Cancer Lett 106: 97–100

Mora Lagares L, Novič M (2022) Recent advances on P-glycoprotein (ABCB1) transporter modelling with in silico methods. Int J Mol Sci 23(23):14804. https://doi.org/10.3390/ijms232314804. PMID: 36499131; PMCID: PMC9740644.

Morris ME, Lee H-J, Predko LM (2003) Gender differences in the membrane transport of endogenous and exogenous compounds. Pharmacol Rev 55:229–240

Nahid NA, Johnson JA (2022) CYP2D6 pharmacogenetics and phenoconversion in personalized medicine. Expert Opin Drug Metab Toxicol 18(11):769–785. https://doi.org/10.1080/17425255.2022.2160317. Epub 2023 Jan 3. PMID: 36597259.

Narendra G, Choudhary S, Raju B, Verma H, Silakari O (2022) Role of genetic polymorphisms in drug-metabolizing enzyme-mediated toxicity and pharmacokinetic resistance to anti-cancer agents: a review on the pharmacogenomics aspect. Clin Pharmacokinet 61(11):1495–1517. https://doi.org/10.1007/s40262-022-01174-7. Epub 2022 Sep 30. PMID: 36180817.

Pardiñas AF, Owen MJ, Walters JTR (2021) Pharmacogenomics: a road ahead for precision medicine in psychiatry. Neuron 109(24):3914–3929. https://doi.org/10.1016/j.neuron.2021.09.011. Epub 2021 Oct 6. PMID: 34619094.] and cancer [Padma VV. An overview of targeted cancer therapy. Biomedicine. 2015;5(4):1–6.

Rodriguez-Antona C, Ingelman-Sundberg M (2006) Cytochrome P450 pharmacogenetics and cancer. Oncogene 25:1679–1691

Seyfinejad B, Ozkan SA, Jouyban A (2021) Recent advances in the determination of unbound concentration and plasma protein binding of drugs: analytical methods. Talanta 225:122052. https://doi.org/10.1016/j.talanta.2020.122052. Epub 2020 Dec 27. PMID: 33592774.

Ueda K, Clark DP, Chen CJ, Roninson IB, Gottesman MM, Pastan I (1987) The human multidrug resistance (mdr1) gene. cDNA cloning and transcription initiation. J Biol Chem 262:505–508

Vree TB, Van Ewijk-Beneken Kolmer EW, Wuis EW, Hekster YA, Broekman MM (1993) Interindividual variation in the capacity-limited renal glucuronidation of probenecid by humans. Pharm World Sci 15:197–202

Wang X, Zhang H, Chen X (2019) Drug resistance and combating drug resistance in cancer. Cancer Drug Resist 2(2):141

Wen HN, Wang CY, Li JM, Jiao Z (2022) Precision cardio-oncology: use of mechanistic pharmacokinetic and pharmacodynamic modeling to predict cardiotoxicities of anti-cancer drugs. Front Oncol 11:814699. https://doi.org/10.3389/fonc.2021.814699. PMID: 35083161; PMCID: PMC8784755.

Wilkinson GR (2005) Drug metabolism and variability among patients in drug response. N Engl J Med 352(21):2211–2221

Yu Y, Zeng D, Ou Q, Liu S, Li A, Chen Y, Lin D, Gao Q, Zhou H, Liao W, Yao H (2019) Association of survival and immune-related biomarkers with immunotherapy in patients with non-small cell lung cancer: a meta-analysis and individual patient-level analysis. JAMA Netw Open 2(7):e196879. https://doi.org/10.1001/jamanetworkopen.2019.6879. PMID: 31290993; PMCID: PMC6625073.

Zhou S-F (2008) Structure, function and regulation of P-glycoprotein and its clinical relevance in drug disposition. Xenobiotica 38:802–832

Biomarkers in Precision Medicine with Special Reference to Oncology

Neha Atale and Vibha Rani

Abbreviations

ADC	Antibody-drug conjugates
ADC	Apparent diffusion coefficient
ALK	Anaplastic lymphoma kinase
ALL	Acute lymphocytic leukemia
BCL2	B cell lymphoma-2
BCMA	B cell maturation antigen
BiTEs	Bispecific T cell engagers
CAR-T	Chimeric antigen receptor-modified T cells
CD	Cluster of differentiation
CDKN2A	Cyclin dependent kinase inhibitor 2A
CLDN	Claudin
CLL	Chronic lymphocytic leukemia
CTLA-4	Cytotoxic T-lymphocyte-associated protein 4
EGFR	Epidermal growth factor receptor
FDA	Food and Drug Administration
FET	18F-Fluoro-ethyl-tyrosine
FISH	Fluorescence *in situ* hybridization
GBM	Glioblastoma multiforme
HER2	Human epidermal growth factor receptor 2
IDH1	Isocitrate dehydrogenase 1

Neha Atale and Vibha Rani contributed equally with all other contributors.

N. Atale
Department of Cardiothoracic Surgery, Division of Lung Transplant and Lung Failure, Thomas E. Starzl Transplantation Institute, University of Pittsburgh, Pittsburgh, PA, USA

V. Rani (✉)
Department of Biotechnology, Jaypee Institute of Information Technology, Noida, Uttar Pradesh, India
e-mail: vibha.rani@jiit.ac.in

H. S. Tuli, M. B. Yerer Aycan (eds.), *Oncology: Genomics, Precision Medicine and Therapeutic Targets*, https://doi.org/10.1007/978-981-99-1529-3_7

IDO1	Indoleamine 2,3-dioxygenase 1
IHC	Immuno-histochemistry
ISH	*In situ* hybridization
JAK	Janus kinase
KRAS	Kirsten rat sarcoma viral oncogene homolog
LAG3	Lymphocyte activating 3
Let-7	Lethal-7
LncRNA	Long non-coding RNA
mAb	Monoclonal antibody
MGMT	Methylguanine-DNA methyltransferase
miRNA	Micro RNA
MRI	Magnetic resonance imaging
mRNA	Messenger RNA
NGS	Next-generation sequencing
NSCLC	Non small cell lung cancer
ORR	Overall response rate
PD-1	Programmed cell death protein 1
PDL1	Programmed death-ligand 1
PET-CT	Positron emission tomography-computed tomography
PPP3CA	Protein phosphatase 3 catalytic subunit-α
qPCR	Quantitative polymerase chain reaction
RB1	Retinoblastoma 1
ROS1	c-ros oncogene 1
ShRNA	Short hairpin RNA
SLAMF7	Signaling lymphocyte activation molecule F7
STK11	Serine/Threonine kinase 11
TCGA	The cancer genome atlas
TIM3	T-cell immunoglobulin and mucin-domain containing-3
TP53	Tumor protein 53

1 Introduction

Cancer is a prime global health issue, which is getting severe day by day. According to GLOBOCAN, an online database, tremendous increase in 19.3 million cases and 10 million deaths were measured due to cancer worldwide (https://www.uicc.org/news/globocan-2020-new-global-cancer-data). Blood, prostate, cervical, and breast cancer are very common cancers in low- and middle-income countries. There are various cancer awareness programs providing information regarding its early detection and predictive measures by biomarkers also taking place across the world. According to US FDA, biomarkers are the effective diagnostic indicators to measure the disease severity (Davis et al. 2020). They are generated either by cancer cells or by other cells in response to tumor (Whiteside 2008). Additionally, cancer biomarkers can be molecules, hormones, enzymes, gene products and cells which can be observed in blood, urine, body fluid and tissues (Rhea and Molinaro 2011). The discovery of these biomarkers set the path to precision medicine for improved patient outcomes through the selective classification of patients on the basis of probable treatment and prognosis (Vargas and Harris 2016).

Biomarkers must be accurate and measurable, with an established real outcome. This needs continuous research, validation, and approving certain tests that analyze the biomarker of interest for drug development. Knowledge of the biomarkers related opportunities among clinicians, patients, and regulators, may create possibilities for improving the current scenario. So, if the professional oncologists and patients become known of the benefits of using biomarkers, which would put pressure on decision committees to prioritize their execution, for development of new advanced cancer drugs.

There are few convenient techniques to identify specific type of cancers, such as cytology-based screening by Papanicolaou (Pap smear) and human papillomavirus (HPV) tests are the most common methods for cervical cancer (CC) detection (Koliopoulos et al. 2017). However, the detection schemes have some limitations such as patient discomfort status, as well as poor sensitivity and specificity. Surgery, radiation and chemotherapy, are the treatment options for cancer along with the severe side effects (Baskar et al. 2012; Bernier et al. 2004). The unsuccessful treatment in the final stage may prone to develop new prognostic, diagnostic, and therapeutic strategies (Piñero et al. 2020; Califf 2018).

Cancer precision medicine denotes the usage of treatments that would hopefully show benefits in cases where cancer progression shows molecular features such as genomic and proteomic alterations (Yates et al. 2018). In some tumors, profiling of relevant molecules may also provide clear prognostic and diagnostic information. Since cancers can have very complex genomics, precision medicine has gone through alteration through new methods in tumor analysis using next-generation sequencing (NGS) and other molecular profiling techniques (Dietz et al. 2016; Malone et al. 2020). Many malignancies including rare ones can today be targeted through precise therapies to achieve successful outcomes in patients.

This chapter reveals the role of cancer biomarkers in precision medicine, emerging approaches, challenges, and proposes potential significance of them in developing therapeutics. The advances in molecular identification techniques like genomics, transcriptomics and proteomics as well as the ability to correlate with drug development towards multi-therapy approaches may amplify the clinical benefits and extent of precision medicine.

2 Types of Biomarkers

Biomarkers are molecules or components which can act as estimators for prognosis or predictor of a disease progression. Biomarkers used in prognosis estimation can provide statistical differences between their quantity in diseased and control groups (Mayeux 2004). These need to be quantified in large selection of control groups such as the positive or negative biomarkers can be identified in patient samples. On the other end, the biomarkers which use for prediction should form an association between their quantity and some drug treatment outcome. They can be used to estimate patients who could benefit from a specific drug. These biomarkers can improve efficiency of clinical trials by allowing selection of possible participants benefited from the trials of different drugs (Boessen et al. 2014). They are mainly used in cases of novel anti-cancer immunotherapies and/or in utilization of programmed death-ligand 1 (PD-L1) testings (Schatz et al. 2020).

3 Major Techniques to Identify Biomarkers

Modern laboratory methods development has greatly influenced medicine. These methods allow for the in-depth study of many disease processes, expanding the scope of what medical laboratories can do beyond direct diagnosis. Several disease processes may now be accurately predicted in advance. Additionally, the results of treatment with specific therapeutic regimens can frequently be quite predictably. This method entails a multi-stage tissue fixation process followed by interpretation of results. It identifies particular antigens in histologic specimens by using target-specific antibodies in conjunction with colorimetric or fluorescent reagents, allowing for *in-situ* protein and phosphoprotein localisation.

Immuno-histochemistry (IHC) is a significant supplementary method utilized in modern labs for a variety of purposes, including examining the pathophysiology of numerous disease processes. It is commonly used to describe a variety of neoplasms, giving rise to an accurate histologic diagnosis and tumor subtype. Additionally, it can help identify benign and malignant cell populations in certain tissues. The distribution and differential expression of certain biomarkers in a tissue samples are also determined using this method. This in turn offers crucial data for assisting in the decision-making process about patient care.

Fluorescent *in situ* hybridization (FISH) uses DNA probes that anneal to the sequences of the target genes. Fluorescent labels on the DNA probes make it possible to identify hybridization using fluorescence microscopy. As a result, particular genomic abnormalities in the sample DNA can be identified. Another technique for detecting interesting RNA sequences is RNA in situ hybridization (RNA ISH) (Dietz et al. 2016; Malone et al. 2020).

While FISH is frequently used to identify genetic illnesses, ISH is a foundational method that is frequently utilized for diagnostic and research purposes. Additionally, FISH involves in the identification of new oncogenes and gene mapping. RNA With the use of ISH, intracellular agents such as messenger RNA (mRNA) and micro RNA (miRNA) can now be detected, which has shown to be important in research aimed at figuring out the pathophysiology of various malignancies.

4 Cancer Biomarkers

Antibody-based cancer biomarkers are commonly used for targeted therapy, which include antibody-drug conjugates (ADCs), monoclonal antibodies (mAbs) and radiolabeled ones, and bispecific T cell engagers (BiTEs) (Kantarjian et al. 2017; Maury et al. 2016). Other markers such as CD19, CD20, CD30, CD38, and SLAMF7 are found to be involved in hematological malignancies (Liu et al. 2017; Liu 2019). CD123, CLL-1, and CD138 are also critical targets in cancer therapeutics (Topp et al. 2015).

In recent years new antibody-drug conjugates such as brentuximab vedotin targeting CD30, inotuzumab ozomagicin targeting CD22 and polatuzumab vedotin targeting CD79b have been used in lymphoid cancer (Kreitman et al. 2018; Palanca-Wessels et al. 2015). Other biomarkers such as CD56, CD70, CD25 have been considered as targets through use of CAR-T cells and ADCs. Various therapeutic drugs such as coltuximab ravtansine, and loncastuximab tesirine (ADCT-402) are under clinical trials (Coiffier et al. 2016; Caimi et al. 2018). Gemtuzumab ozogamicin (GO) is an ADC against CD33 marker and several other biomarkers mainly expressed on myeloid cells (Khan et al. 2017) and given in single or along with chemotherapy (Li et al. 2018).

Solid tumors are another fast-growing cancer types requiring new treatments. They are of several types such as sarcoma, lymphoma and carcinoma. Targeted therapies with CAR T cells marking CD133 have been used against cholangiocarcinoma (Glumac and LeBeau 2018; Barzegar Behrooz et al. 2019). Similarly, CAR-T cells targeting mesothelin was proven to be effective against mesothelioma and lung cancer (Hassan and Ho 2008). In solid tumor immunotherapy CAR-T cells have been used to target epidermal growth factor receptor (EGFR), human epidermal growth factor receptor 2 (HER2), and claudin 18.2 (CLDN 18.2) (Wing et al. 2018; Guo et al. 2016). Figure 1 shows various cancer biomarkers currently used in precision medicine.

4.1 Genomic Biomarkers

Genomic biomarkers are important tools to understand lung adenocarcinoma disease for diagnostic purposes. Carcinoma could be further become drastic due to mutations in the KRAS22 and or EGFR23 genes (Rekhtman et al. 2012). Eventually, the advances in new generation of DNA sequencing techniques are important for elaborated classification of lung cancer genome. The cancer genome atlas (TCGA) network explored the sequencing of approximately 1000 various exomes of lung cancer and identified 15 unique genes that are implicated in enhancing oncogenesis in lung adenocarcinoma. Recently clinical trials were started based on new mutations observed in histone H3K79 methyltransferase DOT1L, protein phosphatase

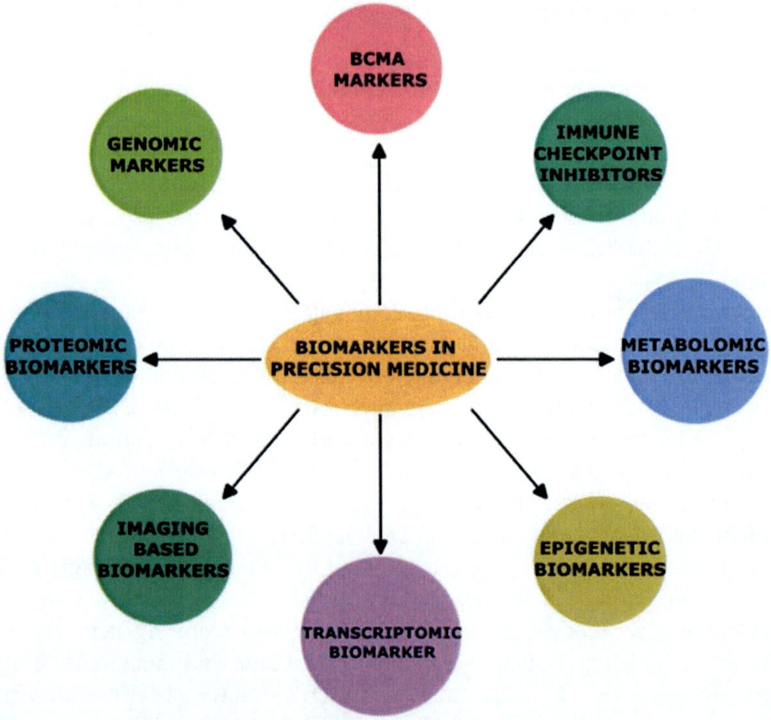

Fig. 1 Various biomarkers in precision medicine

3 catalytic subunit-α (PPP3CA) and FtsJ methyltransferase domain containing 1 (FTSJD1; or CMTR2) (Cancer Genome Atlas Research Network 2014; Campbell et al. 2016).

4.2 *Transcriptomic Biomarkers*

Transcriptomic biomarkers such as mRNA has been observed in association of various types of cancers. In lung cancer, use of microarray technology was used to detect mRNA expression markers (Beer et al. 2002; Kikuchi et al. 2003). Using these techniques studies have detected cohorts of mRNA expression markers to characterize sub groups of lung cancer and their association with prognosis. mRNA sequencing allows for broad sequencing of all mRNA present in sample. As an example, Janus-kinase (JAK) have been identified as a bio-marker as an activator of transcription pathway mRNA (Akagi et al. 2013; Kadara et al. 2011; Govindan et al. 2012).

4.3 *Epigenetic Biomarkers*

Various epigenetic biomarkers such as microRNA (miRNA), long non-coding RNA (lncRNA) and DNA methylation have been considered as biomarkers to function to regulate RNAs (Liloglou et al. 2011; Schmitt and Chang 2016). More specifically, miRNA and lncRNA function as ligands for receptors. DNA methylation on other hand, need to be converted to a detectable form using sequencing analysis through bisulfate conversion or enzyme use. These epigenetic markers have been used in various cancers (Shinjo et al. 2012).

4.4 *Proteomic Biomarkers*

Proteomic biomarkers are also important in exploring various pathological conditions. Recent reports stated that tissue microarrays were found to be novel proteomic markers (Brunnstrom et al. 2013). Proteomic analysis is done by mass spectrophotometry enabling mapping of a peaks to reference libraries to identify different proteins (Zhukov et al. 2003). Additionally, due to their potential role in oncology circulating inflammatory proteins were effective in cancer diagnosis (Alifano et al. 2011).

4.5 *Metabolomic Biomarkers*

Metabolomic markers that can be identified using mass spectroscopy. These are similar to the proteomics and are particularly helpful, as cancer is associated with large changes in metabolism (Hanahan and Weinberg 2011). The metabolites are also usually transported into bodily fluids (blood and urine), and these samples can be easily extracted by non-invasive methods for testing. Different metabolomics have been used to characterize classes of different cancers. These biomarkers require validation against known metabolite samples needing creation of synthetic control samples to make tangible progress in this field (Castle et al. 2006).

4.6 *Imaging Based Biomarkers*

Imaging based biomarkers are characterized as non-invasive biomarkers, indicating disease factors. Quantitative imaging biomarkers (QIB) have major role in precision medicines. This can be used for patient selection and for determining the responders and non-responders (Larue et al. 2017).

Brain tumors can be detected using imaging to evaluate the disease severity. Structural and functional complexity of brain lesions denote their difficulty to diagnose using invasive techniques. This requires a growing focus on noninvasive imaging methods to study these complexities to assess the disease status. Based on their cellular, and molecular characteristics, gliomas are detected mainly through non imaging methods rather than imaging ones. The role of non-imaging biomarkers in gliomas i.e., IDH1 mutation and methylguanine-DNA methyltransferase (MGMT) promoter methylation (Turkalp et al. 2014; Weller et al. 2010).

There are so many biomarkers determined for neurological diseases, such as brain MRI. Multiparametric MRI, having apparent diffusion coefficient (ADC) and perfusion-weighted sequences, are found to be important in the assessment of brain tumors (Lacerda and Law 2009).

Magnetic Resonance Spectroscopy (MRS) is another non-invasive technique to evaluate tissue metabolites and has been used to detect biomarkers such as choline/creatinine ratio found to be lower in cerebral metastases and Cho/NAA ratio has also been useful as biomarker (Tsougos et al. 2012; Server et al. 2010). Glycine amounts have also been found in biopsies in some cases. Other metabolites like lactate have also been detected using MRS in brain tumors. Other non-invasive imaging techniques like PET/CT such as 18F PET are also used in detection of brain tumors. 18F-Fluoro-ethyl-tyrosine (FET)-PET have been used to calculate tumor volume in Glioblastoma multiforme (GBM) has been shown to provide prognostic biomarker. PET targeting 11C-methionine (MET) is another marker in gliomas (Suchorska et al. 2015; De Witte et al. 2001).

4.7 *Immune Checkpoint Inhibition Biomarkers*

Immune checkpoint inhibitors (ICIs) against PD-1, PD-L1 and CTLA-4 have shown a basic epitome change in cancer immunology (Tumeh et al. 2014; Wong et al. 2019). It was observed that the PD-1/PD-L1 checkpoint was specifically found to be manipulated by malignant cells to prevent host immunity. Such chemotherapy is different from conventional one because targeting immune cells instead of cancer cells aim to modulate tumor microenvironment, further leading to restoration of suppressed cancer immunity (Ayers et al. 2017). There are other biomarkers such as IDO, TIM-3, LAG3 and CD47 are supplying the development of selective agents (Jiang et al. 2018; Charoentong et al. 2017; Giraldo et al. 2018).

Checkpoint inhibitors directly block CTLA-4 or PD-1 signaling leading to cytotoxic T-cell activity. This reactivates the effector T-cell populations turned inert by immune response in tumoral, stroma and parenchyma. As a corollary, the number and position of T cells in tumor micro-environment could be used to reliably estimate ICI response. As example an inflated CD8+ and CD4+ T cells are found in tumoral parenchyma in proximity of cancerous cells (Chen and Mellman 2017). This is correlated with better response in patients treated with checkpoint inhibition (Tumeh et al. 2014; Herbst et al. 2014). In early colorectal carcinoma cytotoxic and memory T-cell populations was able to estimate with high confidence the prognosis of survival and probability of recurrence (Galon et al. 2006; Fridman et al. 2012). In patients with melanoma, ICI treatment have also found to have simulated immune-response based score and tumor responding lymphocytes population at tumor micro-environment best correlated with checkpoint inhibitor response.

4.8 B cell Maturation Antigen (BCMA) Biomarkers

B cell maturation antigen (BCMA) expresses in malignant and normal B cells. Based on this knowledge many BCMA targeting CAR-T cell treatments have been developed. Among these bb2121, LCAR-B38M, CT053, and CT103A are currently under testing in many different subtypes of myeloma. Bb2121, specifically, was found to be successful in clinical trials achieving overall response rate (ORR) of 85% (Raje and et al. 2019; Brudno et al. 2018). LCAR-B38M is another BCMA targeting drug which is currently under testing in later stage of clinical trials (Zhao et al. 2018).

5 Biomarkers in Drug Development

The most important role of biomarkers is for development of anti-cancer drugs. Most clinical trials fail due to lack of any efficacy or any biomarker to prove improvement from drug intervention. Table 1 represents the list of biomarkers associated with diseases and targeted drugs. Drugs like gefitinib, crizotinib are some examples of

Table 1 List of biomarkers associated with diseases and targeted drugs

Biomarkers	Diseases	Anticancer drugs	Testing methods
BRCA1, BRCA2	Breast cancer	Olaparib, talazoparib	Sequencing
EGFR	NSCLC	Gefitinib, erlotinib, afatinib	Sequencing
FGFR2, FGFR3	Bladder cancer	Erdafitinib	FISH, sequencing
HER-2	Breast cancer	Trastuzumab, pertuzumab	FISH, IHC
MET	NSCLC	Crizotinib	FISH, sequencing
RAS	Colorectal cancer	Cetuximab, panitumumab	Sequencing
ROS1	NSCLC	Crizotinib	FISH, sequencing
BCR	Chronic myeloid leukemia	Imatinib, dasatinib, nilotinib	IHC, sequencing, PCR
IDH1, IDH2	Chronic myeloid leukemia	Ivosidenib, enasidenib	IHC, sequencing
BRAF	Melanoma, NSCLC	Dabrafenib, trametinib	IHC, sequencing, PCR

success of drug development through use of biomarkers (Gatzemeier et al. 2007; Herbst et al. 2005). Lack of correlative biomarkers are the biggest hurdle today in development of anti-cancer drugs. In case such as EGFR, ALK, ROS1 the activation of these pathways is required for cancer cells and any changes in these factors can directly be used as predictable biomarkers (Herbst et al. 2004; Giaccone et al. 2004). But in many cases, such direct links do not exist as the cancer cells can be killed through indirectly affecting the cancer drivers. However, it has been shown that the PARP inhibitor olaparib can effectively destroy breast cancer early onset (BRCA)-mutant ovarian and breast cancer cells by causing synthetic lethality between the cancer drivers and their synthetic lethal partners (Fong et al. 2009; Audeh et al. 2010). In this situation, olaparib and other PARP inhibitor responses will be predicted by loss-of-function mutations in the BRCA1 and BRCA2 genes or other functional abnormalities in the DNA repair pathways (Fang 2014a). Synthetic lethality is a deadly trait brought on by the interplay of two or more different gene mutations.

New genotype-selective anticancer drugs and/or the discovery of novel therapeutic targets have both lately been made possible due to this notion (Guo and Fang 2012; Fang 2014b). The adoption of this strategy may be advantageous for the indirect targeting of cancer-causing factors such genetic variations in the TP53, STK11, RB1, and RAS genes, which are difficult or impossible to target small molecules. This resulted to the identification of KRAS synthetic lethal genes, depending on the rigor of statistical analysis, indicating the diversity of biological processes or pathways controlled by KRAS (Luo et al. 2009; Brunelli et al. 2014).

Additionally, many KRAS-mutant tumors also include mutations in other cancer-driver genes such TP53, CDKN2A, and STK11. According to research on transgenic mice and animals models with activating mutations in KRAS are predisposed to developing lung adenocarcinomas at an early age (Jackson et al. 2001; Meuwissen et al. 2001). However, neither an invasive nor metastatic characteristics are seen in these malignancies (Kim et al. 2005; Jackson et al. 2005). Finally, because increased RAS activity in human cancers by gene amplification (Filmus and Buick 1985) or an enhanced signals from tyrosine kinase growth factor receptors (Ehrhardt et al. 2004; Buday and Downward 1993), or a reduction in the expression of microRNAs like members of the let-7 family (Barbie et al. 2009).

6 Challenges to Improve Biomarker Based Clinical Trials

Clinical trial-based testing of bio-markers is a challenging endeavor since it does not allow for exhaustive study of these markers to achieve reliable results. In most cases, a very few patients are eligible for clinical trials. This means researchers cannot run an extensive range of tests for different molecular markers in clinical trials as failures would be expensive both in terms of money and time. Furthermore, clinical trials by essence of their testing phase have no accountability towards patients.

This leads us to an urgent need to create efficient testing methods using genomics. Because of that it would be possible to find patients that could be best suited for the corresponding molecular cohort. This way the chances that a drug or bio-marker would have better effect on patient's conditions would increase exponentially. However, this requires analysis of thousands of bio-markers. Projects such as PACT and GENIE have started to work on these particular problems with data now available for over hundreds of thousands of patients. Well designed and coordinated study could lead to precise medical interventions through link of clinical retrospective and proteomic data to clinical outcomes and creating fast and efficient method for biomarker validation, drug modifications and identifying new biomarkers and targets for patients.

7 Conclusion and Future Perspectives

Cancer is a broad disease where the cancerous cell pathways are altered and affecting many individual molecules with overlaps between different cancer subtypes. This broad range of changes to tumor micro-environment requires development of large number of biomarkers and testing them in clinical trials in large cohorts to improve patient lives. The new techniques presented in this chapter have shown more reliable approaches that collate large datasets from molecular measurements and patients. These have given way to tremendous progress in biomarker development. This approach for drug discovery and implementation to perform precise targeting is much faster and better integrated than the broad-spectrum biomarker studies with single biomarker used to estimate progression of many different subtypes of cancers.

Moreover, many clinical trials try to link a given biomarker with a corresponding drug treatment but in reality, several different biomarkers account for different pathways implicated in diseases and eventual prognosis. Some of these are down to regulatory practices which need to be update with these new techniques. Similarly, the prediction or prognosis of exact nature of the cancer subtype using biomarkers is still not reliable as it can give false-positives due to the biomarker measurements being affected by many thousands of molecules. Further advancements in techniques and improved precision of biomarker to molecular measurements would hopefully improve the use of biomarkers to estimate cancer subtypes and prognosis in future.

Acknowledgement We acknowledge University of Pittsburgh and Jaypee Institute of information technology for infrastructural support.

Conflict of Interest Authors declare no conflict of interest.

References

Akagi I et al (2013) Combination of protein coding and noncoding gene expression as a robust prognostic classifier in stage I lung adenocarcinoma. Cancer Res 73:3821–3832

Alifano M et al (2011) Preresection serum C-reactive protein measurement and survival among patients with resectable non-small cell lung cancer. J Thorac Cardiovasc Surg 142:1161–1167

Audeh MW et al (2010) Oral poly(ADP-ribose) polymerase inhibitor olaparib in patients with BRCA1 or BRCA2 mutations and recurrent ovarian cancer: a proof-of-concept trial. Lancet 376:245–251

Ayers M et al (2017) IFN-γ-related mRNA profile predicts clinical response to PD-1 blockade. J Clin Invest 127:2930–2940

Barbie DA et al (2009) Systematic RNA interference reveals that oncogenic KRAS-driven cancers require TBK1. Nature 462:108–112

Barzegar Behrooz A, Syahir A, Ahmad S (2019) CD133: beyond a cancer stem cell biomarker. J Drug Target 27:257–269

Baskar R, Lee KA, Yeo R, Yeoh KW (2012) Cancer and radiation therapy: current advances and future directions. Int J Med Sci 9:193–199

Beer DG et al (2002) Gene-expression profiles predict survival of patients with lung adenocarcinoma. Nat Med 8:816–824

Bernier J, Hall EJ, Giaccia A (2004) Radiation oncology: a century of achievements. Nature 4:737–747

Boessen R, Heerspink HJ, De Zeeuw D et al (2014) Improving clinical trial efficiency by biomarker-guided patient selection. Trials 15:103–114

Brudno JN et al (2018) T cells genetically modified to express an anti-B-cell maturation antigen chimeric antigen receptor cause remissions of poor-prognosis relapsed multiple myeloma. J Clin Oncol 36:267–228

Brunelli L et al (2014) Capturing the metabolomic diversity of KRAS mutants in non-small-cell lung cancer cells. Oncotarget 5:4722–4731

Brunnstrom H et al (2013) Immunohistochemistry in the differential diagnostics of primary lung cancer: an investigation within the Southern Swedish Lung Cancer study. Am J Clin Pathol 140: 37–46

Buday L, Downward J (1993) Epidermal growth factor regulates p21ras through the formation of a complex of receptor, Grb2 adapter protein, and Sos nucleotide exchange factor. Cell 73:611–620

Caimi P, Kahl BS, Hamadani M et al (2018) Safety and efficacy of Adct-402 (Loncastuximab Tesirine), a novel antibody drug conjugate, in relapsed/refractory follicular lymphoma and mantle cell lymphoma: interim results from the phase 1 first-in-human study. Blood 132:2874

Califf RM (2018) Biomarker definitions and their applications. Exp Biol Med 243:213–221

Campbell JD et al (2016) Distinct patterns of somatic genome alternations in lung adenocarcinomas and squamous cell carcinomas. Nat Genet 48:607–616

Cancer Genome Atlas Research Network (2014) Comprehensive molecular profiling of lung adenocarcinoma. A comprehensive analysis of lung adenocarcinoma describing many different molecular subtypes. Nature 511:543–550

Castle AL et al (2006) Metabolomics Standards Workshop and the development of international standards for reporting metabolomics experimental results. Brief Bioinform 7:159–165

Charoentong P et al (2017) Pan-cancer immunogenomic analyses reveal genotype-immunophenotype relationships and predictors of response to checkpoint blockade. Cell Rep 18:248–262

Chen DS, Mellman I (2017) Elements of cancer immunity and the cancer-immune set point. Nature 541:321–330

Coiffier B, Thieblemont C, de Guibert S et al (2016) A phase II, single-arm, multicentre study of coltuximab ravtansine (SAR3419) and rituximab in patients with relapsed or refractory diffuse large B-cell lymphoma. Br J Haematol 173:722–730

Davis KD, Aghaeepour N, Ahn AH et al (2020) Discovery and validation of biomarkers to aid the development of safe and effective pain therapeutics: challenges and opportunities. Nat Rev Neurol 16:381–400

De Witte O et al (2001) Positron emission tomography with injection of methionine as a prognostic factor in glioma. J Neurosurg 95:746–750

Dietz S, Schirmer U, Merce C et al (2016) Low input whole-exome sequencing to determine the representation of the tumor exome in circulating DNA of non-small cell lung cancer patients. PLoS One 11:e0161012

Ehrhardt A et al (2004) Distinct mechanisms determine the patterns of differential activation of H-Ras, N-Ras, K-Ras 4B, and M-Ras by receptors for growth factors or antigen. Mol Cell Biol 24:6311–6323

Fang B (2014a) Development of synthetic lethality anticancer therapeutics. J Med Chem 57:7859–7873

Fang B (2014b) Genetic interactions of STAT3 and anticancer drug development. Cancer 6:494–525

Filmus JE, Buick RN (1985) Stability of c-K-ras amplification during progression in a patient with adenocarcinoma of the ovary. Cancer Res 45:4468–4472

Fong PC et al (2009) Inhibition of poly(ADP-ribose) polymerase in tumors from BRCA mutation carriers. N Engl J Med 361:123–134

Fridman WH et al (2012) The immune contexture in human tumours: impact on clinical outcome. Nat Rev Cancer 12:298–306

Galon J et al (2006) Type, density, and location of immune cells within human colorectal tumors predict clinical outcome. Science 313:1960–1964

Gatzemeier U et al (2007) Phase III study of erlotinib in combination with cisplatin and gemcitabine in advanced non-small-cell lung cancer: the Tarceva Lung Cancer Investigation Trial. J Clin Oncol 25:1545–1552

Giaccone G et al (2004) Gefitinib in combination with gemcitabine and cisplatin in advanced non-small-cell lung cancer: a phase III trial–INTACT 1. J Clin Oncol 22:777–784

Giraldo NA et al (2018) Multidimensional, quantitative assessment of PD-1/PD-L1 expression in patients with Merkel cell carcinoma and association with response to pembrolizumab. J Immunother Cancer 6:99

Glumac PM, LeBeau AM (2018) The role of CD133 in cancer: a concise review. Clin Transl Med 7: 1–42

Govindan R et al (2012) Genomic landscape of non-small cell lung cancer in smokers and never-smokers. Cell 150:1121–1134

Guo Z, Fang B (2012) Synthetic lethality in anticancer drug discovery and target identification. Pharmacol Ther 3:352–361

Guo Y, Wang Y, Han W (2016) Chimeric antigen receptor-modified T cells for solid tumors: challenges and prospects. J Immunol Res 2016:3850839

Hanahan D, Weinberg RA (2011) Hallmarks of cancer: the next generation. A summary of the hallmarks, or characteristics, of cancer. Cell 144:646–674

Hassan R, Ho M (2008) Mesothelin targeted cancer immunotherapy. Eur J Cancer 44:46–53

Herbst RS et al (2004) Gefitinib in combination with paclitaxel and carboplatin in advanced non-small-cell lung cancer: a phase III trial—INTACT 2. J Clin Oncol 22:785–794

Herbst RS et al (2005) TRIBUTE: a phase III trial of erlotinib hydrochloride (OSI-774) combined with carboplatin and paclitaxel chemotherapy in advanced non-small-cell lung cancer. J Clin Oncol 23:5892–5899

Herbst RS et al (2014) Predictive correlates of response to the anti-PD-L1 antibody MPDL3280A in cancer patients. Nature 515:563–567
Jackson EL et al (2001) Analysis of lung tumor initiation and progression using conditional expression of oncogenic K-ras. Genes Dev 15:3243–3248
Jackson EL et al (2005) The differential effects of mutant p53 alleles on advanced murine lung cancer. Cancer Res 65:10280–10288
Jiang P et al (2018) Signatures of T cell dysfunction and exclusion predict cancer immunotherapy response. Nat Med 24:1550–1558
Kadara H et al (2011) A five-gene and corresponding protein signature for stage-I lung adenocarcinoma prognosis. Clin Cancer Res 17:1490–1501
Kantarjian H, Stein A, Gokbuget N et al (2017) Blinatumomab versus chemotherapy for advanced acute lymphoblastic leukemia. N Engl J Med 376:836–847
Khan N, Hills RK, Virgo P et al (2017) Expression of CD33 is a predictive factor for effect of gemtuzumab ozogamicin at different doses in adult acute myeloid leukaemia. Leukemia 31: 1059–1068
Kikuchi T et al (2003) Expression profiles of non-small cell lung cancers on cDNA microarrays: identification of genes for prediction of lymph-node metastasis and sensitivity to anti-cancer drugs. Oncogene 22:2192–2205
Kim CF et al (2005) Mouse models of human non-small-cell lung cancer: raising the bar. Cold Spring Harb Symp Quant Biol 70:241–250
Koliopoulos G, Nyaga VN, Santesso N et al (2017) Cytology versus HPV testing for cervical cancer screening in the general population. Cochrane Database Syst Rev 8:CD008587
Kreitman RJ, Tallman MS, Robak T et al (2018) Minimal residual hairy cell leukemia eradication with moxetumomab pasudotox: phase 1 results and long-term follow-up. Blood 131:2331–2334
Lacerda S, Law M (2009) Magnetic resonance perfusion and permeability imaging in brain tumors. Neuroimaging Clin 19:527–557
Larue RT et al (2017) Quantitative radiomics studies for tissue characterization: a review of technology and methodological procedures. Br J Radiol 90:1–27
Li F, Sutherland MK, Yu C et al (2018) Characterization of SGN-CD123A, a potent CD123-directed antibody-drug conjugate for acute myeloid leukemia. Mol Cancer Ther 17:554–564
Liloglou T et al (2011) Epigenetic biomarkers in lung cancer. Cancer Lett 342:200–212
Liu D (2019) Cancer biomarkers for targeted therapy. Biomark Res 7:25–32
Liu B, Song Y, Liu D (2017) Clinical trials of CAR-T cells in China. J Hematol Oncol 10:166–176
Luo J et al (2009) A genome-wide RNAi screen identifies multiple synthetic lethal interactions with the Ras oncogene. Cell 137:835–848
Malone ER, Oliva M, Sabatini PJB et al (2020) Molecular profiling for precision cancer therapies. Genome Med 12:8–27
Maury S, Chevret S, Thomas X et al (2016) Rituximab in B-lineage adult acute lymphoblastic leukemia. N Engl J Med 375:1044–1053
Mayeux R (2004) Biomarkers: potential uses and limitations. NeuroRx 1:182–188
Meuwissen R et al (2001) Mouse model for lung tumorigenesis through Cre/lox controlled sporadic activation of the K-Ras oncogene. Oncogene 20:6551–6558
Palanca-Wessels MC, Czuczman M, Salles G et al (2015) Safety and activity of the anti-CD79B antibody-drug conjugate polatuzumab vedotin in relapsed or refractory B-cell non-Hodgkin lymphoma and chronic lymphocytic leukaemia: a phase 1 study. Lancet Oncol 16:704–715
Piñero F, Dirchwolf M, Pessôa MG (2020) Biomarkers in hepatocellular carcinoma: diagnosis, prognosis and treatment response assessment. Cell 9:1370
Raje N, et. Al. (2019) Anti-BCMA CAR T-cell therapy bb2121 in relapsed or refractory multiple myeloma. N Engl J Med 380:1726–1737
Rekhtman N, Paik PK, Arcila ME et al (2012) Clarifying the spectrum of driver oncogene mutations in biomarker-verified squamous carcinoma of lung: lack of EGFR/KRAS and presence of PIK3CA/AKT1 mutations. Clin Cancer Res 18:1167–1176

Rhea JM, Molinaro RJ (2011) Cancer biomarkers: Surviving the journey from bench to bedside. MLO 43:10–12, 16, 18; quiz 20, 22.
Schatz S, Falk M, Jóri B et al (2020) Integration of tumor mutation burden and PD-L1 testing in routine laboratory diagnostics in non-small cell lung cancer. Cancer 24:1685–1699
Schmitt AM, Chang HY (2016) Long noncoding RNAs in cancer pathways. Cancer Cell 29:452–463
Server A et al (2010) Proton magnetic resonance spectroscopy in the distinction of high-grade cerebral gliomas from single metastatic brain tumors. Acta Radiol 51:316–325
Shinjo K et al (2012) Integrated analysis of genetic and epigenetic alterations reveals CpG island methylator phenotype associated with distinct clinical characters of lung adenocarcinoma. Carcinogenesis 33:1277–1285
Suchorska B et al (2015) Biological tumor volume in 18FET-PET before radiochemotherapy correlates with survival in GBM. Neurology 84:710–719
Topp MS, Gökbuget N, Stein AS et al (2015) Safety and activity of blinatumomab for adult patients with relapsed or refractory B-precursor acute lymphoblastic leukaemia: a multicentre, single-arm, phase 2 study. Lancet Oncol 16:57–66
Tsougos I et al (2012) Differentiation of glioblastoma multiforme from metastatic brain tumor using proton magnetic resonance spectroscopy, diffusion and perfusion metrics at 3T. Cancer Imaging 12:1–29
Tumeh PC et al (2014) PD-1 blockade induces responses by inhibiting adaptive immune resistance. Nature 515:568–571
Turkalp Z, Karamchandani J, Das S (2014) IDH mutation in glioma: new insights and promises for the future. JAMA Neurol 71:1319–1325
Vargas AJ, Harris CC (2016) Biomarker development in the precision medicine era: lung cancer as a case study. Nat Rev Cancer 16:525–537
Weller M et al (2010) MGMT promoter methylation in malignant gliomas: ready for personalized medicine? Nat Rev Neurol 6:39–51
Whiteside TL (2008) The tumor microenvironment and its role in promoting tumor growth. Oncogene 27:5904–5912
Wing A, Fajardo CA, Posey AD Jr et al (2018) Improving CART-cell therapy of solid tumors with oncolytic virus-driven production of a bispecific T-cell engager. Cancer Immunol Res 6:605–616
Wong PF et al (2019) Multiplex quantitative analysis of tumor-infiltrating lymphocytes and immunotherapy outcome in metastatic melanoma. Clin Cancer Res 25:2442–2449
Yates LR, Seoane J, Le Tourneau C et al (2018) The European society for medical oncology (ESMO) precision medicine glossary. Ann Oncol 29:30–35
Zhao W-H et al (2018) A phase 1, open-label study of LCAR-B38M, a chimeric antigen receptor T cell therapy directed against B cell maturation antigen, in patients with relapsed or refractory multiple myeloma. J Hematol Oncol 11:1–21
Zhukov TA et al (2003) Discovery of distinct protein profiles specific for lung tumors and premalignant lung lesions by SELDI mass spectrometry. Lung Cancer 40:267–279

Bioinformatics and Biostatistics in Precision Medicine

Ahu Cephe, Necla Koçhan, Esma Gamze Aksel, Funda İpekten, Serra İlayda Yerlitaş, Gözde Ertürk Zararsız, and Gökmen Zararsız

1 Introduction

With the recent technological developments in modern medicine and life science areas, the amount of data that can be collected for each patient has increased exponentially. With the appropriate analysis of big data obtained from different sources in the field of genetics, metabolism and imaging; the biological mechanisms that cause the formation of diseases can be characterized, and the information

A. Cephe
Institutional Data Management and Analytics Units, Erciyes University Rectorate, Kayseri, Turkey

Drug Application and Research Center (ERFARMA), Erciyes University, Kayseri, Turkey
e-mail: ahucephe@erciyes.edu.tr

N. Koçhan
İzmir Biomedicine and Genome Center (IBG), İzmir, Turkey
e-mail: necla.kochan@ibg.edu.tr

E. G. Aksel
Drug Application and Research Center (ERFARMA), Erciyes University, Kayseri, Turkey

Department of Genetics, Faculty of Veterinary Medicine, Erciyes University, Kayseri, Turkey
e-mail: gamzeilgar@erciyes.edu.tr

F. İpekten
Department of Biostatistics, Faculty of Medicine, Erciyes University, Kayseri, Turkey

Department of Biostatistics, Faculty of Medicine, Adiyaman University, Adıyaman, Turkey
e-mail: fipekten@adiyaman.edu.tr

S. İ. Yerlitaş · G. E. Zararsız · G. Zararsız (✉)
Drug Application and Research Center (ERFARMA), Erciyes University, Kayseri, Turkey

Department of Biostatistics, Faculty of Medicine, Erciyes University, Kayseri, Turkey
e-mail: serrayerlitas@erciyes.edu.tr; gozdeerturk@erciyes.edu.tr; gokmenzararsiz@erciyes.edu.tr

H. S. Tuli, M. B. Yerer Aycan (eds.), *Oncology: Genomics, Precision Medicine and Therapeutic Targets*, https://doi.org/10.1007/978-981-99-1529-3_8

obtained from this can be used for the prevention and treatment of diseases. Precision medicine is a therapeutic science field that aims to provide the right dose of the right drug to the right patient at the right time by combining fields such as genomics, big data analytics and population health. Precision medicine provides personalized healthcare with a data-driven approach, enabling more efficient spending and better patient outcomes. Precision medicine has replaced the term "personalized medicine" because of the claim that doctors always treat patients on a personalized level (König et al. 2017; Low et al. 2017; MacEachern and Forkert 2021).

Undoubtedly, the rapid development of omics technologies and the increase in the use of machine learning approaches in health studies are among the most important reasons for the rapid progress in precision medicine. Apart from the field of genomics within precision medicine, it also includes other omics fields such as transcriptomics, epigenomics, proteomics, metabolomics and radiomics. Many studies in the current literature are on the integrated solution of problems related to diseases by including multiple omics data with biostatistics and bioinformatics approaches. In this approach, also known as multi-omics, the need for new analytical approaches has emerged in modeling these large and complex data. At the forefront of these approaches are machine learning methods, which involve training mathematical models and are frequently used to make predictions on diseases. Machine learning methods, which are frequently preferred in traditional medicine studies today, will be an indispensable tool for future precision medicine studies (Alyass et al. 2015; Low et al. 2017; MacEachern and Forkert 2021; Quazi 2022).

In this section, we aimed to describe the biostatistics and bioinformatics approaches in precision medicine studies and guide researchers who will work in this field. Particular focus will be given to the analysis of single and multi-omics data and machine-learning approaches.

2 Omics Technologies

Some scientific discoveries in genomics are critical in the development of precision medicine. These are, respectively, the discovery of the DNA double-strand model (in 1953), the discovery of Sanger sequencing (in 1977), and the beginning of the Human Genome Project (in 1990). It took 13 years for the first human genome to be uncovered in this project (Naithani et al. 2021). Although a final version of the project results was reported in 2003, in which approximately 20,500 genes were reported, it was updated with the final version in April 2022. With the development of next-generation sequencing (NGS) analysis, genome sequencing, which takes a long time with the Sanger method, has begun to be obtained in a much shorter time and at a lower cost. After the first results of the Human Genome Project were received, the high-throughput technologies used in medical research today have created a revolution. These technologies, which help to provide the instant content of the biological system in high resolution, form "Omics", a nomenclature applied to the study of the quantification and characterization of biological molecules as a whole (Naithani et al. 2021). The term omic is derived from the Latin suffix -ome, meaning 'many', and represents the 'whole', 'all' of the entity or formation represented by the name to which it is added. Therefore, omics studies do not cover one or a few measurements but the entire matrix examined (Budak and Dönmez 2012). The suffix Ome- was derived from "genome, chromosome" by Hans Winkler in 1920 (Lederberg and Mccray 2001).

With the development of omic technologies, many "omics" terms have emerged from different biomolecules. These are the foundation of the fields of omics research moving from genomics to epigenomics, from transcriptomics to proteomics and metabolomics. Multi-omics research has gained momentum in interactomics research, where the interactions of all these omics technologies are evaluated as a whole. In addition, pharmacogenomics, which is used to create personalized treatments based on individual genetic variations and to predict treatment responses, has also become the focus of attention in precision medicine. In this regard, the correlation between different omic data mentioned in the precision medicine approach should be carefully examined (Hasanzad et al. 2021). By evaluating all the omic data obtained in biostatistics and bioinformatics, it is possible to reveal the relationships between omic levels to determine biomarkers for the diagnosis and treatment of diseases, the detailed disease pathophysiology, and to offer personalized treatment options. Patient cohort classification becomes easier. Analysis of omics data provides preliminary diagnostic data to streamline patient management (Quazi 2022). Most of the available studies examine omics data individually. However, machine-learning data integration is needed in the precision medicine horizon (Hasanzad et al. 2021). Multi-omics data is meaningful if they indicate a disease signature associated with new drug targets during drug development (Hartl et al. 2021).

Therefore, with all its sub-fields, omic is an effective precision medicine tool (Naithani et al. 2021). Molecular markers can be used for all disease management phases (diagnosis, treatment response, and progression) (Sung et al. 2012). Genomic markers are the most important in omic data, but epigenomic, transcriptomic, and metabolomic markers are also important (Naithani et al. 2021). In the integration of multi-omic signatures, omics signatures are affected by the ongoing cross-talk and dynamics in the human body. There are also lifestyle and external environmental factors that affect the omic outcome. Therefore, integrating multiple omic data from an individual is required to generate more meaningful molecular signatures more strongly associated with the disease. This is also called systems biology (Abu-Asab et al. 2011). The primary purpose of systems biology research created by multi-omic data is to develop filtered predictive models with empirical validation to select patients based on personalized multi-omics data, identify who will benefit most from targeted therapies, and classify them (Chakraborty et al. 2018; GuhaThakurta et al. 2013).

3 Genomics

The beginning of precision medicine was genomic research. The term genome describes an organism's complete set of DNA (where, when, and how much). In humans, genetic material is present in the cell nucleus. However, mitochondria have their genome (Zeggini et al. 2019). Genomic studies are valuable for gene mapping detection of genetic variants associated with single genes and multifactorial diseases (Hasin et al. 2017). In addition to diagnosing the disease, medical research also focuses on identifying genetic variants associated with the response to treatments and the disease (McGuire et al. 2020). One of the main goals of genomics is to reveal the essential function of the gene by establishing relationships between the existing genotype and phenotype, and it which is critical in the formation of predictive models and the development of precision medicine. At this stage, deep learning can be used to perform genome-wide association analyses accurately and quickly (Eraslan et al. 2019; Ho et al. 2019). Genome-wide association (GWAS) and whole-exome sequencing (WES) methods are currently being studied in the field of genomics for the detection of variants associated with common multifactorial diseases (Hasanzad et al. 2021). It has been reported that next-generation sequencing (NGS)-based technology in genomics is the closest to clinical research among the technologies presented in various omics fields (Tebani et al. 2016). However, one of the challenging aspects of NGS is its clinical validity. Also another problem is the analysis of the large amount of data generated (a sample whole genome sequenced 200 Gb). In this respect, bioinformatics has always been at the center of genetic data analysis (Pettini et al. 2021). For NGS data analysis, data were stored in the FASTQ format, and various alignment tools were used for sequencing. Some popular aligners (using SAM or BAM format) are BWA (Li and Durbin 2009), Bowtie (Langmead et al. 2009), Bowtie2 (Langmead and Salzberg 2012) and, SNAP (Zaharia et al. 2011; Pettini et al. 2021). Databases with data on the human genome are NCBI (Brown et al. 2015), ENSEMBL, GOA (Dimmer et al. 2011) and, ENCODE (ENCODE 2004).

4 Epigenomics

Our stable genome at birth can undergo chemical changes throughout our lives due to diet, environment, and lifestyle (Chadwick et al. 2015). Chemical modifications of DNA, nuclear RNA, histone, and nonhistone chromatin proteins (Bird 2007), microRNA (miRNA) expression, processing, and chromatin compaction (Sadakierska-Chudy and Filip 2015) define the epigenome. This can lead to transcriptional repression or activation. Thus, it affects the synthesis of the corresponding protein. In the mentioned changes, the expression of the gene is affected without changing the DNA sequence. Epigenetic markers can be transient or inherited (Bird 2007). These modifications are quantified and characterized using

Epigenomics (Chadwick et al. 2015). Epigenomic changes depend on time and environmental factors. Thus, they may be tissue-specific in response to environmental factors and disease-related variables. The comprehensive mapping of epigenetic structure has been reported in many tissues and cell types (Kundaje et al. 2015). Various strategies have been adopted to evaluate the epigenome (Mensaert et al. 2013). Studies have shown that interindividual variations may be associated with epigenetic modifications in some diseases. The most widely used epigenomic methods are methylation analysis by bisulfite sequencing (BSSeq) or whole-genome bisulfite sequencing (WGBS), and NGS (ChIP-seq) combined with chromatin immunoprecipitation (ChIP) assays (Chakraborty et al. 2018; Tebani et al. 2016). In addition, the method known as DNase-seq, which identifies regulatory regions in the genome combines the digestion of chromatin with DNase I enzyme with high-throughput sequencing (Yaragatti et al. 2008). The ATAC-seq method also allows mapping of chromatin accessibility across the genome (Buenrostro et al. 2015; Tebani et al. 2016). Taking advantage of the developments in the field of NGS, it is now possible to detect methylated cytosine bases on genomic DNA (Tebani et al. 2016). Data obtained by array-based methods can be analyzed using special packages such as methylation Array Analysis (Maksimovic et al. 2016). SICER2 (Zang et al. 2009), PeakRanger (Feng et al. 2011), GEM (Guo et al. 2012), MUSIC (Harmanci et al. 2014), PePr (Zhang et al. 2014), DFilter (Kumar et al. 2013), and MACS (Zhang et al. 2008) are some of the packages used for ChIP-seq data processing (Pettini et al. 2021). NCBI epigenomics (Bernstein et al. 2010), 4DGenome (Teng et al. 2015), HEA, and MethylomeDB (Xin et al. 2012) are popular epigenomic databases for humans.

5 Transcriptomics

The transcriptome describes all the mRNAs produced in an organism at a "specific location and time". A transcriptome is the whole of the coded (messenger RNA) and non-coding RNA (soluble RNA, transfer RNA, siRNA, etc.) in a cell, tissue, or sample. A comprehensive analysis of these data is called as transcriptomics (Mele et al. 2015). Quantitative analysis of the transcriptome can be performed by RNA sequencing (RNA-seq) or microarrays (microchips). Next-generation sequencing-based transcriptome analysis (RNA-seq) (Mortazavi et al. 2008) provides quantitative and qualitative gene expression profiles. Microarrays can also be used to measure the relative abundance of mRNAs for large numbers of genes. However, with the advent of NGS, it has become possible to determine the abundance and presence of genome-wide RNA transcripts. Furthermore, RNA-seq analyses are not related to existing transcript-specific probes in the microarray technique, therefore they can unbiasedly and efficiently detect fewer new transcripts with high specificity and sensitivity (Pettini et al. 2021). Evaluation of the data obtained in the transcriptomic phase is critical for determining the pathophysiology of the disease and shaping treatments in precision medicine. It also plays a vital role in determining biomarkers because of its integration with other omic data. For bioinformatics analysis, RNA-seq has been used in alignment-based methods such as STAR (Dobin et al. 2013), Bowtie (Langmead et al. 2009), Hisat (Kim et al. 2015a), Salmon (Patro et al. 2017), and includes non-alignment methods such as Kallisto (Bray et al. 2016). Some of the transcriptomic databases for human being are NCBI GEO (SRA) (Edgar 2002), expression atlas (Petryszak et al. 2014), CMAP (Lamb 2006), CoxPRESdb (Okamura et al. 2014), and GeneFriends (van Dam et al. 2015).

6 Proteomics

All of the proteins expressed in a cell are called the proteome, and its work in this context is called proteomics (Duarte and Spencer 2016; Shapiro 2009). Proteomics is used to simultaneously determine the structure and functions of all proteins synthesized by the genome in a particular biological system, tissue, or organism under certain conditions (Carbonaro 2004). In general, proteomics, which characterizes different protein combinations according to their constituents, is analyzed in three steps: obtaining protein information via mass spectrometry (MS), protein separation by two-dimensional gel electrophoresis, and using bioinformatic databases (Graves and Haystead 2002; Beranova-Giorgianni 2003). Unlike the genome, the proteome is more dynamic. It can vary significantly depending on the cell type and the cell's physiological state (Aebersold and Mann 2003). Therefore, proteomic studies are more complex than genomic studies (Amiour et al. 2002). Genomic studies are insufficient to characterize protein modifications, discover drug targets, and reveal all the information necessary for cell genotype and phenotype association analyses (Graves and Haystead 2002). Thus, proteomics is essential in precision medicine and proteomic technologies that require a correlation between the discovery of protein biomarkers and the diagnosis and treatment of disease (Beranova-Giorgianni 2003; Jain 2004). In addition, understanding protein biomarkers can aid the discovery of new drugs and drug targets (Hale 2003). State-of-the-art high-resolution Orbitrap mass spectrometry instruments have simplified the genome-wide detection of all proteins expressed in human cells and tissues with powerful computational tools, namely Perseus (Tyanova et al. 2016), and MaxQuant (Cox and Mann 2008). Some of the proteomic databases for the human species are UniProt (UniProt 2014), NeXtProt (Gaudet et al. 2015), HPA (Uhlen et al. 2015), IntAct (Kerrien et al. 2011), RCSB PBD (Berman 2000), BioGrid (Chatr-aryamontri et al. 2015), STRING (Szklarczyk et al. 2014) and I2D (Brown and Jurisica 2005).

7 Metabolomics

The metabolome comprises all metabolites present simultaneously in a cell, fluid, or tissue (Oliver 1998). Metabolomics is an omic approach that reveals changes in the biochemical characterization of metabolites due to genetic (internal) and environmental (external) factors (Nicholson et al. 1999). Nuclear magnetic resonance (NMR) and MS are techniques used in metabolomics (Alonso et al. 2015). In complex diseases such as cancer, metabolite levels may vary over time during disease stages, indicating abnormal metabolic functions. Therefore, cancer metabolomics is a valuable research area in oncology (Armitage and Ciborowski 2017). Metabolomics is closer to inclusion in clinical practice than proteomics. Metabolite analyses using MS routinely performed are in clinical laboratories for IEM screening and drug monitoring (Tebani et al. 2016). Recent studies have suggested that this omic domain is fundamental in determining the phenotypes of various diseases by highlighting changes in the concentrations of individual metabolites and biochemical pathways (Puchades-Carrasco and Pineda-Lucena 2017; Baraldi et al. 2009). This allows us to identify new biomarkers of metabolomic diagnosis (Puchades-Carrasco and Pineda-Lucena 2017; Baraldi et al. 2009; Bekri 2016). There is the metaRbolomics package for identifying biomarkers (Stanstrup et al. 2019) or the metaBolomics package (Aggio et al. 2011) for metabolomics derived from gas chromatography-MS data. The SMPD (Jewison et al. 2014), KEGG pathway (Ogata et al. 1999), HMDB (Wishart et al. 2007), Reactome (Croft et al. 2014), UniCarbKB (Campbell et al. 2013), SugarBindDB (Shakhsheer et al. 2013), KEGG glycan (Hashimoto et al. 2006) are some of the metabolomic databases for humans.

8 Pharmacogenomics

Variations in genes encoding pharmacodynamic components affect the efficacy and toxicity of a drug due to its pharmacokinetics (Roden et al. 2019). While pharmacogenetics examines variations in a single gene related to normal drug response or adverse drug effects, pharmacogenomics (PGx) deals with the relationship between these effects and many genes. Pharmacogenomics aims to increase the efficacy and safety of treatments by identifying drugs suitable for an individual's genetic structure (Charlab and Zhang 2013). Therefore, studies in this field have provided evidence for the use of precision medicine. These results pave the way for the faster and more cost-effective determination of drug targets to be used in the future, especially in the field of cancer, and a radical change in drug development and clinical trial processes (Nakatani and Nobori 2013; Hockings et al. 2020).

Furthermore, it has been reported that deep learning and machine-learning methods are used in this area, which allow information about personalized drug design and mechanical prediction of drug design (Kalinin et al. 2018). For example, it has been reported that the genomic sequences of pancreatic tumors can be investigated using the bioinformatics diagnostic tool called as TREATMENTMAP, which uses a Pharmacogenomic marker panel. According to the results obtained, it has been reported that known driver mutations were identified as well as identifying biomarkers for effective treatment (Malgerud et al. 2017). The STRING (Szklarczyk et al. 2014), SIDER (Kuhn et al. 2015), PharmGKB (Whirl-Carrillo et al. 2012), CTD (Davis et al. 2022), COSMIC (Tate et al. 2019) and, GDSC (Yang et al. 2013) are some of the pharmacogenomic bioinformatics online databases and algorithms.

9 Management and Processing Omics Data

More recently, high-dimensional data sets have been generated using omics technologies. Omics datasets, such as DNA/RNA-seq, protein sequence data and metabolomics, are stored in a variety of file formats due to the high dimensionality of the data. Omics datasets can be stored in a variety of formats, including FASTA (DNA and protein sequences), FASTQ (an extension of the FASTA file format and Illumina readouts), EMBL (European Molecular Biology Laboratory), PHYLIP (mainly used for phylogeny extraction), PAUP-NEXUS, CLUSTAL (a self-explanatory alignment format), STRUCTURE (Structure's proprietary data storage and analysis format), ARLEQUIN (a free population genetics software with an integrated graphical user interface), SAM and BAM file formats. SAM files are text file formats that store the alignment information of different sequences that have been aligned against reference sequences. Similarly, BAM files contain the same information as SAM files, except that they are in binary file format.

The data generated simultaneously by each of the omics technologies (genomics, transcriptomics, proteomics, metabolomics), the collection of the information associated with these data and their storage in an easily accessible environment have led to the creation of many databases containing omics data. Comprehensive omics analysis is a multifaceted process and requires more than one software. Therefore, the conversion of data formats from one to another has become mandatory. Some of the desktop programs used for this purpose are PGD Spider (Lischer and Excoffier 2012), Picard Tools, formatomatic (Manoukis 2007), and WidgetCon (Aydın et al. 2019), and some of the programming languages used for this purpose are R, Python, Perl, Java, and C#.

Once the raw data has been collected from the various tools, the quality of the sequencing data is assessed. Quality assessment, one of the pre-analysis steps, is an important step because it can affect the results of subsequent analyses. Data quality assessment is also important for those performing assembly/reassembly operations and those attempting to use assemblies as maps in sub-analyses. NGS platforms themselves provide quality assessment results. The quality assessment process should consider the identification of problems in sequencing and other experimental processes, such as the initiation of library material. For example, FASTQC

(Andrews 2010) is a widely used software to assess the quality of NGS reads (Zararsız 2015). HTSeq (Anders et al. 2015), PRINSEQ (Schmieder and Edwards 2011) and R ShortRead (Morgan et al. 2009) are quality assessment tools for RNA-Seq data; GenomeQC (Manchanda et al. 2020), QUAST, (Gurevich et al. 2013), SQUAT (Yang et al. 2019) are for genomic data and DOGMA (Dohmen et al. 2016), for proteomic data Metriculator (Taylor et al. 2013), QCQuan (Van Houtven et al. 2019), PTXQC (Bielow et al. 2016), QC-ART (Stanfill et al. 2018) are some of the tools for transcriptome data.

The next step after quality assessment is to pre-process the data to obtain high quality and clean reads. This pre-processing includes quality filtering, quality clipping, removal of sequencing adapters, and masking nucleotides with 'N' (Zararsız 2015). The FASTX (Hannon Lab 2015) tool is a collection of these and some other modules. Furthermore, QTrim (Shrestha et al. 2014) is another popular tool used for this purpose.

After quality assessment and filtering, the reads are aligned to a reference genome or transcriptome. Reference genome-based approaches include Bowtie 2 (Langmead and Salzberg 2012), Bowtie (Langmead et al. 2009), Tophat2 (Kim et al. 2013), Star (Dobin et al. 2013), Burrows-Wheeler Aligner (BWA) (Li and Durbin 2009), MapSplice (Wang et al. 2010). SOAPdenovo-Trans (Xie et al. 2014), SPAdes (Bankevich et al. 2012), SOAPdenovo2 (Luo et al. 2012) Oases (Schulz et al. 2012), Trans-ABySS (Wang and Gribskov 2017), Trinity (Hass et al. 2013) are de novo-based approaches.

The number of matched reads has been reported to correlate with transcript abundance. Therefore, the number of matched reads for each transcript must be counted to obtain expression levels. This step can be achieved using the featureCounts (Liao et al. 2014), bedtools (Quinlan and Hall 2010) and HTSeq (Anders et al. 2015) tools.

Normalization and transformation should be applied to the data we obtained with the match numbers. Normalization options: Trimmed mean of M-values normalization (Robinson and Smyth 2008), Deseq median ratio normalization (Anders and Huber 2012).

With metabolomic data, there are processes that depend on the type of data. MS data handling & processing: Centroiding, Direct infusion MS, Hyphenated MS, Targeted MS, Additional dimensionality, Isotope labeling using MS, Data structure & meta-data, Feature grouping and Feature annotation. NMR data handling & processing: data pre-processing (Fourier transformation, Apodization, Zero filling, Phase and baseline correction, Bucketing) and Annotation & identification. UV data handling & processing: Alignment, Baseline correction and Componentization (Stanstrup et al. 2019).

The first step in metabolomics studies is to process the raw data obtained. For this, chromatographic, spectroscopic data must be converted into a data matrix in accordance with data analysis. When working with LC-MS data, feature matching is performed between the samples and then feature detection (or peak picking) is performed on the samples. When working with spectroscopic data, the spectra are aligned and separated into buckets. The matrix obtained after these steps will be

2-dimensional. There will be observations in one dimension and features (spectral bins or unique chromatographic features) in the other dimension. When studying NMR-based metabolomics, several steps need to be taken, such as staging spectra and baseline correction. These steps are necessary to improve quality. In NMR-based metabolomics, once these steps have been completed, the peaks are aligned and then the data are split or peak fitting is performed, depending on the method used.

After preprocessing the data, different statistical approaches are used to find interesting features (e.g. to separate patients from healthy individuals) in the experimental design. In untargeted metabolomics, the selected features include only those from the measurement. For example, m/z, chemical shift, retention time and intensity. This means that chemical structure is not included. If untargeted metabolomics is to be used, some approaches range from breaking down models obtained from databases.

Tools that can be used in the preprocessing phase: CROP (Kouřil et al. 2020), TidyMS (Riquelme et al. 2020), ncGTW (Wu et al. 2020), AutoTuner (McLean and Kujawinski 2020), hRUV (Kim et al. 2020), MetumpX (Wajid et al. 2020).

The LC-MS data preprocessing step is essential in the analysis of LC-MS-based proteomics (Tsai et al. 2016). Preprocessing steps of LC-MC data: including steps of noise filtering, deisotoping, peak detection, normalization, retention time alignment, and peak matching. In general, these steps produce a list of peaks characterised by the m/z values of the RTs. The first step is noise filtering. LC-MS data contains electronic and chemical noise. The filtering step can be done using software tools such as MZmine 2 (Pluskal et al. 2010). Gaussian filter and Savitzky-Golay filter (Savitzky and Golay 1964) can be used as Smoothing filters. In the next step, deisotoping combines the ion peaks and recapitulates them by their monoisotopic mass. These ion peaks occur in the same peptide. The deisotoping step reduces the complexity of the analysis and makes the data easier to interpret. DeconTools (Jaitly et al. 2009) is one of the tools used for this purpose. The next step, the Peak Detection step, detects the presence of the peak in a given range of Retention Time and m/z values and measures its intensity. The MaxQuant method (Cox and Mann 2008) is one of the tools developed for peak detection. After this step, the density measurements should be appropriately normalized to remove the effect of systematic bias and variability on the data (Tsai et al. 2016). Among the normalization approaches, it is common to set a reference for ion concentrations and adjust based on this reference.

The next step is Peak Alignment. The Peak Alignment step groups consensus peaks across multiple LC-MS runs prior to analysis. Peak alignment is also essential for possible extensions of the normalization and peak detection steps. The final step is Retention Time (RT) Alignment. LC-MS preprocessing pipelines are msInspect (Bellew et al. 2006), OpenMS (Sturm et al. 2008), and MZmine 2 (Pluskal et al. 2010). These pipelines integrate the RT variability estimation into the highest matching step. RT alignment approaches are grouped in 2 ways: profile-based and feature-based. The PEPPer platform is available for RT Alignment (Vandenbogaert et al. 2008; Robinson and Smyth 2008).

10 Predictive Modeling with Machine-Learning Algorithms

10.1 Machine-Learning for Precision Medicine

Machine-learning (ML) is a branch of artificial intelligence, which seeks to uncover hidden and complex patterns in the data that can be used to predict or classify new observations. Machine-learning analysis in precision medicine enables researchers to analyze large and complex datasets and ultimately a better understanding of many diseases. ML approaches are increasingly being used to identify individuals at risk, to determine which patient-specific prevention strategies will be most effective, to identify different subtypes of diseases, and to discover new drugs. They have also been implemented effectively in disease diagnosis and prognosis.

Recently, it has become easy to access large amounts of patient data, including clinical data such as age, gender, laboratory results and omics data such as genomics, proteomics, metabolomics (Dash et al. 2019). Omics data, which are large and complex even for a single patient, grows enormously and becomes even more complex when collected from a large number of patients. The statistical models created with such datasets are critical for making effective and accurate medical decisions for further treatments in personalized medicine (Ahmed 2020). However, there are many challenges, such as high-dimensionality, heterogeneity, and highly correlated genomic features to overcome. To process such large datasets, identify molecular biomarkers, and derive meaningful results, traditional statistical methods are insufficient at the stages of data storage, processing, and analysis. Thus, machine-learning algorithms that deal with data storage, acquisition, and powerful statistical methods are critical in precision medicine to ensure that each patient receives the best possible treatment.

There are various machine-learning algorithms used for various purposes, which can be mainly classified as supervised and unsupervised learning. Supervised learning makes inferences from labeled data, whereas unsupervised learning draws conclusions from unlabeled data. Two major supervised learning algorithms in the field of precision medicine are classification and regression, which are used to identify biomarkers and predict disease outcomes and survival of patients with various diseases. Clustering, on the other hand, is an unsupervised learning method in the field used to identify disease subtypes. A comprehensive review of the literature on ML in precision medicine can be found in (Quazi 2022).

In this section of the chapter, we will cover a number of topics that are crucial to precision medicine, such as discovery of disease subtypes, biomarker discovery analysis, predicting disease outcomes, patient survival analysis, and drug discovery analysis.

11 Discovery of Disease Subtypes

One of the main goals of precision medicine is personalized treatment, in which patients receive customized care rather than a one-size-fits-all therapy. Personalized treatment is becoming extremely important in many diseases that are heterogeneous, such as cancer, cardiovascular and autoimmune diseases, where patient specific signature differences can be overlooked given the same diagnosis (Peng et al. 2021). An efficient way of providing personalized treatment is to correctly identify subtypes of a disease or cancer based on the individual's unique signatures, since many diseases progress differently from patient to patient.

Discovery of disease subtypes, a.k.a. disease subtyping, is the process of categorizing patients into subgroups based on their genomic, epigenomic, transcriptomic, and clinical data. The main objective in disease subtyping is to generate more accurate prognosis predictions of patients that can help clinicians improve treatment strategies (Golub et al. 1999; Perou et al. 2000). Disease subtyping (Saria and Goldenberg 2015) has been beneficial in treatment of various diseases (Gligorijević et al. 2016) such as Parkinson's, cardiovascular diseases, and cancers including colon and rectal cancer (The Cancer Genome Atlas Network 2012a), breast cancer (The Cancer Genome Atlas Network 2012b), ovarian cancer (The Cancer Genome Atlas Network 2011).

To identify disease subtypes, unsupervised ML algorithms have been modified and/or applied to datasets generated by omics technologies. For instance, k-means clustering, consensus clustering, density-based clustering, neural network-based clustering, hierarchical clustering, Support Vector Machines (SVMs), Bayesian approaches have been used to discover meaningful subgroups from gene expression profiles (Tibshirani et al. 2002; Monti et al. 2003; Vaske et al. 2010; Witten 2011; Shen et al. 2012; Kim et al. 2015b; Wang et al. 2015; Nguyen et al. 2017; De Souto et al. 2008; Huang et al. 2018; Durmuşçelebi 2019; Kaissis et al. 2019; Rintala et al. 2021). Numerous studies have also employed unsupervised ML algorithms in the fields of metabolomics (Alakwaa et al. 2018; Azad and Shulaev 2019; Neumann et al. 2022) and proteomics (Mallavarapu et al. 2019; Tijms et al. 2020; Sinkala et al. 2020; Thakur and Singh 2021) with the aim of disease subtyping. Moreover, many packages and tools, which use ML algorithms in disease subtype discovery, have been implemented in the field of precision medicine. Table 1 provides a list of some of these packages and tools.

12 Biomarker Discovery Analysis

Another ML application in personalized medicine is the discovery of novel biomarkers for specific diseases that are beneficial for early disease detection, advancing therapeutic management, prediction of treatment response and disease prognosis. Recently, it has been shown that ML-driven biomarkers, whether prognostic or

Table 1 The packages and tools for discovery of disease subtypes

Package/tools	Function/algorithm	Description	Address	Platform
ConsensusCluster (Seiler et al. 2010)		ConsensusCluster is a software tool, which implements the consensus clustering algorithm and principal component analysis to nucleotide polymorphism (SNP) and microarray data.	https://code.google.com/archive/p/consensus-cluster/	Python
ConsensusClusterPlus (Wilkerson and Hayes 2010)	ConsensusClusterPlus()	ConsensusClusterPlus extends Concensus Clusterwith new features and graphical outputs that can aid users in class discovery.	https://bioconductor.org/packages/release/bioc/html/ConsensusClusterPlus.html	R/ BIOCONDUCTOR
geNetClassifier (Aibar et al. 2015)	geNetClassifier()	The package aims to discover subsets of genes that unequivocally differentiate and classify different disease subtypes. geNetClassifier() function returns a parameter that measures the importance the classifier attaches to each gene. This package provides a list of the most important genes for each cluster that have been associated with certain diseases.	https://www.bioconductor.org/packages/release/bioc/vignettes/geNetClassifier/inst/doc/geNetClassifier-vignette.pdf	R/ BIOCONDUCTOR
CoINcIDE (Planey and Gevaert 2016)	clusterMatrixKmeansGap(), clusterMatrixHclustGap(), consensusClusterMatrix()	CoINcIDE package implements a set of novel methods for the discovery of patient subtypes across multiple datasets.	https://github.com/kplaney/CoINcIDE	R/ BIOCONDUCTOR
PINSPlus (Nguyen et al. 2018)	PerturbationClustering()	PerturbationClustering() function automatically calculates the optimal number of clusters and assigns each observation to its corresponding cluster.	https://cran.r-project.org/web/packages/PINSPlus/index.html	R/CRAN

(continued)

Table 1 (continued)

Package/tools	Function/algorithm	Description	Address	Platform
voomCLUSTER (Durmuşçelebi 2019)	voomQW and voomPW	It is a comprehensive web application which can perform clustering analysis of any kinds of RNA-seq data including preprocessing, normalization and transformation techniques. There are two clustering approaches that use precision weights or quality weights depending strategy to be implemented in the clustering step, where k-means, k-medoids, and hierarchical clustering algorithms were applied.	http://biosoft.erciyes.edu.tr/apps/voomCLUSTER/	R Shiny web-tool
SNFtool (Wang et al. 2022)	SNF()	After Similarity Network Fusion (SNF) finds a sample similarity networks from each of the data types separately, it iteratively integrates these networks using a novel network fusion method.	https://cran.r-project.org/web/packages/SNFtool/index.html	R/CRAN
iClusterPlus (Mo and Shen 2023)	iClusterPlus()	iClusterPlus is used for integrative clustering analysis of multi-type genomic data. iClusterPlus() function fits a regularized latent model based clustering. This clustering creates an integrated cluster assignment based on joint inference across data types.	https://bioconductor.org/packages/release/bioc/html/iClusterPlus.html	R/ BIOCONDUCTOR

diagnostic, were critical in the discovery and development of new therapeutic drugs (Li et al. 2015; Van Gool et al. 2017; Kraus 2018).

To discover biomarkers for complex diseases, many feature selection methods in ML algorithms such as Support Vector Machine Recursive Feature Elimination (SVM-RFE), partial least squares (PLS) based feature selection approaches, Random Forest-Recursive Feature Elimination algorithm (RF-RFE), Rule-guided Iterative Feature Elimination (RGIFE), have been implemented to omics data (Swan et al. 2015; Zararsiz et al. 2017; Azad and Shulaev 2019; Bravo-Merodio et al. 2019; Long et al. 2019; Kaur et al. 2021; Quazi 2022). Zheng et al. (2016), for instance, implemented Self-Organizing Maps (SOM) to metabolomics data composed of renal cell carcinoma patients to discover biomarker metabolites for early renal cell carcinoma diagnosis. Hamzeh and Rueda (2019) proposed an ML algorithm that incorporates filter-based and wrapper-based feature selection techniques into the model to find cancer biomarkers. On the other hand, Bravo-Merodio et al. (2019) employed elastic-net and lasso methods as feature selection techniques in their ML pipeline to discover biomarkers using high-dimensional omics and clinical dataset. Additionally, several packages and tools have been published in the literature that use ML algorithms to discover disease subtypes. A few of these packages and tools are included in Table 2.

13 Prediction of Disease Outcomes

Prediction of any disease is the process of identifying a disease from its signs and symptoms, such as medical history, imaging tests, blood tests, biopsy, physical examination, as well as omics data. Early detection and prediction of diseases, particularly cancer, is extremely critical for providing patients with the best treatment possible. Specialists struggle to accurately predict disease outcomes because they identify or predict diseases based on specific symptoms that may also be symptoms of other diseases, given that diseases are more complex and heterogeneous. The heterogeneity of many diseases is being resolved with the help of omics technologies, but these technologies have resulted in a large amount of datasets from which is not an easy task to extract the significant and biological information. To overcome this problem, ML algorithms have been implemented and developed in this field. Supervised ML algorithms, such as classification and regression approaches, have been employed and developed for disease prediction. SVM, RF, artificial neural network, Bayesian approaches, k-nearest neighbors, regularized logistic regressions such as lasso and ridge; linear discriminant analysis, gradient boosting, extreme gradient boosting and many other approaches have been applied in order to make predictions (Haberman et al. 2014; Cai et al. 2015; Anaissi et al. 2016; Fatima and Pasha 2017; Zararsiz et al. 2017; Alakwaa et al. 2018; Cho et al. 2019; Goksuluk et al. 2019; Koçhan et al. 2019; Kochan et al. 2021; Romagnoni et al. 2019; Sigin et al. 2020; De Souza et al. 2021; Ahsan et al. 2022). There are

Table 2 The packages and tools for biomarker discovery analysis

Package/tools	Function/algorithm	Description	Address	Platform
WGCNA (Langfelder and Horvath 2008)	chooseOneHubInEachModule() and chooseTopHubInEachModule()	WGCNA is a comprehensive package for identifying candidate biomarkers using network based gene screening methods or therapeutic targets as well as clustering and its detailed analysis module.	https://cran.r-project.org/web/packages/WGCNA/index.html	R/CRAN
netClass (Cun and Frohlich 2014)	classify.stsvm()	netClass uses network and data integration for biomarker signature discovery. It includes five approaches (average expression profile of pathways, pathway activity classification, classification based on differential expression of hub genes and correlated partners, filtering of genes according to a modified Google PageRank algorithm, kernel-based smoothing of t-statistics over a network structure) for incorporating network information into gene selection.	https://cran.r-project.org/web/packages/netClass/index.html	R/CRAN
PAA (Turewicz et al. 2016)	selectFeatures()	The package, which involves three variations of recursive feature elimination methods, is used for protein microarrays. It is a comprehensive package that includes preprocessing, univariate feature preselection and multivariate feature selection steps.	https://bioconductor.org/packages/release/bioc/html/PAA.html	R/BIOCONDUCTOR
PharmacoGx (Smirnov et al. 2015)	drugSensitivitySig()	The goal of the PharmacoGx package is to discover new genomic biomarkers of drug response so that new targeted therapies can be developed.	https://bioconductor.org/packages/release/bioc/html/PharmacoGx.html	R/BIOCONDUCTOR

MetaboAnalyst (Pang et al. 2021)	Biomarker Analysis Module	MetaboAnalyst is a comprehensive a web-based interface for analyzing metabolomics data. MetaboAnalyst identify potential biomarker creating the receiver operating characteristic (ROC) curve. It provides users with several options for single (classical) or multiple biomarker analysis, as well as for predictive biomarker model creation and evaluation.	https://www.metaboanalyst.ca/	MetaboAnalyst 5.0
INDEED (Zuo et al. 2016)	non_partial_cor() partial_cor()	The INDEED algorithm was developed by integrating differential expression and differential network based on omic data for cancer biomarker discovery. non_partial_cor() function is a one step function to perform the analysis based on typical correlation analysis, no pre-processing step required. partial_cor() function is the second step of partial correlation calculation after getting the result from select_rho_partial() function.	https://bioconductor.org/packages/release/bioc/html/INDEED.html	R/ BIOCONDUCTOR

numerous packages and tools available that use ML algorithms to predict disease outcomes. Table 3 outlines a list of the most used packages and tools in this field.

14 Survival Analysis

Survival analysis is an important statistical method, particularly in cancer research, where accurate prediction of patients' survival has been an interesting and challenging problem for decades. Survival analysis mainly focuses on the time to event such as death, relapse or diagnosis of a disease. The common characteristic of survival analysis is to contain the censored data, i.e., the event did not occur during the study period. That is, the time to event cannot be fully observed but only the lower bound of the time to event.

Numerous conventional statistical models, such as Cox proportional hazards model, have been successfully used to analyze survival data with censored observations, but such models are only applicable to small datasets. High-throughput technologies, on the other hand, have generated high-dimensional omics datasets, necessitating the development of predictive survival models for high-dimensional omic datasets using more sophisticated methods, such as regularization. Finding a correlation between the time to event and the gene expression profile or identifying prognostic biomarkers associated with the progression of diseases can lead to more accurate prognosis forecasting and, as a result, better treatment strategies. Due to this, there has recently been a lot of interest in predicting cancer patient survival using high-dimensional omics data, such as proteomics, gene expression, and transcriptomics. One of the difficulties in survival analysis of high-dimensional omics data is the "curse of dimensionality", which occurs when the number of features exceeds the number of patients. This leads to overfitting and overfitting leads to inaccurate predictions. Another difficulty, known as high collinearity, is the high correlation between genomic features, such as highly correlated genes. Recently, in order to handle high-dimensional and censored data, as well as data with high collinearity problems, several ML approaches have been modified for survival analysis.

In the survival analysis of large-scale data, first, variable subset selection methods including principal component regression (PCR), partial least squares (PLS), ridge regression and penalized Cox regression methods such as lasso penalized function were implemented. Then, ML methods such as survival trees, ensemble methods, SVMs, Cox boosting, bagging survival trees, random survival forests, Artificial Neural Network (ANN) approaches were applied (Bøvelstad et al. 2009). Moreover, classical survival methods or regularization methods, with different penalty functions, as well as machine-learning algorithms adapted to survival analysis of high-dimensional omics datasets, can be found in (Ching et al. 2018; Lee and Lim 2019; Qiu et al. 2020; Fan et al. 2022; Zhang et al. 2022). ML methods used for survival analysis are also employed in a variety of packages and tools. Table 4 lists the most frequently used packages and tools in this field.

Table 3 The packages and tools for prediction of disease outcomes

Package/tools	Function/ algorithm	Description	Address	Platform
MLInterface (Mar et al. 2008)	MLearn()	The MLInterface package includes supervised methods such as RPART, random forest, linear discriminant, neural network, SVM, as well as unsupervised approaches such as clustering and principal component.	https://www.bioconductor.org/packages/release/bioc/html/MLInterfaces.html	R/ BIOCONDUCTOR
caret (Kuhn 2008)	train()	The package is developed for complex regression and classification problems. train() function can resample and use tuning parameters of model, choose the best model across these parameters and estimate model performance from a training set.	https://cran.r-project.org/web/packages/caret/index.html	R/CRAN
ClassifyR (Strbenac et al. 2015)	ClassifyResult ()	The package is used for classification and has four stages: data transformation, feature selection, classifier training, and prediction.	https://www.bioconductor.org/packages/release/bioc/html/ClassifyR.html	R/ BIOCONDUCTOR
mlr (Bischl et al. 2016)	classif.cv_glmnet(), classif.glmnet (), classif.kknn (), classif.lda(),	The mlr package includes supervised methods like classification,	https://mlr.mlr-org.com/	R/CRAN

(continued)

Table 3 (continued)

Package/tools	Function/ algorithm	Description	Address	Platform
	classif.log_reg (), classif. multinom(), classif. naive_bayes(), classif.nnet(), classif.qda(), classif.ranger (), classif.svm (), classif. xgboost()	regression, and survival analysis, as well as unsupervised approaches such as clustering. This package can carry out all processes in detail, from data pre-processing to reporting.		
MLSeq (Goksuluk et al. 2019)	classify()	MLSeq is a comprehensive package for application of various machine-learning algorithms in classification of next-generation RNA-seq data. The MLSeq package contains 90+ microarray-based classifiers besides discrete-based classifiers, such as Poisson linear discriminant analysis and negative binomial linear discriminant analysis. Also, this package provide opportunities for the researchers preprocessing approaches such as normalization, filtering, transformation.	https://www.bioconductor.org/packages/release/bioc/html/MLSeq.html	R/ BIOCONDUCTOR

(continued)

Table 3 (continued)

Package/tools	Function/ algorithm	Description	Address	Platform
sampleClassifier (El Amrani 2022)	classifyProfile ()	The package is used for classification micro-array and RNA-seq gene expression profiles.	https://www.bioconductor.org/packages/devel/bioc/manuals/sampleClassifier/man/sampleClassifier.pdf	R/ BIOCONDUCTOR

15 Drug Discovery Analysis

Personalized medicine is defined by the US Food and Drug Administration (FDA) as "providing care for the right patient with the right drug at the right dose at the right time." Due to the genetic differences between people, the same drug may have different effects on each patient. For instance, one patient may have a genetic variation that causes the medication to stay in the body longer than usual, resulting in serious side effects, while another patient may have a variation that reduces the medication's effectiveness (Vogenberg et al. 2010). With the development of omics technologies, variable genes that affect the response to drugs in patients can be identified, and thus disease-susceptible genes that represent potential new drug targets can be discovered (Mancinelli et al. 2000). Biomarkers that can be used to personalize the use of many drugs have also recently been identified (Mayo Clinic 2018).

Drug discovery, the process by which new candidate medications are discovered, is critical for pharmaceutical industries. Drug discovery and drug development is still an expensive and labor-intensive process that requires labor-intensive cycles of medicinal chemistry synthesis and analysis, including Phase I, II, and III for clinical trials. Machine-learning has recently become one of the most important subjects in drug discovery, with the aim of discovering and designing new pharmaceuticals and vaccines with superior biological functions. ML approaches have the potential to continue growing in popularity since they can speed up the research process, lower the risk and expense of clinical trials.

One of the most common implementations of machine-learning in drug discovery is to assist researchers in understanding and interpreting correlations between chemical compounds or structures and their biological activities (Ali et al. 1997). For instance, one may be interested in knowing how to optimize a medicine's chemical structure in order to increase its binding affinity (the ability to bind to the receptors), biological reactions and physiochemical characteristics. Although a number of physiochemical aspects of pharmaceuticals, such as toxicity, metabolism, drug-drug interactions, and carcinogenesis, have been successfully modeled using quantitative structure-property relationships (QSAR) approaches, they have the linearity assumption in modeling, which cannot be used in modeling nonlinear datasets

Table 4 The packages and tools for survival analysis

Package/ tools	Function/ algorithm	Description	Address	Platform
RTNsurvival (Groeneveld et al. 2019)	tnsCox(), tnsCoxIntereaction ()	RTNsurvival is a tool generated by the RTN package based on survival information. It has two main survival analysis workflows including a Cox Proportional Hazards approach used to model regulons as predictors of survival time, and a Kaplan-Meier analysis assessing the stratification of a cohort based on the regulon activity.	https://bioconductor.org/packages/release/bioc/html/RTNsurvival.html	R/ BIOCONDUCTOR
mlr3proba (Sonabend et al. 2021)	surv.akritas(), surv.blackboost(), surv.cforest(),surv.coxboost(), surv.coxph(), surv.coxtime(), surv.ctree(),surv.cv_coxboost(), surv.cv_glmnet(), surv.deephit(), surv.deepsurv(), surv.dnnsurv(), surv.flexible(), surv.gamboost(), surv.gbm(), surv.glmboost(),surv.glmnet(), surv.kaplan(), surv.loghaz(), surv.mboost(), surv.nelson(),surv.obliqueRSF(), surv.parametric(), surv.pchazard(), surv.penalized(), surv.ranger(), surv.rfsrc(), surv.rpart(), surv.svm(), surv.xgboost()	mlr3proba is a machine-learning toolkit for making probabilistic survival predictions within the mlr/mlr3 package.	https://mlr3proba.mlr-org.com/	R/CRAN

(Cherkasov et al. 2014). Therefore, advanced ML techniques are of great importance in the field of drug discovery.

Both supervised and unsupervised ML algorithms, particularly deep learning approaches, have been used in drug discovery, which includes various subfields such as biomarker discovery, target validation, drug repurposing, and so on. Numerous studies have been conducted on the application of ML algorithms to drug discovery (Lo et al. 2018; Stephenson et al. 2019; Keshavarzi Arshadi et al. 2020; Dara et al. 2022; Priya et al. 2022). A more recent study, for example, proposed ML approaches for drug repositioning against COVID19, targeting SARS-CoV-2 and its cellular processes in the host (Santos et al. 2022). According to their findings, the top anticipated broad-spectrum antivirals include drugs approved for use in COVID-19 patients. The packages or tools developed for drug discovery analyses are given in Table 5.

16 Integration of Multi-Omics Data

High-dimensional data can be used in precision medicine for the prediction and prevention of diseases and for personalized treatments. It is known that using a single omic technology is common in analyzing such high-dimensional data. However, in the first studies, it is suggested that using a single omics technology is unlikely to provide an overview of omics (metabolite, proteomics, transcriptomics, etc.) produced by biological systems in a short period (Beale et al. 2018; Pinu et al. 2019). Therefore, it is important to accurately detect and digitize many omics simultaneously, and it is recommended to use a combination of different omics (Pinu et al. 2019). At the same time, cellular metabolism and disease etiologies can be examined in more detail using multiple omic platforms. Additionally, multi-omics technologies enable more powerful and easily interpretable results in assessing the molecular complexity of disease etiology. It can also be determined what metabolites or genes may affect diseases. However, since the metabolite or gene on a single pathway will be considered with a single omic, other metabolites associated with the disease are not considered (Marco-Ramell et al. 2018). Therefore, there is an increasing need for developing data integration methods that can respond to biological problems and challenges where these omic data are lacking (Pruitt et al. 2011).

The primary purposes of the developed data integration methodology are to obtain information that cannot be obtained from any single data set, analyze all data simultaneously, create common models based on all their data, and identify vital biological factors that can explain a biological mechanism and help researchers predict disease risk or a particular phenotypic outcome (Leung et al. 2016). In general, data integration methods can be examined under three different headings, each of which is explained separately (see Fig. 1).

The first method, the concatenated-based method (a), different omic data types are brought together using a multidimensional matrix before the model is created.

Table 5 The packages and tools for drug discovery analysis

Package/tools	Function/algorithm	Description	Address	Platform
CACTVS (Ihlenfeldt et al. 1994)		It is a chemoinformatics toolkit for chemistry data analysis on web-based application.	http://www.xemistry.com/academic	Web
WOMBAT (Olah et al. 2007)		WOMBAT is a bioactivity database and applies software solutions and services for lead and drug discovery.	http://www.chemaxon.com/	Web
ChemmineR (Cao et al. 2008)	cmp.search()	The package is the first open-source framework for drug discovery and chemical genomics.	http://manuals.bioinformatics.ucr.edu/home/chemminer	R/BIOCONDUCTOR
Rcpi (Cao et al. 2014)	calcDrugFPSim(), calcDrugMCSSim(), searchDrug(), calcTwoProtSeqSim(), calcParProtSeqSim(), calcTwoProtGOSim(), calcParProtGOSim()	The package is the comprehensive integration of bioinformatics and chemoinformatics into a molecular informatics platform for drug discovery.	https://bioconductor.org/packages/Rcpi	R/BIOCONDUCTOR
cogena (Jia et al. 2016)	coExp()	The cogena is developed for drug discovery and drug repositioning. When identifying candidate drugs, predictions are made using disease-related gene expression.	https://www.bioconductor.org/packages/release/bioc/html/cogena.html	R/BIOCONDUCTOR
DeepPINCS (Jung 2022)	fit_cpi()	It is very important that the compound-protein interaction is correct for an effective drug discovery process.	https://www.bioconductor.org/packages/release/bioc/html/DeepPINCS.html	R/BIOCONDUCTOR

(continued)

Table 5 (continued)

Package/tools	Function/algorithm	Description	Address	Platform
		To find unknown compound-protein interactions, the package uses prediction methods based on machine-learning and deep learning.		
MolEngine		MolEngine is a web-tool, which enables .NET developers to quickly build chemistry desktop. Researchers can perform analysis without using complicated Cheminformatics algorithms.	http://www.scilligence.com/	Web
OpenBabel		OpenBabel is an open-source application, which is written in C++ but accessible from Python, Ruby, Perl, Shell scripts. It enables read, write and convert over 110 different chemical file formats.	http://openbabel.org/	Web
RDKit		RDKit is an open-source cheminformatics software. It is developed for cheminformatics and machine learning. It is written in C++ and Python.	http://www.rdkit.org	Web
Daylight toolkit		The toolkit has chemical information processing, substructure pattern searching and fingerprinting.	http://www.daylight.com/products/toolkit.html	Web

(continued)

Table 5 (continued)

Package/tools	Function/algorithm	Description	Address	Platform
Chemistry Development Kit (CDK)		CDK, which is a library for bio-, cheminformatics and computational chemistry written in Java, can read and write chemical data formats, render chemical structures, algorithms for chemical graph theory and QSAR descriptors.	http://sourceforge.net/projects/cdk/	Web
Indigo toolkit		Indigo is a universal organic chemistry toolkit including cheminformatics algorithms. The open-source chemical search engine Bingo is developed on top of the Indigo library.	http://lifescience.opensource.epam.com/indigo	Web
Chemf		It is a purely functional cheminformatics toolkit written in Scala.	https://github.com/stefan-hoeck/chemf	Web
OEChem		OEChem TK is a programming library for chemistry and cheminformatics for leading discovery, leading optimization, and chemical information processing facility.	http://eyesopen.com/	Web

The advantage of this matrix created the use of various machine-learning methods to analyze quantitative or qualitative data. Integrating different data types with different dimensions causes dimensionality problems in the analyses. Let us consider a concatenated matrix comprised of metabolite and transcript data. Gene data will dominate any integration model since there are 100 metabolites and 10,000

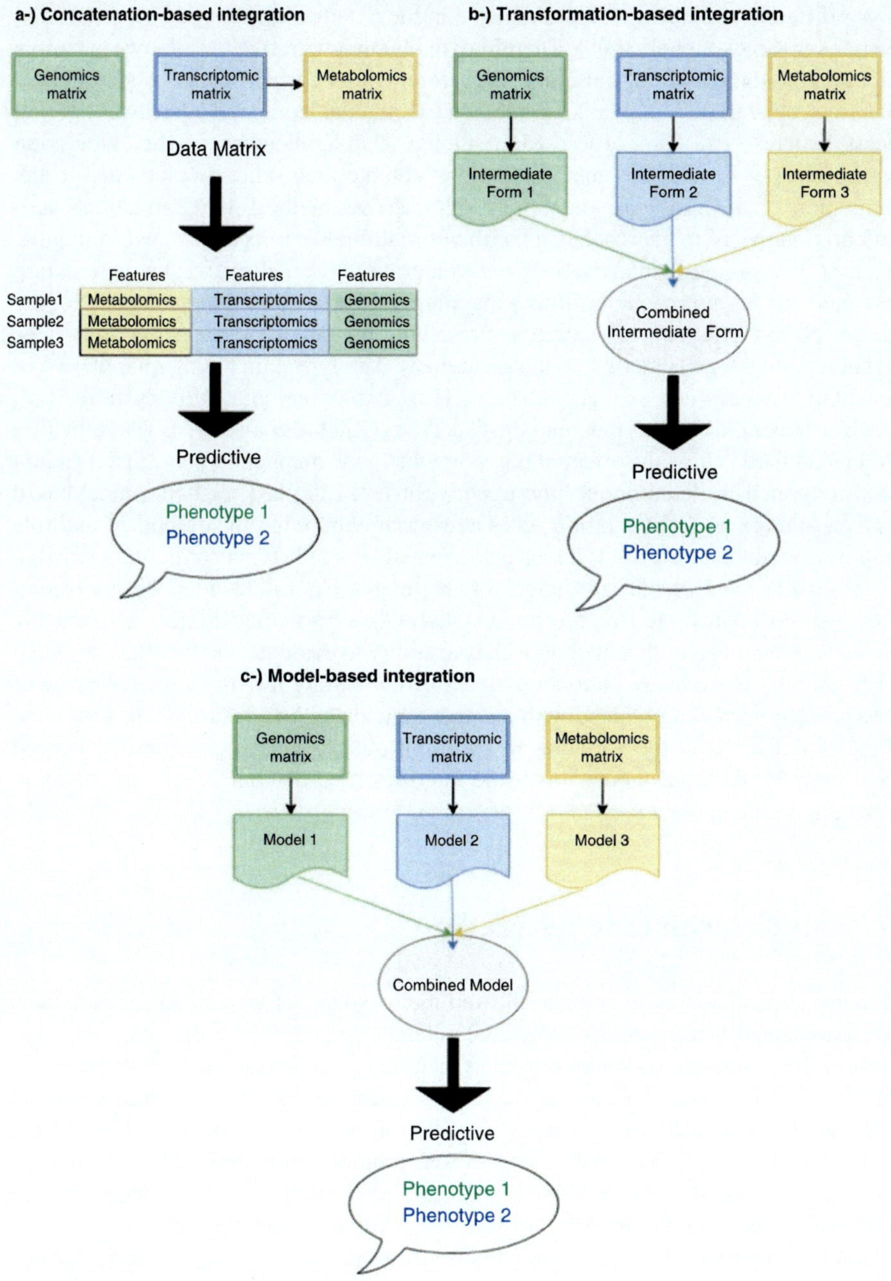

Fig. 1 Flowchart for the integrated methods

transcripts in this matrix. The model constructed will be incorrect due to this unbalanced condition, which may be resolved by giving equal weight to each omic data in the model. In addition to the high-dimensionality problem, it is obtained from different technologies that the noise in the data has different distributions and variances (Spicker et al. 2008). Therefore, to obtain integrative links between simple integration data and to minimize the aforementioned problems, the combined datasets by variance-maximizing methods (Principal component analysis, partial least squares, etc.) are processed. Although the application of the integration approaches is simple, this method is more suitable than other methods due to the limitation of the problems mentioned. The second method, transformation-based integration (b), is an approach that combines multiple datasets after transforming the dataset into an intermediate form, such as a graph or kernel matrix, before creating the model. This approach performs a mapping or data transformation suitable for the data type and offers an advantage by preserving the data type's specific properties. Another advantage is that it can integrate many data types, including quantitative or qualitative values and sequence data, as long as they contain different properties, such as patient identifiers that bind the data types (Ritchie et al. 2015). The limitation of this method is how the effect of the individual gene on phenotype, is that it cannot be interpreted in detail about how effective it is. In the last method, model-based integration (c), a model is produced as a result of combining the outputs of multiple models obtained from the training data created using different omic data (Ritchie et al. 2015). One of the advantages of this method is to consider the interaction between different omic structures associated with a particular disease or genotype. Another advantage of this method is the capability to evaluate whether the combined data can produce a better prediction model by comparing it with the performance of the prediction model obtained with a single omic data. As a result, it can obtain the best model from all the variables in the multi-omic data and provides a balanced accuracy for the final model. It can be interpreted easily, but it has limitations in creating a holistic data view.

17 Studies with Data Integration

Emerging precision medicine has allowed the personalization of medical decisions, treatments and healthcare for individual patients. In addition to transcriptomic and genomic biomarkers, other multi-omics data have an important role in precision medicine. For example, Tong et al. (2021) conducted a study to combine multi-omic datasets to predict the overall survival of breast cancer patients. Zhu et al. (2017) used somatic mutations, mRNA expression, protein expression, DNA copy numbers, DNA methylation, miRNA expression, and combinations to predict overall survival. They created core/similarity matrices for all molecular datasets to understand the underlying biology leading to disease progression among patients' cancers. They stated that its predictive power varies between cancer types and molecular profiles. They also stated that a single data type did not show superiority among all cancers (Zhu et al. 2017). Chaudhary et al. (2018), in their study, used miRNA, RNA-seq, and methylation data from patients with hepatocellular carcinoma to identify survival subpopulations. In addition, they used the Cox Proportional-

Hazards feature selection strategy and the k-means clustering method (Chaudhary et al. 2018). Their study combined RNA-seq and miRNA data with SVM classifier to identify prognostic biomarkers for pancreatic ductal adenocarcinoma (Kwon et al. 2015). Lin and Lane (2017) provided a detailed perspective on approaching multi-omic data integration (Lin and Lane 2017). Tong et al. (2021) considering this study, stated that they chose to use model-based integration using methylation, miRNA, and RNA-seq data. Biological integrity is provided by combining and analyzing different omics data. Genomic, epigenomic, and proteomic data were integrated to improve drug sensitivity in breast cancer using the Multiple Kernel Learning (MKL) method (Costello et al. 2014). Furthermore, ML-based approaches used for these analyzes offer new mechanistic insights into the phenotype of interest and facilitate progress for precision medicine (Joyce and Palsson 2006).

Some of the tools used to integrate omics data are given in Table 6. First, omics data integration is used in INMEX pathway analysis. Besides MixOmics, OmicsPLS packages, and SIMCA data integration, chemometric analysis is also performed. Second, PathVisio3 is used to create interactive pathways. Finally, unlike others, MetScape2 performs merging by pulling data from the database (Beale et al. 2016; Fondi and Liò 2015; El Bouhaddani et al. 2018).

18 Discussion

High-throughput omics technologies are capable of generating big data that can be exploited in precision medicine. Big data in healthcare is often a combination of clinical and omic data. The definition of "big data" related to complexity, diversity and rich content includes 5v dimensions, i.e., volume, variety, velocity, veracity, and value (Higdon et al. 2013). Volume refers to amount of data, velocity refers to how quickly data is generated and how quickly that data moves, variety shows that how many different data sources are used to obtain data, veracity is relevant to quality and accuracy of data, value is related to the value of the collected data and includes the concept of what kind of information can be extracted from this collected data. The amount of structured and unstructured data produced for precision medicine creates a mechanism that 5v can use.

Big data poses significant challenges for data integration, storage and analysis. Various difficulties are encountered when using computational approaches in omics technology studies, such as the discovery of disease biomarkers: (1) *experimental challenges* encountered when acquiring a universal sample collection and preparation protocol when generating multi-omics data; (2) *data preprocessing challenge*, which includes processes such as different types of information in different omics datasets require the use of very different standardization approaches, processing of each omics dataset independently or over a single combined dataset before statistical analysis; (3) *integration issues* that increase the difficulty of accounting for false positives in merged datasets, (4) *data storage issues* involving the creation of new omic platforms for storing omics data for reproducibility, (5) *biological knowledge*

Table 6 Tools used in multi-omics integration

Software tool	Omics integrated	Description	References	License
OmicsPLS	Metagenomics	It can work with low and high-dimensional data. The data can be used in visualization.	El Bouhaddani et al. (2018)	Open
	Transcriptomics			
	Proteomics			
	Metabolomics			
PathVisio 3	Transcriptomics	Biological pathways can be drawn and edited. It can also be performed in analysis.	Kutmon et al. (2015)	Open
	Proteomics			
	Metabolomics			
mixOmics (R package)	Metagenomics	Various methods are used to statistically integrate several datasets simultaneously to investigate relationships between heterogeneous omics datasets.	Günther et al. (2014)	Open
	Transcriptomics			
	Proteomics			
	Metabolomics			
INMEX	Transcriptomics	It has three different modules.	Xie et al. (2014)	Open
	Metabolomics	Data preparation module		
		Statistical analysis module		
		Functional analysis module		
SIMCA	Metagenomics	It offers a simplified and customized interface for omic data. Multivariate methods are used.	Wheelock and Wheelock (2013)	Commercial
	Transcriptomics			
	Proteomics			
	Metabolomics			
MetScape2	Transcriptomics	It is a method that combines data from KEGG and EHMN databases. It is used in the visualization of gene and metabolite networks.	Karnovsky et al. (2012)	Open
	Metabolomics			
PaintOmics	Transcriptomics	Data visualization and pathway analyzes are used for multi-omics data.	García-Alcalde et al. (2011)	Open
	Metabolomics			
IOMA	Proteomics	Quantitative coupling is done with genome-scale metabolic models.	Yizhak et al. (2010)	Open
	Metabolomics			
VitisNet	Metagenomics	It uses manually annotated networks in SBML or XML format, enabling the integration of large datasets, facilitating functional biological processing, and improving the understanding of dynamic processes in systems biology experiments.	Grimplet et al. (2009)	Open
	Transcriptomics			
	Proteomics			
	Metabolomics			

INMEX integrative meta-analysis of expression data, *IOMA* integrative omics-metabolic analysis

challenge about having detailed information about the relevant biological system for interpretation of analysis results (Pettini et al. 2021); (6) *ethical, social and legal challenge*, where the privacy of the people participating in the study, the confidentiality of personal and health information is very important. Before participating in the study, participants must go through the consent process with information about the risks and benefits of the study; (7) *cost issues* in realizing technologies such as sequencing large amounts of DNA and cost of drug development process for treatments based on molecular and genetic variation.

Despite the above-mentioned difficulties encountered in precision medicine, it offers more accurate and more effective treatments by eliminating the monotype approach applied to everyone in the treatment of diseases. When the patient is treated with the appropriate drug by solving the biology of a disease, the likelihood of the patient responding to it increases. With this type of treatment, patients will not be exposed to the side/negative effects of unnecessary drugs. Precision medicine will also help the country's economy by reducing the amount of drugs and services used in the field of health carelessly.

With the increase in the number, type and size of data that can be collected from patients and the development of omics technologies, the need to analyze large and complex data has arisen. Such data can be analyzed with machine learning methods. Statistics-based machine learning approaches developed today offer significant advantages to solve the problems about analysis stages of omics data such as preprocessing, analysis, and data merging. Many machine learning algorithms help achieve accurate results, especially in cancer therapy, drug discovery and development. These algorithms provide a significant contribution to the field of precision medicine by reducing costs and time and assisting physicians in a variety of settings. They also serve as crucial supporting mechanisms in diagnosis, prognosis, and treatment.

The development of reliable machine learning algorithms is very important for understanding the biology of diseases and, accordingly, the development of treatments for precision medicine. Therefore, these algorithms used to analyze this big data and extract useful information will be of great importance in the future. As existing algorithms are improved and new ones are developed, continuous progress will be made towards solving many problems in the healthcare field.

19 Conclusion

The advancement of precision medicine and the emergence of artificial intelligence in healthcare are shifting disease control from a community-based approach to an individualistic one. The high-efficiency technologies used in medical research with the human genome project have revolutionized this issue. In recent years, it has been emphasized that these data should be evaluated as a whole, together with the separate examination of omic technologies that help to provide instant content of the biological system in high resolution. For this purpose, existing biostatistics and bioinformatics approaches continue to be developed day by day. Omics data, which is large and complex, even for a single patient, becomes enormously complex when collected from a large number of patients. Therefore, statistical models created with large datasets are important for making effective and accurate medical decisions for advanced treatments in personalized medicine. However, traditional statistical methods are insufficient due to high dimensionality, heterogeneity and highly correlated genomic features in data storage, processing and analysis stages for processing large data sets, identifying relevant molecular biomarkers and obtaining meaningful results. Therefore, machine learning algorithms that deal with data management, organization, data storage, collection and powerful statistical methods are critical in precision medicine to ensure that every patient receives the best possible treatment. In this section, information is given on the compilation, storage, filtering, analysis of data to be obtained from omics technologies for the development of precision medicine, the use and analysis of artificial intelligence and machine learning techniques through appropriate statistical methods and models. In addition, topics that are very important for precision medicine such as the discovery of disease subtypes, biomarker discovery analysis, predicting disease outcomes, patient survival analysis and drug discovery analysis are covered. Looking to the future, the sharing of data in a secure environment and the integration of heterogeneous data sources will also play a major role in precision medicine. The development and application of highly efficient tools to access corporate multi-source big data in medical decision-making will also take personalized medicine to a new level.

References

Abu-Asab MS, Chaouchi M, Alesci S, Galli S, Laassri M, Cheema AK, Atouf F, VanMeter J, Amri H (2011) Biomarkers in the age of omics: time for a systems biology approach. OMICS 15(3): 105–112. https://doi.org/10.1089/omi.2010.0023

Aebersold R, Mann M (2003) Mass spectrometry-based proteomics. Nature 422(6928):198–207. https://doi.org/10.1038/nature01511

Aggio R, Villas-Boas SG, Ruggiero K (2011) Metab: an R package for high-throughput analysis of metabolomics data generated by GC-MS. Bioinformatics 27(16):2316–2318. https://doi.org/10.1093/bioinformatics/btr379

Ahmed Z (2020) Practicing precision medicine with intelligently integrative clinical and multi-omics data analysis. Hum Genom 14(1):35. https://doi.org/10.1186/s40246-020-00287-z

Ahsan MM, Luna SA, Siddique Z (2022) Machine-learning-based disease diagnosis: a comprehensive review. Healthcare 10(3):541. https://doi.org/10.3390/healthcare10030541

Aibar S, Fontanillo C, Droste C, Roson-Burgo B, Campos-Laborie FJ, Hernandez-Rivas JM, De Las Rivas J (2015) Analyse multiple disease subtypes and build associated gene networks using genome-wide expression profiles. BMC Genomics 16(S5):S3. https://doi.org/10.1186/1471-2164-16-s5-s3

Alakwaa FM, Chaudhary K, Garmire LX (2018) Deep learning accurately predicts estrogen receptor status in breast cancer metabolomics data. J Proteome Res 17(1):337–347. https://doi.org/10.1021/acs.jproteome.7b00595

Ali SM, Hoemann MZ, Aubé J, Georg GI, Mitscher LA, Jayasinghe LR (1997) Butitaxel analogues: synthesis and structure–activity relationships. J Med Chem 40(2):236–241. https://doi.org/10.1021/jm960505t

Alonso A, Marsal S, JuliÃ A (2015) Analytical methods in untargeted metabolomics: state of the art in 2015. Front Bioeng Biotechnol 3:23. https://doi.org/10.3389/fbioe.2015.00023

Alyass A, Turcotte M, Meyre D (2015) From big data analysis to personalized medicine for all: challenges and opportunities. BMC Med Genom 8:33

Amiour N, Merlino M, Leroy P, Branlard G (2002) Proteomic analysis of amphiphilic proteins of hexaploid wheat kernels. Proteomics 2(6):632–641. https://doi.org/10.1002/1615-9861(200206)2:6<632::AID-PROT632>3.0.CO;2-M

Anaissi A, Goyal M, Catchpoole DR, Braytee A, Kennedy PJ (2016) Ensemble feature learning of genomic data using support vector machine. PLoS One 11(6):e0157330. https://doi.org/10.1371/journal.pone.0157330

Anders S, Huber W (2012) Differential expression of RNA-Seq data at the gene level–the DESeq package. European Molecular Biology Laboratory (EMBL), Heidelberg

Anders S, Pyl PT, Huber W (2015) HTSeq—a Python framework to work with high-throughput sequencing data. Bioinformatics 31(2):166–169

Andrews S (2010) FastQC: a quality control tool for high throughput sequence data

Armitage EG, Ciborowski M (2017) Applications of metabolomics in cancer studies. Adv Exp Med Biol 965:209–234. https://doi.org/10.1007/978-3-319-47656-8_9

Aydın M, Kryvoruchko IS, Şakiroğlu M (2019) widgetcon: a website and program for quick conversion among common population genetic data formats. Mol Ecol Resour 19(5): 1374–1377

Azad RK, Shulaev V (2019) Metabolomics technology and bioinformatics for precision medicine. Brief Bioinform 20(6):1957–1971. https://doi.org/10.1093/bib/bbx170

Bankevich A, Nurk S, Antipov D, Gurevich AA, Dvorkin M, Kulikov AS, Lesin VM, Nikolenko SI, Pham S, Prjibelski AD, Pyshkin AV, Sirotkin AV, Vyahhi N, Tesler G, Alekseyev MA, Pevzner PA (2012) SPAdes: a new genome assembly algorithm and its applications to single-cell sequencing. J Comput Biol 19(5):455–477

Baraldi E, Carraro S, Giordano G, Reniero F, Perilongo G, Zacchello F (2009) Metabolomics: moving towards personalized medicine. Ital J Pediatr 35(1):30. https://doi.org/10.1186/1824-7288-35-30

Beale DJ, Karpe AV, Ahmed W (2016) Beyond metabolomics: a review of multi-omics-based approaches. In: Beale D, Kouremenos K, Palombo E (eds) Microbial metabolomics. Springer, Cham, pp 289–312

Beale DJ, Pinu FR, Kouremenos KA, Poojary MM, Narayana VK, Boughton BA, Kanojia K, Dayalan S, Jones OAH, Dias DA (2018) Review of recent developments in GC–MS approaches to metabolomics-based research. Metabolomics 14(11):152. https://doi.org/10.1007/s11306-018-1449-2

Bekri S (2016) The role of metabolomics in precision medicine. Exp Rev Prec Med Drug Dev. https://doi.org/10.1080/23808993.2016.1273067

Bellew M, Coram M, Fitzgibbon M, Igra M, Randolph T, Wang P, May D, Eng J, Fang R, Lin C, Chen J, Goodlett D, Whiteaker J, Paulovich A, McIntosh M (2006) A suite of algorithms for the comprehensive analysis of complex protein mixtures using high-resolution LC-MS. Bioinformatics 22(15):1902–1909

Beranova-Giorgianni S (2003) Proteome analysis by two-dimensional gel electrophoresis and mass spectrometry: strengths and limitations. TrAC Trends Anal Chem 22(5):273–281. https://doi.org/10.1016/s0165-9936(03)00508-9

Berman HM (2000) The protein data bank. Nucleic Acids Res 28(1):235–242. https://doi.org/10.1093/nar/28.1.235

Bernstein BE, Stamatoyannopoulos JA, Costello JF, Ren B, Milosavljevic A, Meissner A, Kellis M, Marra MA, Beaudet AL, Ecker JR, Farnham PJ, Hirst M, Lander ES, Mikkelsen TS, Thomson JA (2010) The NIH Roadmap Epigenomics Mapping Consortium. Nat Biotechnol 28(10): 1045–1048. https://doi.org/10.1038/nbt1010-1045

Bielow C, Mastrobuoni G, Kempa S (2016) Proteomics quality control: quality control software for MaxQuant results. J Proteome Res 15(3):777–787

Bird A (2007) Perceptions of epigenetics. Nature 447(7143):396–398. https://doi.org/10.1038/nature05913

Bischl B, Lang M, Kotthoff L, Schiffner J, Richter J, Studerus E, Casalicchio G, Jones ZM (2016) mlr: Machine learning in R. J Mach Learn Res 17:1–5

Bøvelstad HM, Nygård S, Borgan Ø (2009) Survival prediction from clinico-genomic models - a comparative study. BMC Bioinform 10(1):413. https://doi.org/10.1186/1471-2105-10-413

Bravo-Merodio L, Williams JA, Gkoutos GV, Acharjee A (2019) Omics biomarker identification pipeline for translational medicine. J Transl Med 17(1):155. https://doi.org/10.1186/s12967-019-1912-5

Bray NL, Pimentel H, Melsted P, Pachter L (2016) Erratum: near-optimal probabilistic RNA-seq quantification. Nat Biotechnol 34(8):888. https://doi.org/10.1038/nbt0816-888d. Erratum for: Nat Biotechnol. 34(5):525-527. doi:10.1038/nbt0816-888d

Brown KR, Jurisica I (2005) Online predicted human interaction database. Bioinformatics 21(9): 2076–2082. https://doi.org/10.1093/bioinformatics/bti273

Brown GR, Hem V, Katz KS, Ovetsky M, Wallin C, Ermolaeva O, Tolstoy I, Tatusova T, Pruitt KD, Maglott DR, Murphy TD (2015) Gene: a gene-centered information resource at NCBI. Nucleic Acids Res 43(Database Issue):D36–D42. https://doi.org/10.1093/nar/gku1055

Budak ŞÖ, Dönmez S (2012) Novel omics technologies in food science. J Food 37(3):173–179

Buenrostro JD, Wu B, Chang HY, Greenleaf WJ (2015) ATAC-seq: a method for assaying chromatin accessibility genome-wide. Curr Protoc Mol Biol 109:21.29.1–21.29.9. https://doi.org/10.1002/0471142727.mb2129s109

Cai Z, Xu D, Zhang Q, Zhang J, Ngai S-M, Shao J (2015) Classification of lung cancer using ensemble-based feature selection and machine learning methods. Mol BioSyst 11(3):791–800. https://doi.org/10.1039/c4mb00659c

Campbell MP, Peterson R, Mariethoz J, Gasteiger E, Akune Y, Aoki-Kinoshita KF, Lisacek F, Packer NH (2013) UniCarbKB: building a knowledge platform for glycoproteomics. Nucleic Acids Res 42(D1):D215–D221. https://doi.org/10.1093/nar/gkt1128

Cao Y, Charisi A, Cheng LC, Jiang T, Girke T (2008) ChemmineR: a compound mining framework for R. Bioinformatics 24(15):1733–1734. https://doi.org/10.1093/bioinformatics/btn307

Cao DS, Xiao N, Xu QS, Chen AF (2014) Rcpi: R/Bioconductor package to generate various descriptors of proteins, compounds and their interactions. Bioinformatics 31(2):279–281. https://doi.org/10.1093/bioinformatics/btu624

Carbonaro M (2004) Proteomics: present and future in food quality evaluation. Trends Food Sci Technol 15(3–4):209–216. https://doi.org/10.1016/j.tifs.2003.09.020

Chadwick LH, Sawa A, Yang IV, Baccarelli A, Breakefield XO, Deng H-W, Dolinoy DC, Fallin MD, Holland NT, Houseman EA, Lomvardas S, Rao M, Satterlee JS, Tyson FL, Vijayanand P, Greally JM (2015) New insights and updated guidelines for epigenome-wide association studies. Neuroepigenetics 1:14–19. https://doi.org/10.1016/j.nepig.2014.10.004

Chakraborty S, Hosen MI, Ahmed M, Shekhar HU (2018) Onco-multi-OMICS approach: a new frontier in cancer research. Biomed Res Int 2018:1–14. https://doi.org/10.1155/2018/9836256

Charlab R, Zhang L (2013) Pharmacogenomics: historical perspective and current status. Methods Mol Biol 1015:3–22. https://doi.org/10.1007/978-1-62703-435-7_1

Chatr-aryamontri A, Breitkreutz B-J, Oughtred R, Boucher L, Heinicke S, Chen D, Stark C, Breitkreutz A, Kolas N, O'Donnell L, Reguly T, Nixon J, Ramage L, Winter A, Sellam A, Chang C, Hirschman J, Theesfeld C, Rust J, Livstone MS (2015) The BioGRID interaction database: 2015 update. Nucleic Acids Res 43(D1):D470–D478. https://doi.org/10.1093/nar/gku1204

Chaudhary K, Poirion OB, Lu L, Garmire LX (2018) Deep learning–based multi-omics integration robustly predicts survival in liver cancer. Clin Cancer Res 24(6):1248–1259. https://doi.org/10.1158/1078-0432.ccr-17-0853

Cherkasov A, Muratov EN, Fourches D, Varnek A, Baskin II, Cronin M, Dearden J, Gramatica P, Martin YC, Todeschini R, Consonni V, Kuźmin VE, Cramer R, Benigni R, Yang C, Rathman J, Terfloth L, Gasteiger J, Richard A, Tropsha A (2014) QSAR modeling: where have you been? Where are you going to? J Med Chem 57(12):4977–5010. https://doi.org/10.1021/jm4004285

Ching T, Zhu X, Garmire LX (2018) Cox-nnet: an artificial neural network method for prognosis prediction of high-throughput omics data. PLoS Comput Biol 14(4):e1006076. https://doi.org/10.1371/journal.pcbi.1006076

Cho G, Yim J, Choi Y, Ko J, Lee SH (2019) Review of machine learning algorithms for diagnosing mental illness. Psychiatry Investig 16(4):262–269. https://doi.org/10.30773/pi.2018.12.21.2

Costello JC, Heiser LM, Georgii E, Gönen M, Menden MP, Wang NJ, Bansal M, Ammad-ud-din M, Hintsanen P, Khan SA, Mpindi J-P, Kallioniemi O, Honkela A, Aittokallio T, Wennerberg K, Collins JJ, Gallahan D, Singer D, Saez-Rodriguez J, Kaski S (2014) A community effort to assess and improve drug sensitivity prediction algorithms. Nat Biotechnol 32(12):1202–1212. https://doi.org/10.1038/nbt.2877

Cox J, Mann M (2008) MaxQuant enables high peptide identification rates, individualized p.p.b.-range mass accuracies and proteome-wide protein quantification. Nat Biotechnol 26(12): 1367–1372. https://doi.org/10.1038/nbt.1511

Croft D, Mundo AF, Haw R, Milacic M, Weiser J, Wu G, Caudy M, Garapati P, Gillespie M, Kamdar MR, Jassal B, Jupe S, Matthews L, May B, Palatnik S, Rothfels K, Shamovsky V, Song H, Williams M, Birney E (2014) The Reactome pathway knowledgebase. Nucleic Acids Res 42 (D1):D472–D477. https://doi.org/10.1093/nar/gkt1102

Cun Y, Frohlich H (2014) netClass: an R-package for network based, integrative biomarker signature discovery. Bioinformatics 30(9):1325–1326. https://doi.org/10.1093/bioinformatics/btu025

van Dam S, Craig T, de Magalhães JP (2015) GeneFriends: a human RNA-seq-based gene and transcript co-expression database. Nucleic Acids Res 43(D1):D1124–D1132. https://doi.org/10.1093/nar/gku1042

Dara S, Dhamercherla S, Jadav SS, Babu CM, Ahsan MJ (2022) Machine learning in drug discovery: a review. Artif Intell Rev 55:1947–1999. https://doi.org/10.1007/s10462-021-10058-4

Dash S, Shakyawar SK, Sharma M, Kaushik S (2019) Big data in healthcare: management, analysis and future prospects. J Big Data 6(1):1. https://doi.org/10.1186/s40537-019-0217-0

Davis AP, Wiegers TC, Johnson RJ, Sciaky D, Wiegers J, Mattingly C (2022) Comparative Toxicogenomics Database (CTD): update 2023. Nucleic Acids Res 51:D1257. https://doi.org/10.1093/nar/gkac833

De Souto MCP, Costa IG, de Araujo DSA, Ludermir TB, Schliep A (2008) Clustering cancer gene expression data: a comparative study. BMC Bioinform 9:497. https://doi.org/10.1186/1471-2105-9-497

De Souza FSH, Hojo-Souza NS, Dos Santos EB, Da Silva CM, Guidoni DL (2021) Predicting the disease outcome in COVID-19 positive patients through machine learning: a retrospective cohort study with Brazilian data. Front Artif Intell 4:579931. https://doi.org/10.3389/frai.2021.579931

Dimmer EC, Huntley RP, Alam-Faruque Y, Sawford T, O'Donovan C, Martin MJ, Bely B, Browne P, Mun Chan W, Eberhardt R, Gardner M, Laiho K, Legge D, Magrane M, Pichler K, Poggioli D, Sehra H, Auchincloss A, Axelsen K, Blatter MC, Boutet E, Braconi-Quintaje S, Breuza L, Bridge A, Coudert E, Estreicher A, Famiglietti L, Ferro-Rojas S, Feuermann M, Gos A, Gruaz-Gumowski N, Hinz U, Hulo C, James J, Jimenez S, Jungo F, Keller G, Lemercier P, Lieberherr D, Masson P, Moinat M, Pedruzzi I, Poux S, Rivoire C, Roechert B, Schneider M, Stutz A, Sundaram S, Tognolli M, Bougueleret L, Argoud-Puy G, Cusin I, Duek-Roggli P, Xenarios I, Apweiler R (2011) The UniProt-GO annotation database in 2011. Nucleic Acids Res 40(D1):D565–D570. https://doi.org/10.1093/nar/gkr1048

Dobin A, Davis CA, Schlesinger F, Drenkow J, Zaleski C, Jha S, Batut P, Chaisson M, Gingeras TR (2013) STAR: ultrafast universal RNA-seq aligner. Bioinformatics 29(1):15–21. https://doi.org/10.1093/bioinformatics/bts635

Dohmen E, Kremer LP, Bornberg-Bauer E, Kemena C (2016) DOGMA: domain-based transcriptome and proteome quality assessment. Bioinformatics 32(17):2577–2581

Duarte T, Spencer C (2016) Personalized proteomics: the future of precision medicine. Proteomes 4(4):29. https://doi.org/10.3390/proteomes4040029

Durmuşçelebi A (2019) Novel statistical approaches in clustering RNA-sequencing data, Erciyes University School of Medicine, Kayseri, Turkey. https://tez.yok.gov.tr/UlusalTezMerkezi/tezDetay.jsp?id=2pZmRB_VOVad_nKYWE9hbA&no=y2v9vd_e5TfI_1JzX78ang

Edgar R (2002) Gene expression omnibus: NCBI gene expression and hybridization array data repository. Nucleic Acids Res 30(1):207–210. https://doi.org/10.1093/nar/30.1.207

El Amrani K (2022) sampleClassifier: Sample Classifier. R package version 1.22.0

El Bouhaddani S, Uh H-W, Jongbloed G, Hayward C, Klarić L, Kiełbasa SM, Houwing-Duistermaat J (2018) Integrating omics datasets with the OmicsPLS package. BMC Bioinform 19(1):371. https://doi.org/10.1186/s12859-018-2371-3

ENCODE Project Consortium (2004) The encode (encyclopedia of DNA elements) project. Science 306:636–640

Eraslan G, Avsec Ž, Gagneur J, Theis FJ (2019) Deep learning: new computational modelling techniques for genomics. Nat Rev Genet 20(7):389–403. https://doi.org/10.1038/s41576-019-0122-6

Fan Y, Zhang S, Ma S (2022) Survival analysis with high-dimensional omics data using a threshold gradient descent regularization-based neural network approach. Genes 13(9):1674. https://doi.org/10.3390/genes13091674

Fatima M, Pasha M (2017) Survey of machine learning algorithms for disease diagnostic. J Intell Learn Syst Appl 9(1):1–16. https://doi.org/10.4236/jilsa.2017.91001

Feng X, Grossman R, Stein L (2011) PeakRanger: a cloud-enabled peak caller for ChIP-seq data. BMC Bioinform 12(1):139. https://doi.org/10.1186/1471-2105-12-139

Fondi M, Liò P (2015) Multi -omics and metabolic modelling pipelines: challenges and tools for systems microbiology. Microbiol Res 171:52–64. https://doi.org/10.1016/j.micres.2015.01.003

García-Alcalde F, García-López F, Dopazo J, Conesa A (2011) Paintomics: a web based tool for the joint visualization of transcriptomics and metabolomics data. Bioinformatics 27(1):137–139. https://doi.org/10.1093/bioinformatics/btq594

Gaudet P, Michel P-A, Zahn-Zabal M, Cusin I, Duek PD, Evalet O, Gateau A, Gleizes A, Pereira M, Teixeira D, Zhang Y, Lane L, Bairoch A (2015) The neXtProt knowledgebase on human proteins: current status. Nucleic Acids Res 43(D1):D764–D770. https://doi.org/10.1093/nar/gku1178

Gligorijević V, Malod-Dognin N, Pržulj N (2016) Integrative methods for analyzing big data in precision medicine. Proteomics 16(5):741–758. https://doi.org/10.1002/pmic.201500396

Goksuluk D, Zararsiz G, Korkmaz S, Eldem V, Zararsiz GE, Ozcetin E, Ozturk A, Karaagaoglu AE (2019) MLSeq: machine learning interface for RNA-sequencing data. Comput Methods Prog Biomed 175:223–231. https://doi.org/10.1016/j.cmpb.2019.04.007

Golub TR, Slonim DK, Tamayo P, Huard C, Gaasenbeek M, Mesirov JP, Coller H, Loh ML, Downing JR, Caligiuri MA, Bloomfield CD, Lander ES (1999) Molecular classification of cancer: class discovery and class prediction by gene expression monitoring. Science 286(5439): 531–537. https://doi.org/10.1126/science.286.5439.531

van Gool AJ, Bietrix F, Caldenhoven E, Zatloukal K, Scherer A, Litton J-E, Meijer G, Blomberg N, Smith A, Mons B, Heringa J, Koot W-J, Smit MJ, Hajduch M, Rijnders T, Ussi A (2017) Bridging the translational innovation gap through good biomarker practice. Nat Rev Drug Discov 16(9):587–588. https://doi.org/10.1038/nrd.2017.72

Graves PR, Haystead TAJ (2002) Molecular biologist's guide to proteomics. Microbiol Mol Biol Rev 66(1):39–63. https://doi.org/10.1128/mmbr.66.1.39-63.2002

Grimplet J, Cramer GR, Dickerson JA, Mathiason K, Van Hemert J, Fennell AY (2009) VitisNet: "Omics" integration through grapevine molecular networks. PLoS One 4(12):e8365. https://doi.org/10.1371/journal.pone.0008365

Groeneveld CS, Chagas VS, Jones SJM, Robertson AG, Ponder BAJ, Meyer KB, Castro MAA (2019) RTNsurvival: an R/Bioconductor package for regulatory network survival analysis. Bioinformatics 35(21):4488–4489. https://doi.org/10.1093/bioinformatics/btz229

GuhaThakurta D, Sheikh NA, Meagher TC, Letarte S, Trager JB (2013) Applications of systems biology in cancer immunotherapy: from target discovery to biomarkers of clinical outcome. Expert Rev Clin Pharmacol 6(4):387–401. https://doi.org/10.1586/17512433.2013.811814

Günther OP, Shin H, Ng RT, McMaster WR, McManus BM, Keown PA, Tebbutt SJ, Lê Cao K-A (2014) Novel multivariate methods for integration of genomics and proteomics data: applications in a kidney transplant rejection study. Omics: J Integr Biol 18(11):682–695

Guo Y, Mahony S, Gifford DK (2012) High resolution genome wide binding event finding and motif discovery reveals transcription factor spatial binding constraints. PLoS Comput Biol 8(8): e1002638. https://doi.org/10.1371/journal.pcbi.1002638

Gurevich A, Saveliev V, Vyahhi N, Tesler G (2013) QUAST: quality assessment tool for genome assemblies. Bioinformatics 29(8):1072–1075

Haas BJ, Papanicolaou A, Yassour M, Grabherr M, Blood PD, Bowden J, Couger MB, Eccles D, Li B, Lieber M, MacManes MD, Ott M, Orvis J, Pochet N, Strozzi F, Weeks N, Westerman R, William T, Dewey CN, Henschel R, LeDuc RD, Friedman N, Regev A (2013) De novo transcript sequence reconstruction from RNA-seq using the Trinity platform for reference generation and analysis. Nat Protoc 8(8):1494–1512

Haberman Y, Tickle TL, Dexheimer PJ, Kim M-O, Tang D, Karns R, Baldassano RN, Noe JD, Rosh J, Markowitz J, Heyman MB, Griffiths AM, Crandall WV, Mack DR, Baker SS, Huttenhower C, Keljo DJ, Hyams JS, Kugathasan S, Walters TD (2014) Pediatric Crohn disease patients exhibit specific ileal transcriptome and microbiome signature. J Clin Invest 124(8): 3617–3633. https://doi.org/10.1172/JCI75436

Hale EJ (2003) Application of proteomics for discovery of protein biomarkers. Brief Funct Genom Proteom 2(3):185–193. https://doi.org/10.1093/bfgp/2.3.185

Hamzeh O, Rueda L (2019) A gene-disease-based machine learning approach to identify prostate cancer biomarkers. In: Proceedings of the 10th ACM international conference on bioinformatics, computational biology and health informatics. Association for Computing Machinery, New York, NY, pp 633–638

Hannon Lab (2015) http://hannonlab.cshl.edu/fastx_toolkit/index.html. Accessed 3 Jun 2015

Harmanci A, Rozowsky J, Gerstein M (2014) MUSIC: identification of enriched regions in ChIP-Seq experiments using a mappability-corrected multiscale signal processing framework. Genome Biol 15(10):474. https://doi.org/10.1186/s13059-014-0474-3

Hartl D, de Luca V, Kostikova A, Laramie J, Kennedy S, Ferrero E, Siegel R, Fink M, Ahmed S, Millholland J, Schuhmacher A, Hinder M, Piali L, Roth A (2021) Translational precision medicine: an industry perspective. J Transl Med 19(1):245. https://doi.org/10.1186/s12967-021-02910-6

Hasanzad M, Sarhangi N, Ehsani Chimeh S, Ayati N, Afzali M, Khatami F, Nikfar S, Aghaei Meybodi HR (2021) Precision medicine journey through omics approach. J Diab Metab Disord 21(1):881–888. https://doi.org/10.1007/s40200-021-00913-0

Hashimoto K, Goto S, Kawano S, Aoki-Kinoshita KF, Ueda N, Hamajima M, Kawasaki T, Kanehisa M (2006) KEGG as a glycome informatics resource. Glycobiology 16(5):63R–70R. https://doi.org/10.1093/glycob/cwj010

Hasin Y, Seldin M, Lusis A (2017) Multi-omics approaches to disease. Genome Biol 18(1):83. https://doi.org/10.1186/s13059-017-1215-1

Higdon R, Haynes W, Stanberry L, Stewart E, Yandl G, Howard C, Broomall W, Kolker N, Kolker E (2013) Unraveling the complexities of life sciences data. Big Data 1(1):42–50. https://doi.org/10.1089/big.2012.1505. Epub 2012 Nov 7. PMID: 27447037

Ho DSW, Schierding W, Wake M, Saffery R, O'Sullivan J (2019) Machine learning SNP based prediction for precision medicine. Front Genet 10:267. https://doi.org/10.3389/fgene.2019.00267

Hockings JK, Pasternak AL, Erwin AL, Mason NT, Eng C, Hicks JK (2020) Pharmacogenomics: an evolving clinical tool for precision medicine. Cleve Clin J Med 87(2):91–99. https://doi.org/10.3949/ccjm.87a.19073

Huang S, Cai N, Pacheco PP, Narrandes S, Wang Y, Xu W (2018) Applications of support vector machine (SVM) learning in cancer genomics. Cancer Genomics Proteomics 15(1):41–51. https://doi.org/10.21873/cgp.20063. PMID: 29275361; PMCID: PMC5822181

Ihlenfeldt WD, Takahashi Y, Abe H, Sasaki S (1994) Computation and management of chemical properties in CACTVS: an extensible networked approach toward modularity and compatibility. J Chem Inf Model 34(1):109–116. https://doi.org/10.1021/ci00017a013

Jain KK (2004) Role of oncoproteomics in the personalized management of cancer. Exp Rev Proteom 1(1):49–55. https://doi.org/10.1586/14789450.1.1.49

Jaitly N, Mayampurath A, Littlefield K, Adkins JN, Anderson GA, Smith RD (2009) Decon2LS: an open-source software package for automated processing and visualization of high resolution mass spectrometry data. BMC Bioinform 10(1):1–15

Jewison T, Su Y, Disfany FM, Liang Y, Knox C, Maciejewski A, Poelzer J, Huynh J, Zhou Y, Arndt D, Djoumbou Y, Liu Y, Deng L, Guo AC, Han B, Pon A, Wilson M, Rafatnia S, Liu P, Wishart DS (2014) SMPDB 2.0: big improvements to the small molecule pathway database. Nucleic Acids Res 42(D1):D478–D484. https://doi.org/10.1093/nar/gkt1067

Jia Z, Liu Y, Guan N, Bo X, Luo Z, Barnes MR (2016) Cogena, a novel tool for co-expressed gene-set enrichment analysis, applied to drug repositioning and drug mode of action discovery. BMC Genomics 17(1):414. https://doi.org/10.1186/s12864-016-2737-8

Joyce AR, Palsson BØ (2006) The model organism as a system: integrating 'omics' data sets. Nat Rev Mol Cell Biol 7(3):198–210. https://doi.org/10.1038/nrm1857

Jung D (2022) DeepPINCS: protein Interactions and Networks with Compounds based on Sequences using Deep Learning. R package version 1.6.0

Kaissis G, Ziegelmayer S, Lohöfer F, Steiger K, Algül H, Muckenhuber A, Yen H-Y, Rummeny E, Friess H, Schmid R, Weichert W, Siveke JT, Braren R (2019) A machine learning algorithm predicts molecular subtypes in pancreatic ductal adenocarcinoma with differential response to gemcitabine-based versus FOLFIRINOX chemotherapy. PLoS One 14(10):e0218642. https://doi.org/10.1371/journal.pone.0218642

Kalinin AA, Higgins GA, Reamaroon N, Soroushmehr S, Allyn-Feuer A, Dinov ID, Najarian K, Athey BD (2018) Deep learning in pharmacogenomics: from gene regulation to patient stratification. Pharmacogenomics 19(7):629–650. https://doi.org/10.2217/pgs-2018-0008

Karnovsky A, Weymouth T, Hull T, Tarcea VG, Scardoni G, Laudanna C, Sartor MA, Stringer KA, Jagadish HV, Burant C, Athey B, Omenn GS (2012) Metscape 2 bioinformatics tool for the analysis and visualization of metabolomics and gene expression data. Bioinformatics 28(3):373–380. https://doi.org/10.1093/bioinformatics/btr661

Kaur P, Singh A, Chana I (2021) Computational techniques and tools for omics data analysis: state-of-the-art, challenges, and future directions. Archiv Comput Methods Eng 28:4595–4631. https://doi.org/10.1007/s11831-021-09547-0

Kerrien S, Aranda B, Breuza L, Bridge A, Broackes-Carter F, Chen C, Duesbury M, Dumousseau M, Feuermann M, Hinz U, Jandrasits C, Jimenez RC, Khadake J, Mahadevan U, Masson P, Pedruzzi I, Pfeiffenberger E, Porras P, Raghunath A, Roechert B (2011) The IntAct molecular interaction database in 2012. Nucleic Acids Res 40(D1):D841–D846. https://doi.org/10.1093/nar/gkr1088

Keshavarzi Arshadi A, Webb J, Salem M, Cruz E, Calad-Thomson S, Ghadirian N, Collins J, Diez-Cecilia E, Kelly B, Goodarzi H, Yuan JS (2020) Artificial intelligence for COVID-19 drug discovery and vaccine development. Front Artif Intell 3:65. https://doi.org/10.3389/frai.2020.00065

Kim D, Pertea G, Trapnell C, Pimentel H, Kelley R, Salzberg SL (2013) TopHat2: accurate alignment of transcriptomes in the presence of insertions, deletions and gene fusions. Genome Biol 14(4):1–13

Kim D, Langmead B, Salzberg SL (2015a) HISAT: a fast spliced aligner with low memory requirements. Nat Methods 12(4):357–360. https://doi.org/10.1038/nmeth.3317

Kim S, Herazo-Maya JD, Kang DD, Juan-Guardela BM, Tedrow J, Martinez FJ, Sciurba FC, Tseng GC, Kaminski N (2015b) Integrative phenotyping framework (iPF): integrative clustering of multiple omics data identifies novel lung disease subphenotypes. BMC Genomics 16(1):924. https://doi.org/10.1186/s12864-015-2170-4

Kim T, Tang O, Vernon ST, Kott KA, Koay YC, Park J, James DE, Grieve SM, Speed TP, Yang P, Figtree GA, O'Sullivan JF, Yang JYH (2020) hRUV: hierarchical approach to removal of unwanted variation for large-scale metabolomics data. bioRxiv

Koçhan N, Tutuncu GY, Smyth GK, Gandolfo LC, Giner G (2019) qtQDA: quantile transformed quadratic discriminant analysis for high-dimensional RNA-seq data. PeerJ 7:e8260. https://doi.org/10.7717/peerj.8260

Kochan N, Tütüncü GY, Giner G (2021) A new local covariance matrix estimation for the classification of gene expression profiles in high dimensional RNA-Seq data. Expert Syst Appl 167:114200. https://doi.org/10.1016/j.eswa.2020.114200

König IR, Fuchs O, Hansen G, von Mutius E, Kopp MV (2017) What is precision medicine? Eur Respir J 50(4):1700391. https://doi.org/10.1183/13993003.00391

Kouřil Š, de Sousa J, Václavík J, Friedecký D, Adam T (2020) CROP: correlation-based reduction of feature multiplicities in untargeted metabolomic data. Bioinformatics 36(9):2941–2942

Kraus VB (2018) Biomarkers as drug development tools: discovery, validation, qualification and use. Nat Rev Rheumatol 14(6):354–362. https://doi.org/10.1038/s41584-018-0005-9

Kuhn M (2008) Building predictive models in R using the caret package. J Stat Softw 28:1–26

Kuhn M, Letunic I, Jensen LJ, Bork P (2015) The SIDER database of drugs and side effects. Nucleic Acids Res 44(D1):D1075–D1079. https://doi.org/10.1093/nar/gkv1075

Kumar V, Muratani M, Rayan NA, Kraus P, Lufkin T, Ng HH, Prabhakar S (2013) Uniform, optimal signal processing of mapped deep-sequencing data. Nat Biotechnol 31(7):615–622. https://doi.org/10.1038/nbt.2596

Kundaje A, Meuleman W, Ernst J, Bilenky M, Yen A, Heravi-Moussavi A, Kheradpour P, Zhang Z, Wang J, Ziller MJ, Amin V, Whitaker JW, Schultz MD, Ward LD, Sarkar A, Quon G, Sandstrom RS, Eaton ML, Wu Y-C, Pfenning AR (2015) Integrative analysis of 111 reference human epigenomes. Nature 518(7539):317–330. https://doi.org/10.1038/nature14248

Kutmon M, van Iersel MP, Bohler A, Kelder T, Nunes N, Pico AR, Evelo CT (2015) PathVisio 3: an extendable pathway analysis toolbox. PLoS Comput Biol 11(2):e1004085. https://doi.org/10.1371/journal.pcbi.1004085

Kwon MS, Kim Y, Lee S, Namkung J, Yun T, Yi SG, Han S, Kang M, Kim SW, Jang JY, Park T (2015) Integrative analysis of multi-omics data for identifying multi-markers for diagnosing pancreatic cancer. BMC Genomics 16(9):1–10

Lamb J (2006) The connectivity map: using gene-expression signatures to connect small molecules, genes, and disease. Science 313(5795):1929–1935. https://doi.org/10.1126/science.1132939

Langfelder P, Horvath S (2008) WGCNA: an R package for weighted correlation network analysis. BMC Bioinform 9(1):559. https://doi.org/10.1186/1471-2105-9-559

Langmead B, Salzberg SL (2012) Fast gapped-read alignment with Bowtie 2. Nat Methods 9(4): 357–359. https://doi.org/10.1038/nmeth.1923

Langmead B, Trapnell C, Pop M, Salzberg SL (2009) Ultrafast and memory-efficient alignment of short DNA sequences to the human genome. Genome Biol 10(3):R25. https://doi.org/10.1186/gb-2009-10-3-r25

Lederberg J, Mccray AT (2001) Ome sweet 'omics--a genealogical treasury of word. Scientist 15(7):8

Lee S, Lim H (2019) Review of statistical methods for survival analysis using genomic data. Genom Inform 17(4):e41. https://doi.org/10.5808/GI.2019.17.4.e41

Leung MKK, Delong A, Alipanahi B, Frey BJ (2016) Machine learning in genomic medicine: a review of computational problems and data sets. Proc IEEE 104(1):176–197. https://doi.org/10.1109/jproc.2015.2494198

Li H, Durbin R (2009) Fast and accurate short read alignment with Burrows-Wheeler transform. Bioinformatics 25(14):1754–1760. https://doi.org/10.1093/bioinformatics/btp324

Li B, Shin H, Gulbekyan G, Pustovalova O, Nikolsky Y, Hope A, Trepicchio WL (2015) Develop a drug-response modelling framework to identify cell line-derived translational biomarkers that can predict treatment outcomes to erlotinib or sorafenib. PLoS One 10(6):e0130700

Liao Y, Smyth GK, Shi W (2014) featureCounts: an efficient general purpose program for assigning sequence reads to genomic features. Bioinformatics 30(7):923–930

Lin E, Lane H-Y (2017) Machine learning and systems genomics approaches for multi-omics data. Biomark Res 5(1):2. https://doi.org/10.1186/s40364-017-0082-y

Lischer HE, Excoffier L (2012) PGDSpider: an automated data conversion tool for connecting population genetics and genomics programs. Bioinformatics 28(2):298–299

Lo YC, Rensi SE, Torng W, Altman RB (2018) Machine learning in chemoinformatics and drug discovery. Drug Discov Today 23(8):1538–1546. https://doi.org/10.1016/j.drudis.2018.05.010

Long N, Park S, Anh N, Nghi T, Yoon S, Park J, Lim J, Kwon S (2019) High-throughput omics and statistical learning integration for the discovery and validation of novel diagnostic signatures in colorectal cancer. Int J Mol Sci 20(2):296. https://doi.org/10.3390/ijms20020296

Low SK, Zembutsu H, Nakamura Y (2017) Breast cancer: the translation of big genomic data to cancer precision medicine. Cancer Sci 109:497–506

Luo R, Liu B, Xie Y, Li Z, Huang W, Yuan J, He G, Chen Y, Pan Q, Liu Y, Tang J, Wu G, Zhang H, Shi Y, Liu Y, Yu C, Wang B, Lu Y, Han C, Cheung DW, Yiu SM, Peng S, Xiaoqian Z, Liu G, Liao X, Li Y, Yang H, Wang J, Lam TW, Wang J (2012) SOAPdenovo2: an empirically improved memory-efficient short-read de novo assembler. GigaScience 1(1): 2047–217X

MacEachern SJ, Forkert ND (2021) Machine learning for precision medicine. Genome 64:416–425. https://doi.org/10.1139/gen-2020-0131

Maksimovic J, Phipson B, Oshlack A (2016) A cross-package Bioconductor workflow for analysing methylation array data. F1000Research 5:1281. https://doi.org/10.12688/f1000research.8839.1

Malgerud L, Lindberg J, Wirta V, Gustafsson-Liljefors M, Karimi M, Moro CF, Stecker K, Picker A, Huelsewig C, Stein M, Bohnert R, Del Chiaro M, Haas SL, Heuchel RL, Permert J, Maeurer MJ, Brock S, Verbeke CS, Engstrand L, Jackson DB (2017) Bioinformatory-assisted analysis of next-generation sequencing data for precision medicine in pancreatic cancer. Mol Oncol 11(10):1413–1429. https://doi.org/10.1002/1878-0261.12108

Mallavarapu T, Hao J, Kim Y, Oh JH, Kang M (2019) Pathway-based deep clustering for molecular subtyping of cancer. Methods 173:24–31. https://doi.org/10.1016/j.ymeth.2019.06.017

Manchanda N, Portwood JL, Woodhouse MR, Seetharam AS, Lawrence-Dill CJ, Andorf CM, Hufford MB (2020) GenomeQC: a quality assessment tool for genome assemblies and gene structure annotations. BMC Genomics 21(1):1–9

Mancinelli L, Cronin M, Sadée W (2000) Pharmacogenomics: the promise of personalized medicine. AAPS Pharm Sci 2(1):29–41. https://doi.org/10.1208/ps020104

Manoukis NC (2007) FORMATOMATIC: a program for converting diploid allelic data between common formats for population genetic analysis. Mol Ecol Notes 7(4):592–593

Mar J, Gentleman R, Carey V (2008) MLInterfaces: uniform interfaces to R machine learning procedures for data in Bioconductor containers. R package version. 1.24.0. http://www.bioconductor.org

Marco-Ramell A, Palau-Rodriguez M, Alay A, Tulipani S, Urpi-Sarda M, Sanchez-Pla A, Andres-Lacueva C (2018) Evaluation and comparison of bioinformatic tools for the enrichment analysis of metabolomics data. BMC Bioinform 19(1):1. https://doi.org/10.1186/s12859-017-2006-0

Mayo Clinic (2018) Precision medicine and pharmacogenomics. https://www.mayoclinic.org/healthy-lifestyle/consumer-health/indepth/personalized-medicine/art-20044300. Accessed 5 Oct 2022

McGuire AL, Gabriel S, Tishkoff SA, Wonkam A, Chakravarti A, Furlong EEM, Treutlein B, Meissner A, Chang HY, López-Bigas N, Segal E, Kim J-S (2020) The road ahead in genetics and genomics. Nat Rev Genet 21(10):581–596. https://doi.org/10.1038/s41576-020-0272-6

McLean C, Kujawinski EB (2020) AutoTuner: high fidelity and robust parameter selection for metabolomics data processing. Anal Chem 92(8):5724–5732

Mele M, Ferreira PG, Reverter F, DeLuca DS, Monlong J, Sammeth M, Young TR, Goldmann JM, Pervouchine DD, Sullivan TJ, Johnson R, Segre AV, Djebali S, Niarchou A, Consortium TG, Wright FA, Lappalainen T, Calvo M, Getz G, Dermitzakis ET (2015) The human transcriptome across tissues and individuals. Science 348(6235):660–665. https://doi.org/10.1126/science.aaa0355

Mensaert K, Denil S, Trooskens G, Van Criekinge W, Thas O, De Meyer T (2013) Next-generation technologies and data analytical approaches for epigenomics. Environ Mol Mutagen 55(3): 155–170. https://doi.org/10.1002/em.21841

Mo Q, Shen R (2023) iClusterPlus: integrative clustering of multi-type genomic data. R package version 1.34.3

Monti S, Tamayo P, Mesirov J, Golub T (2003) Consensus clustering: a resampling-based method for class discovery and visualization of gene expression microarray data. Mach Learn 52:91–118

Morgan M, Anders S, Lawrence M, Aboyoun P, Pages H, Gentleman R (2009) ShortRead: a bioconductor package for input, quality assessment and exploration of high-throughput sequence data. Bioinformatics 25(19):2607–2608

Mortazavi A, Williams BA, McCue K, Schaeffer L, Wold B (2008) Mapping and quantifying mammalian transcriptomes by RNA-Seq. Nat Methods 5(7):621–628. https://doi.org/10.1038/nmeth.1226

Naithani N, Sinha S, Misra P, Vasudevan B, Sahu R (2021) Precision medicine: concept and tools. Med J Arm Forc India 77(3):249–257. https://doi.org/10.1016/j.mjafi.2021.06.021

Nakatani K, Nobori T (2013) Pharmacogenomics. Rinsho Byori 61(11):1018–1025

Neumann JM, Freitag H, Hartmann JS, Niehaus K, Galanis M, Griesshammer M, Kellner U, Bednarz H (2022) Subtyping non-small cell lung cancer by histology-guided spatial metabolomics. J Cancer Res Clin Oncol 148(2):351–360. https://doi.org/10.1007/s00432-021-03834-w

Nguyen T, Tagett R, Diaz D, Draghici S (2017) A novel approach for data integration and disease subtyping. Genome Res 27(12):2025–2039. https://doi.org/10.1101/gr.215129.116

Nguyen H, Shrestha S, Draghici S, Nguyen T (2018) PINSPlus: a tool for tumor subtype discovery in integrated genomic data. Bioinformatics 35(16):2843–2846. https://doi.org/10.1093/bioinformatics/bty1049

Nicholson JK, Lindon JC, Holmes E (1999) 'Metabonomics': understanding the metabolic responses of living systems to pathophysiological stimuli via multivariate statistical analysis of biological NMR spectroscopic data. Xenobiotica 29(11):1181–1189. https://doi.org/10.1080/004982599238047

Ogata H, Goto S, Sato K, Fujibuchi W, Bono H, Kanehisa M (1999) KEGG: Kyoto Encyclopedia of genes and genomes. Nucleic Acids Res 27(1):29–34. https://doi.org/10.1093/nar/27.1.29

Okamura Y, Aoki Y, Obayashi T, Tadaka S, Ito S, Narise T, Kinoshita K (2014) COXPRESdb in 2015: coexpression database for animal species by DNA-microarray and RNAseq-based expression data with multiple quality assessment systems. Nucleic Acids Res 43(D1):D82–D86. https://doi.org/10.1093/nar/gku1163

Olah M, Rad R, Ostopovici L et al (2007) WOMBAT and WOMBATPK: bioactivity databases for lead and drug discovery. Chem Biol Small Mol Syst Biol Drug Des 1:760–786

Oliver S (1998) Systematic functional analysis of the yeast genome. Trends Biotechnol 16(9): 373–378. https://doi.org/10.1016/s0167-7799(98)01214-1

Pang Z, Chong J, Zhou G, de Lima Morais DA, Chang L, Barrette M, Gauthier C, Jacques P-É, Li S, Xia J (2021) MetaboAnalyst 5.0: narrowing the gap between raw spectra and functional insights. Nucleic Acids Res 49:W388. https://doi.org/10.1093/nar/gkab382

Patro R, Duggal G, Love MI, Irizarry RA, Kingsford C (2017) Salmon provides fast and bias-aware quantification of transcript expression. Nat Methods 14(4):417–419. https://doi.org/10.1038/nmeth.4197

Peng J, Jury EC, Dönnes P, Ciurtin C (2021) Machine learning techniques for personalised medicine approaches in immune-mediated chronic inflammatory diseases: applications and challenges. Front Pharmacol 12:720694. https://doi.org/10.3389/fphar.2021.720694

Perou C, Sørlie T, Eisen M et al (2000) Molecular portraits of human breast tumours. Nature 406: 747–752. https://doi.org/10.1038/35021093

Petryszak R, Burdett T, Fiorelli B, Fonseca NA, Gonzalez-Porta M, Hastings E, Huber W, Jupp S, Keays M, Kryvych N, McMurry J, Marioni JC, Malone J, Megy K, Rustici G, Tang AY, Taubert J, Williams E, Mannion O, Parkinson HE (2014) Expression Atlas update—a database of gene and transcript expression from microarray- and sequencing-based functional genomics experiments. Nucleic Acids Res 42(D1):D926–D932. https://doi.org/10.1093/nar/gkt1270

Pettini F, Visibelli A, Cicaloni V, Iovinelli D, Spiga O (2021) Multi-omics model applied to cancer genetics. Int J Mol Sci 22(11):5751. https://doi.org/10.3390/ijms22115751

Planey CR, Gevaert O (2016) CoINcIDE: a framework for discovery of patient subtypes across multiple datasets. Genome Med 8(1). https://doi.org/10.1186/s13073-016-0281-4

Pluskal T, Castillo S, Villar-Briones A, Orešič M (2010) MZmine 2: modular framework for processing, visualizing, and analyzing mass spectrometry-based molecular profile data. BMC Bioinform 11(1):1–11

Pinu FR, Beale DJ, Paten AM, Kouremenos K, Swarup S, Schirra HJ, Wishart D (2019) Systems biology and multi-omics integration: viewpoints from the metabolomics research community. Metabolites 9(4):76. https://doi.org/10.3390/metabo9040076

Priya S, Kumar A, Singh DB, Jain P, Tripathi G (2022) Machine learning approaches and their applications in drug discovery and design. Chem Biol Drug Des 100:136–153. https://doi.org/10.1111/cbdd.14057

Pruitt KD, Tatusova T, Brown GR, Maglott DR (2011) NCBI Reference Sequences (RefSeq): current status, new features and genome annotation policy. Nucleic Acids Res 40(D1):D130–D135. https://doi.org/10.1093/nar/gkr1079

Puchades-Carrasco L, Pineda-Lucena A (2017) Metabolomics applications in precision medicine: an oncological perspective. Curr Top Med Chem 17(24):2740. https://doi.org/10.2174/1568026617666170707120034

Qiu YL, Zheng H, Devos A, Selby H, Gevaert O (2020) A meta-learning approach for genomic survival analysis. Nat Commun 11(1):6350. https://doi.org/10.1038/s41467-020-20167-3

Quazi S (2022) Artificial intelligence and machine learning in precision and genomic medicine. Med Oncol 39(8):120. https://doi.org/10.1007/s12032-022-01711-1

Quinlan AR, Hall IM (2010) BEDTools: a flexible suite of utilities for comparing genomic features. Bioinformatics 26(6):841–842

Rintala TJ, Federico A, Latonen L, Greco D, Fortino V (2021) A systematic comparison of data- and knowledge-driven approaches to disease subtype discovery. Brief Bioinform 22(6): bbab314. https://doi.org/10.1093/bib/bbab314

Riquelme G, Zabalegui N, Marchi P, Jones CM, Monge ME (2020) A python-based pipeline for preprocessing LC–MS data for untargeted metabolomics workflows. Metabolites 10(10):416

Ritchie MD, Holzinger ER, Li R, Pendergrass SA, Kim D (2015) Methods of integrating data to uncover genotype–phenotype interactions. Nat Rev Genet 16(2):85–97. https://doi.org/10.1038/nrg3868

Robinson MD, Smyth GK (2008) Small-sample estimation of negative binomial dispersion, with applications to SAGE data. Biostatistics 9(2):321–332

Roden DM, McLeod HL, Relling MV, Williams MS, Mensah GA, Peterson JF, Van Driest SL (2019) Pharmacogenomics. Lancet 394(10197):521–532. https://doi.org/10.1016/s0140-6736(19)31276-0

Romagnoni A, Jégou S, Van Steen K, Wainrib G, Hugot J-P, International Inflammatory Bowel Disease Genetics Consortium (IIBDGC) (2019) Comparative performances of machine learning methods for classifying Crohn Disease patients using genome-wide genotyping data. Sci Rep 9(1):10351. https://doi.org/10.1038/s41598-019-46649-z

Sadakierska-Chudy A, Filip M (2015) A comprehensive view of the epigenetic landscape. Part II: Histone post-translational modification, nucleosome level, and chromatin regulation by ncRNAs. Neurotox Res 27(2):172–197. https://doi.org/10.1007/s12640-014-9508-6

Santos SS, Torres M, Galeano D, Sánchez MDM, Cernuzzi L, Paccanaro A (2022) Machine learning and network medicine approaches for drug repositioning for COVID-19. Patterns 3(1):100396. https://doi.org/10.1016/j.patter.2021.100396

Saria S, Goldenberg A (2015) Subtyping: what it is and its role in precision medicine. IEEE Intell Syst 30(4):70–75. https://doi.org/10.1109/mis.2015.60

Savitzky A, Golay MJ (1964) Smoothing and differentiation of data by simplified least squares procedures. Anal Chem 36(8):1627–1639

Schmieder R, Edwards R (2011) Quality control and preprocessing of metagenomic datasets. Bioinformatics 27(6):863–864

Schulz MH, Zerbino DR, Vingron M, Birney E (2012) Oases: robust de novo RNA-seq assembly across the dynamic range of expression levels. Bioinformatics 28(8):1086–1092

Seiler M, Huang CC, Szalma S, Bhanot G (2010) ConsensusCluster: a software tool for unsupervised cluster discovery in numerical data. OMICS: J Integr Biol 14(1):109–113. https://doi.org/10.1089/omi.2009.0083

Shakhsheer B, Anderson M, Khatib K, Tadoori L, Joshi L, Lisacek F, Hirschman L, Mullen E (2013) SugarBind database (SugarBindDB): a resource of pathogen lectins and corresponding glycan targets. J Mol Recognit 26(9):426–431. https://doi.org/10.1002/jmr.2285

Shapiro JA (2009) Revisiting the Central Dogma in the 21st century. Ann N Y Acad Sci 1178(1): 6–28. https://doi.org/10.1111/j.1749-6632.2009.04990.x

Shen R, Mo Q, Schultz N, Seshan VE, Olshen AB, Huse J, Ladanyi M, Sander C (2012) Integrative subtype discovery in glioblastoma using iCluster. PLoS One 7(4):e35236. https://doi.org/10.1371/journal.pone.0035236

Shrestha RK, Lubinsky B, Bansode VB, Moinz MB, McCormack GP, Travers SA (2014) QTrim: a novel tool for the quality trimming of sequence reads generated using the Roche/454 sequencing platform. BMC Bioinform 15(1):1–6

Sigin VO, Kalinkin AI, Kuznetsova EB, Simonova OA, Chesnokova GG, Litviakov NV, Slonimskaya EM, Tsyganov MM, Ibragimova MK, Volodin IV, Vinogradov II, Vinogradov MI, Vinogradov IY, Kutsev SI, Strelnikov VV, Zaletaev DV, Tanas AS (2020) DNA methylation markers panel can improve prediction of response to neoadjuvant chemotherapy in luminal B breast cancer. Sci Rep 10(1):9239. https://doi.org/10.1038/s41598-020-66197-1

Sinkala M, Mulder N, Martin D (2020) Machine learning and network analyses reveal disease subtypes of pancreatic cancer and their molecular characteristics. Sci Rep 10(1):1. https://doi.org/10.1038/s41598-020-58290-2

Smirnov P, Safikhani Z, El-Hachem N, Wang D, She A, Olsen C, Freeman M, Selby H, Gendoo DMA, Grossmann P, Beck AH, Aerts HJWL, Lupien M, Goldenberg A, Haibe-Kains B (2015) PharmacoGx: an R package for analysis of large pharmacogenomic datasets. Bioinformatics 32(8):1244–1246. https://doi.org/10.1093/bioinformatics/btv723

Sonabend R, Király FJ, Bender A, Bischl B, Lang M (2021) mlr3proba: an R package for machine learning in survival analysis. Bioinformatics 37(17):2789–2791. https://doi.org/10.1093/bioinformatics/btab039

Spicker JS, Brunak S, Frederiksen KS, Toft H (2008) Integration of clinical chemistry, expression, and metabolite data leads to better toxicological class separation. Toxicol Sci 102(2):444–454. https://doi.org/10.1093/toxsci/kfn001

Stanfill BA, Nakayasu ES, Bramer LM, Thompson AM, Ansong CK, Clauss TR, Gritsenko MA, Monroe ME, Moore RJ, Orton DJ, Piehowski PD, Schepmoes AA, Smith RD, Webb-Robertson BM, Metz TO (2018) Quality control analysis in real-time (QC-ART): a tool for real-time quality control assessment of mass spectrometry-based proteomics data. Mol Cell Proteomics 17(9):1824–1836

Stanstrup J, Broeckling CD, Helmus R, Hoffmann N, Mathé E, Naake T, Nicolotti L, Peters K, Rainer J, Salek RM, Schulze T, Schymanski EL, Stravs MA, Thévenot EA, Treutler H, Weber RJM, Willighagen E, Witting M, Neumann S (2019) The metaRbolomics toolbox in bioconductor and beyond. Metabolites 9(10):200. https://doi.org/10.3390/metabo9100200

Stephenson N, Shane E, Chase J, Rowland J, Ries D, Justice N, Zhang J, Chan L, Cao R (2019) Survey of machine learning techniques in drug discovery. Curr Drug Metab 20(3):185–193. https://doi.org/10.2174/1389200219666180820112457

Strbenac D, Mann GJ, Ormerod JT, Yang JYH (2015) ClassifyR: an R package for performance assessment of classification with applications to transcriptomics. Bioinformatics 31:1851–1853

Sturm M, Bertsch A, Gröpl C, Hildebrandt A, Hussong R, Lange E, Pfeifer N, Schulz-Trieglaff O, Zerck A, Reinert K, Kohlbacher O (2008) OpenMS–an open-source software framework for mass spectrometry. BMC Bioinform 9(1):1–11

Sung J, Wang Y, Chandrasekaran S, Witten DM, Price ND (2012) Molecular signatures from omics data: from chaos to consensus. Biotechnol J 7(8):946–957. https://doi.org/10.1002/biot.201100305

Swan AL, Stekel DJ, Hodgman C, Allaway D, Alqahtani MH, Mobasheri A, Bacardit J (2015) A machine learning heuristic to identify biologically relevant and minimal biomarker panels from omics data. BMC Genomics 16(S1):S2. https://doi.org/10.1186/1471-2164-16-s1-s2

Szklarczyk D, Franceschini A, Wyder S, Forslund K, Heller D, Huerta-Cepas J, Simonovic M, Roth A, Santos A, Tsafou KP, Kuhn M, Bork P, Jensen LJ, von Mering C (2014) STRING v10: protein-protein interaction networks, integrated over the tree of life. Nucleic Acids Res 43 (Database Issue):D447–D452. https://doi.org/10.1093/nar/gku1003

Tate JG, Bamford S, Jubb HC, Sondka Z, Beare DM, Bindal N, Boutselakis H, Cole CG, Creatore C, Dawson E, Fish P, Harsha B, Hathaway C, Jupe SC, Kok CY, Noble K, Ponting L, Ramshaw CC, Rye CE, Speedy HE (2019) COSMIC: the catalogue of somatic mutations in cancer. Nucleic Acids Res 47(D1):D941–D947. https://doi.org/10.1093/nar/gky1015

Taylor RM, Dance J, Taylor RJ, Prince JT (2013) Metriculator: quality assessment for mass spectrometry-based proteomics. Bioinformatics 29(22):2948–2949

Tebani A, Afonso C, Marret S, Bekri S (2016) Omics-based strategies in precision medicine: toward a paradigm shift in inborn errors of metabolism investigations. Int J Mol Sci 17(9):1555. https://doi.org/10.3390/ijms17091555

Teng L, He B, Wang J, Tan K (2015) 4DGenome: a comprehensive database of chromatin interactions. Bioinformatics 32(17):2727–2727. https://doi.org/10.1093/bioinformatics/btw375

Thakur R, Singh PK (2021) Molecular subtypes of pancreatic cancer: a proteomics approach. Clin Cancer Res 27(12):3272–3274. https://doi.org/10.1158/1078-0432.ccr-21-0640

The Cancer Genome Atlas Network (2011) Integrated genomic analyses of ovarian carcinoma. Nature 474:609–615

The Cancer Genome Atlas Network (2012a) Comprehensive molecular characterization of human colon and rectal cancer. Nature 487(7407):330–337. https://doi.org/10.1038/nature11252

The Cancer Genome Atlas Network (2012b) Comprehensive molecular portraits of human breast tumours. Nature 490:61–70

The UniProt Consortium (2014) UniProt: a hub for protein information. Nucleic Acids Res 43(D1): D204–D212. https://doi.org/10.1093/nar/gku989

Tibshirani R, Hastie T, Narasimhan B, Chu G (2002) Diagnosis of multiple cancer types by shrunken centroids of gene expression. Proc Natl Acad Sci 99(10):6567–6572. https://doi.org/10.1073/pnas.082099299

Tijms BM, Gobom J, Reus L, Jansen I, Hong S, Dobricic V, Kilpert F, Ten Kate M, Barkhof F, Tsolaki M, Verhey FRJ, Popp J, Martinez-Lage P, Vandenberghe R, Lleó A, Molinuevo JL, Engelborghs S, Bertram L, Lovestone S, Streffer J, Vos S, Bos I, Alzheimer's Disease Neuroimaging Initiative (ADNI), Blennow K, Scheltens P, Teunissen CE, Zetterberg H, Visser PJ (2020) Pathophysiological subtypes of Alzheimer's disease based on cerebrospinal fluid proteomics. Brain 143(12):3776–3792. https://doi.org/10.1093/brain/awaa325

Tong L, Wu H, Wang MD (2021) Integrating multi-omics data by learning modality invariant representations for improved prediction of overall survival of cancer. Methods 189:74–85

Tsai TH, Wang M, Ressom HW (2016) Preprocessing and analysis of LC-MS-based proteomic data. In: Statistical Analysis in Proteomics. Humana Press, New York, NY, pp 63–76

Turewicz M, Ahrens M, May C, Marcus K, Eisenacher M (2016) PAA: an R/bioconductor package for biomarker discovery with protein microarrays. Bioinformatics 32(10):1577–1579. https://doi.org/10.1093/bioinformatics/btw037

Tyanova S, Temu T, Sinitcyn P, Carlson A, Hein MY, Geiger T, Mann M, Cox J (2016) The Perseus computational platform for comprehensive analysis of (prote)omics data. Nat Methods 13(9):731–740. https://doi.org/10.1038/nmeth.3901

Uhlen M, Fagerberg L, Hallstrom BM, Lindskog C, Oksvold P, Mardinoglu A, Sivertsson A, Kampf C, Sjostedt E, Asplund A, Olsson I, Edlund K, Lundberg E, Navani S, Szigyarto CA-K, Odeberg J, Djureinovic D, Takanen JO, Hober S, Alm T (2015) Tissue-based map of the human proteome. Science 347(6220):1260419. https://doi.org/10.1126/science.1260419

Van Houtven J, Agten A, Boonen K, Baggerman G, Hooyberghs J, Laukens K, Valkenborg D (2019) Qcquan: a web tool for the automated assessment of protein expression and data quality of labeled mass spectrometry experiments. J Proteome Res 18(5):2221–2227

Vandenbogaert M, Li-Thiao-Té S, Kaltenbach HM, Zhang R, Aittokallio T, Schwikowski B (2008) Alignment of LC-MS images, with applications to biomarker discovery and protein identification. Proteomics 8(4):650–672

Vaske CJ, Benz SC, Sanborn JZ, Earl D, Szeto C, Zhu J, Haussler D, Stuart JM (2010) Inference of patient-specific pathway activities from multi-dimensional cancer genomics data using PARADIGM. Bioinformatics 26(12):i237–i245. https://doi.org/10.1093/bioinformatics/btq182

Vogenberg FR, Carol IB, Michael P (2010) Personalized medicine: Part 1: Evolution and development into theranostics. Pharm Therapeut 35(10):560–576

Wajid B, Iqbal H, Jamil M, Rafique H, Anwar F (2020) MetumpX—a metabolomics support package for untargeted mass spectrometry. Bioinformatics 36(5):1647–1648

Wang S, Gribskov M (2017) Comprehensive evaluation of de novo transcriptome assembly programs and their effects on differential gene expression analysis. Bioinformatics 33(3):327–333

Wang K, Singh D, Zeng Z, Coleman SJ, Huang Y, Savich GL, He X, Mieczkowski P, Grimm SA, Perou CM, JN ML, Chiang DY, Prins JF, Liu J (2010) MapSplice: accurate mapping of RNA-seq reads for splice junction discovery. Nucleic Acids Res 38(18):e178

Wang L, Li F, Sheng J, Wong ST (2015) A computational method for clinically relevant cancer stratification and driver mutation module discovery using personal genomics profiles. BMC Genomics 16(Suppl 7):S6. https://doi.org/10.1186/1471-2164-16-s7-s6

Wang B, Mezlini AM, Demir F, Fiume M, Tu Z, Brudno M, Haibe-Kains B, Goldenberg A (2022) SNFtool:Similarity Network Fusion. R package version 2.3.1

Wheelock ÅM, Wheelock CE (2013) Trials and tribulations of 'omics data analysis: assessing quality of SIMCA-based multivariate models using examples from pulmonary medicine. Mol BioSyst 9(11):2589–2596. https://doi.org/10.1039/c3mb70194h

Whirl-Carrillo M, McDonagh EM, Hebert JM, Gong L, Sangkuhl K, Thorn CF, Altman RB, Klein TE (2012) Pharmacogenomics knowledge for personalized medicine. Clin Pharmacol Ther 92(4):414–417. https://doi.org/10.1038/clpt.2012.96

Wilkerson MD, Neil Hayes D (2010) ConsensusClusterPlus: a class discovery tool with confidence assessments and item tracking. Bioinformatics 26(12):1572–1573. https://doi.org/10.1093/bioinformatics/btq170

Wishart DS, Tzur D, Knox C, Eisner R, Guo AC, Young N, Cheng D, Jewell K, Arndt D, Sawhney S, Fung C, Nikolai L, Lewis M, Coutouly MA, Forsythe I, Tang P, Shrivastava S, Jeroncic K, Stothard P, Amegbey G et al (2007) HMDB: the human metabolome database. Nucleic Acids Res 35:521–526. https://doi.org/10.1093/nar/gkl923

Witten DM (2011) Classification and clustering of sequencing data using a Poisson model. Ann Appl Stat 5(4):2493–2518. https://doi.org/10.1214/11-aoas493

Wu CT, Wang Y, Wang Y, Ebbels T, Karaman I, Graça G, Pinto R, Herrington DM, Wang Y, Yu G (2020) Targeted realignment of LC-MS profiles by neighbor-wise compound-specific graphical time warping with misalignment detection. Bioinformatics 36(9):2862–2871

Xie Y, Wu G, Tang J, Luo R, Patterson J, Liu S, Huang W, He G, Gu S, Li S, Zhou X, Lam TW, Li Y, Xu X, Wong GK, Wang J (2014) SOAPdenovo-Trans: de novo transcriptome assembly with short RNA-Seq reads. Bioinformatics 30(12):1660–1666

Xin Y, Chanrion B, O'Donnell AH, Milekic M, Costa R, Ge Y, Haghighi FG (2012) MethylomeDB: a database of DNA methylation profiles of the brain. Nucleic Acids Res 40 (D1):D1245–D1249

Yang W, Soares J, Greninger P, Edelman EJ, Lightfoot H, Forbes S, Bindal N, Beare D, Smith JA, Thompson IR, Ramaswamy S, Futreal PA, Haber DA, Stratton MR, Benes C, McDermott U, Garnett MJ (2013) Genomics of Drug Sensitivity in Cancer (GDSC): a resource for therapeutic biomarker discovery in cancer cells. Nucleic Acids Res 41:955–961. https://doi.org/10.1093/nar/gks1111

Yang LA, Chang YJ, Chen SH, Lin CY, Ho JM (2019) SQUAT: a sequencing quality assessment tool for data quality assessments of genome assemblies. BMC Genomics 19(9):1–12

Yaragatti M, Basilico C, Dailey L (2008) Identification of active transcriptional regulatory modules by the functional assay of DNA from nucleosome-free regions. Genome Res 18(6):930–938. https://doi.org/10.1101/gr.073460.107

Yizhak K, Benyamini T, Liebermeister W, Ruppin E, Shlomi T (2010) Integrating quantitative proteomics and metabolomics with a genome-scale metabolic network model. Bioinformatics 26(12):i255–i260. https://doi.org/10.1093/bioinformatics/btq183

Zaharia M, Bolosky WJ, Curtis K, Fox A, Patterson D, Shenker S, Stoica I, Karp RM, Sittler T (2011) Faster and more accurate sequence alignment with SNAP. arXiv:arXiv:1111.5572

Zang C, Schones DE, Zeng C, Cui K, Zhao K, Peng W (2009) A clustering approach for identification of enriched domains from histone modification ChIP-Seq data. Bioinformatics 25(15):1952–1958. https://doi.org/10.1093/bioinformatics/btp340

Zararsız G (2015) Development and application of novel machine learning approaches for RNA-seq data classification. Hacettepe University School of Medicine, Hacettepe, Turkey. https://tez.yok.gov.tr/UlusalTezMerkezi/tezDetay.jsp?id=FUuBLcFdKB0WqIKVAVL-vA&no=o75jI9oxbTyeMIQZYeveyQ

Zararsiz G, Goksuluk D, Klaus B, Korkmaz S, Eldem V, Karabulut E, Ozturk A (2017) voomDDA: discovery of diagnostic biomarkers and classification of RNA-seq data. PeerJ 5:e3890. https://doi.org/10.7717/peerj.3890

Zeggini E, Gloyn AL, Barton AC, Wain LV (2019) Translational genomics and precision medicine: moving from the lab to the clinic. Science 365(6460):1409–1413. https://doi.org/10.1126/science.aax4588

Zhang Y, Liu T, Meyer CA, Eeckhoute J, Johnson DS, Bernstein BE, Nussbaum C, Myers RM, Brown M, Li W, Liu XS (2008) Model-based Analysis of ChIP-Seq (MACS). Genome Biol 9(9):R137. https://doi.org/10.1186/gb-2008-9-9-r137

Zhang Y, Lin Y-H, Johnson TD, Rozek LS, Sartor MA (2014) PePr: a peak-calling prioritization pipeline to identify consistent or differential peaks from replicated ChIP-Seq data. Bioinformatics 30(18):2568–2575. https://doi.org/10.1093/bioinformatics/btu372

Zhang Y, Wong G, Mann G, Muller S, Yang JYH (2022) SurvBenchmark: comprehensive benchmarking study of survival analysis methods using both omics data and clinical data. GigaScience 11:giac071. https://doi.org/10.1093/gigascience/giac071

Zheng H, Ji J, Zhao L, Chen M, Shi A, Pan L, Huang Y, Zhang H, Dong B, Gao H (2016) Prediction and diagnosis of renal cell carcinoma using nuclear magnetic resonance-based serum metabolomics and self-organizing maps. Oncotarget 7(37):59189–59198. https://doi.org/10.18632/oncotarget.10830. PMID: 27463020; PMCID: PMC5312304

Zhu B, Song N, Shen R, Arora A, Machiela MJ, Song L, Landi MT, Ghosh D, Chatterjee N, Baladandayuthapani V, Zhao H (2017) Integrating clinical and multiple omics data for prognostic assessment across human cancers. Sci Rep 7(1):1. https://doi.org/10.1038/s41598-017-17031-8

Zuo Y, Cui Y, Di Poto C, Varghese RS, Yu G, Li R, Ressom HW (2016) INDEED: integrated differential expression and differential network analysis of omic data for biomarker discovery. Methods 111:12–20. https://doi.org/10.1016/j.ymeth.2016.08.015

Omics in Precision Medicine

Ozgecan Kayalar, Hadi Rajabi, Deniz Mortazavi, and Hasan Bayram

1 Introduction

The significance of patient-to-patient differences in disease characteristics in the assessment of treatments became better understood in the latter part of the twentieth century (Sørensen 1996; Longford and Nelder 1999). This variability revealed the need for individualization of therapy within the framework of evidence-based medicine (Kravitz et al. 2004). Modern precision medicine is a notion that has emerged in the century we currently live in. In this field, patient heterogeneity is more sensitively disclosed using methods based on multi-omics data to enhance treatment decisions and provide the right care to the right patient at the right time. Since President Obama introduced the Precision Medicine Initiative in his 2015 State of the Union Speech (White House), precision medicine has gained national attention (White House Office of The Press Secretary 2015). The terms "precision medicine" and "personalized medicine" are conceptually equivalent (Kosorok and Laber 2019; Lonergan et al. 2017). In recent years, with the precision medicine approach, the term of precision public health has started to gain importance in terms of screening individuals in the society within the concept of public health, early diagnosis and reaching the right treatment. Genes, diet, and environmental exposures are some of the factors that influence health. The primary goal of precision health is to protect your health by measuring and acting on these factors. It aims to ensure that all treatment approaches can be applied individually rather than using the same approach for everyone (CDC 2020).

O. Kayalar (✉) · H. Rajabi · D. Mortazavi · H. Bayram
Koc University Research Centre for Translational Medicine (KUTTAM), School of Medicine, Koc University, Istanbul, Turkey
e-mail: okayalar@ku.edu.tr; hrajabi20@ku.edu.tr; dmortazavi18@ku.edu.tr; habayram@ku.edu.tr

H. S. Tuli, M. B. Yerer Aycan (eds.), *Oncology: Genomics, Precision Medicine and Therapeutic Targets*, https://doi.org/10.1007/978-981-99-1529-3_9

Cancer is a complex and heterogeneous disease consisting of many genetic and epigenetic changes. Cancer is a multidimensional disease arising from the disruption of genetic, epigenetic, and biochemical processes responsible for the continuity of physiological conditions in the cell (Srivastava 2022). The main contributing factors to disease-related deaths are the difficulty to get the right diagnosis, the delays encountered during this process, and the failure to administer the patient with the appropriate treatment regimen. In the worst-case scenario, the disease has a poor prognosis due to factors related to the disease process, including the intensity and duration of the disease, sensitivity to medications, and the emergence of drug resistance, in addition to all these negative events. Because of its high morbidity and mortality, cancer is one of the most serious diseases that threaten human health globally and places a significant financial and health burden on society (Ferlay et al. 2021). There were 19.3 million new cases of cancer and 9.6 million deaths occurred in 2020, according to GLOBOCAN 2020 statistics, which are updated on a regular basis by the World Health Organization (WHO)'s International Agency for Research on Cancer (IARC). The results of the statistics have also indicated that, by 2040, the incidence and mortality rates of cancer will rise by 49% and 65%, respectively (Fig. 1) (Ferlay et al. 2020a, b). On the other hand, factors such as age, gender, weight gain, smoking and alcohol consumption, an irregular diet, exposure to occupational and environmental pollutants and stressors, sleep disturbances, and exposure to pathogens are among the important causes of cancer. However, in recent years, it has been thought that even if they have one or more of these risk factors, the biological processes that trigger cancer may differ from patient to patient. It is also emphasized that it is extremely important to consider the treatment processes to be applied with this approach. Recent advances in omics technologies have led to better identification of variations in deoxyribonucleic acid (DNA), epigenetic changes on genes and chromatin, changes in gene expression, post-translational modifications, metabolic processes, and the exosome that indicate the patient's environmental and lifestyle exposure.

Studies have shown that ten different cancer hallmarks are effective in understanding the nature of cancer and giving the appropriate treatment option so far. These hallmarks include genome instability and mutation, resistance to cell death, dysregulation of cellular metabolism, maintenance of proliferative signalling, evasion of growth suppressors, avoidance of immune destruction, provision of replicative immortality, inflammation that promotes tumour growth, activation of invasion and metastasis, and induction or access to the vasculature. Among these, the concepts of the eight hallmarks reveal that cancers and their constituent cells play a prominent role in supporting their development, progression, and persistence (Hanahan 2022). Molecular profiling of these biological processes gains meaning because of the analysis of big data obtained from multi-omics studies to understand the underlying processes from the early establishment of the disease to its progression. These data and their processing offer an opportunity to define precision oncology in a new way (Biswas and Chakrabarti 2020). Omics studies help to understand the complexity of cancer, from the molecular processes involved to its clinical subtypes. Multi-omics data provides insight into the evaluations before and

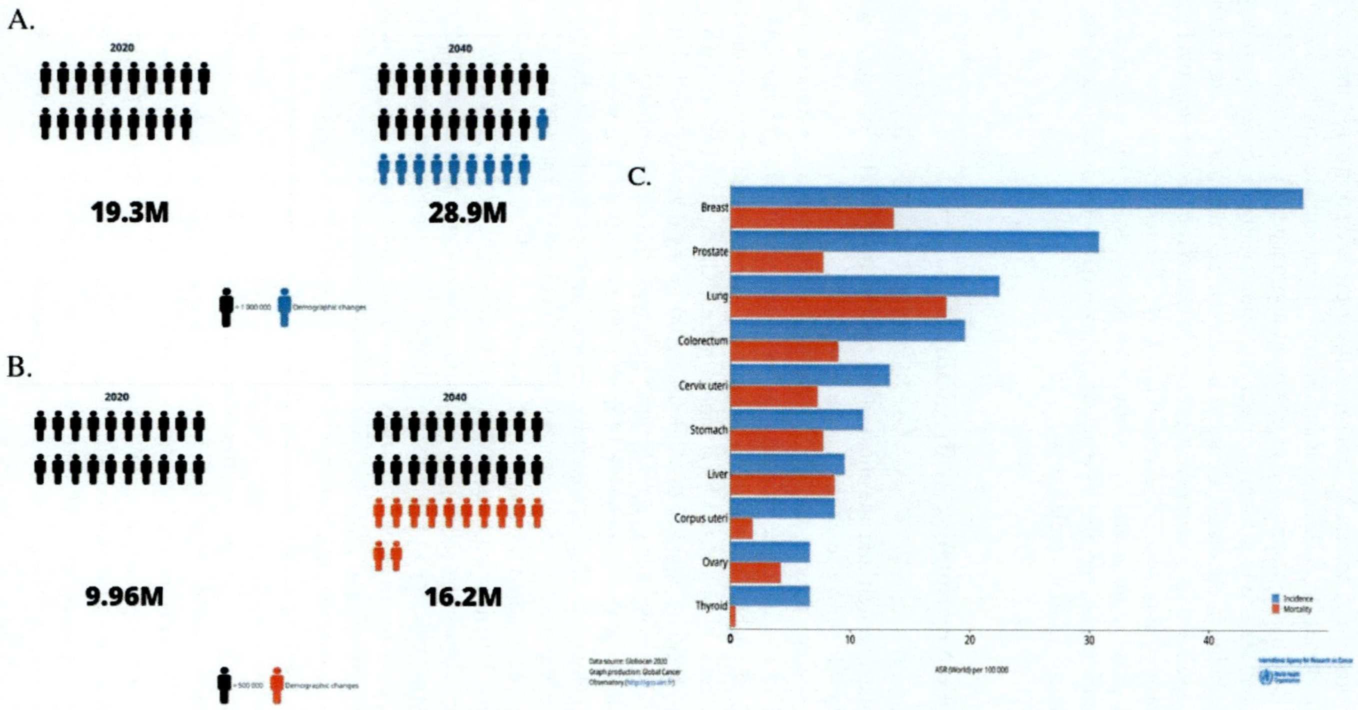

Fig. 1 (**a**) The predictive number of new cancer cases in both sexes and ages (0–85+) from 2020 to 2040. (**b**) The estimated number of deaths cases in both sexes and ages (0–85+) from 2020 to 2040. (**c**) The age standardized world-wide incidence and mortality rates in 2020. (Data retracted from WHO-GLOBOCAN 2020 Database)

after the formation of cancer, as well as information about the structure and heterogeneity of the tumour. The results emerging from these data provide meaningful information for the emergence of biomarkers and therapeutic targets required for sensitive oncology-based diagnosis and treatment approaches.

In this chapter, we aim to present a current viewpoint on multi-omics approaches to precision medicine in oncology. First, we have reviewed the most recent research on omics concepts used in cancer investigations. We also focused on improvements in omics technology for cancer research. Finally, we discussed the advances, challenges, and urgent future needs in discovering genes, epigenetic modifications, transcripts, and metabolites that are used or potentially used for diagnosis or treatment using multi-omics techniques.

2 Genomics Approaches in Precision Oncology

Perhaps the first research and the most development in the fields of omics were thanks to the developments and innovations in genomics. Ever since the first results of the human genome project were announced, it was in everyone's mind that one day everyone would have the opportunity to access their own genome library. Genomic studies enable identification of genetic variants implicated in diseases that are multifactorial or caused by a single gene (Khodadadian et al. 2020). A human haploid genome consists of 20,000–30,000 genes that are encoded within three billion DNA base pairs (bp). Human genomes remain essentially constant throughout history, with less than 1% of their sequence differing between individuals (Vora and Norton 2022). As abnormalities and errors accumulate in the genome and epigenome, cancer evolves as a disease (Gray et al. 2015; Yates et al. 2018). A large-scale cancer genome sequencing efforts such as The Cancer Genome Atlas (TCGA) and the International Cancer Genome Consortium (ICGC), by world-wide next-generation sequencers (NGSs) revealed that cancer genome contained millions and various type-nucleic acid sequence alterations. Furthermore, advances in NGS have enabled precision medicine to identify key drivers of tumor development. NGS technology is increasingly replacing conventional methods in routine molecular pathology, such as single gene testing, due to the increasing of targetable mutations and predictive biomarkers. By analyzing multiple genes simultaneously, NGS can identify different types of genomic alterations including mutations, fusion genes, CNVs and also translocations. Comparatively to conventional techniques, it offers more efficiency, cost-effectiveness, and tissue preservation. According to the type of application, NGS can be divided into three groups (Yip et al. 2019; Moorcraft et al. 2015). Whole Genome Sequencing (WGS) is the study of entire genome including not only coding but also the intronic regions. This analysis method provides extensive information covering all regions of intronic regions, untranslated regions (UTR), promoters, regulatory elements, non-coding functional RNA, repetitive regions, and mitochondrial genomes, including coding regions. This technique is an important tool given that structural variations in cancer genomes, including major deletions/insertions, inversion, duplication, translocation, and pathogen (virus) integration, remain largely unexplored. However, even today, costs per sample remain high. Whole exome sequencing (WES) is the most common targeted sequencing method. This method covers less than 2% of the human genome (coding part), but because most of the pathogenic variants are located in the protein coding region or so-called exon, this method is much more economical than WGS. Another type of NGS is where a group of genes or a specific genomic region is isolated and sequenced. A wide variety of sequencing kits are available for this method that focus on a specific area, for example genes involved in a specific cancer and other diseases. Compared to WGS and WES, this technique has better and higher coverage, therefore it has the ability to find pathogenic variants more accurately (Lim et al. 2016).

Cancer is, in one respect, a genome disease that occurs and worsens through epigenetic changes with accumulations of various mutations, including structural variants such as copy number variations (CNV), point mutations, deletions, translocations, inversions, insertions, and frameshift mutations (Zhang et al. 2009; Pritchard et al. 2014). Familial segregation studies and loss of heterozygosity studies identified several somatic mutations and copy number changes in tumour tissues included tumour suppressor genes such as RB1, TP53, APC, and proto-oncogenes such as KRAS, BRAS, HER2/ERBB2 and MYC (Dudek et al. 2020). Some of these genome changes have begun to be used in molecular cancer diagnosis and treatments and are the most widely used personalized medicine facts today as important biomarkers of susceptibility to treatment, prognosis, and re-observation of the disease. Although there are a variety of sample materials used to assist precision medicine, solid biopsy samples containing tumor tissue and liquid biopsy samples comprising blood and urine are the most often utilized and trustworthy samples. Especially, liquid biopsy is one of the most important sources of biomarkers used in the personalized medicine approach because it can be obtained with faster and less invasive methods. Thanks to liquid biopsy, important information about the processes of the disease can be obtained. One of the important advantages is that it can be made with a small number of samples. However, there is still a need for the perfection of the technique with the development of new techniques. Tissue collecting from cancer patients face many challenges and limitations in words of invasiveness, tumor heterogeneity and alteration probability of tissues during preparing of tumor samples for testing. As an emerging field in cancer genomics, patients may benefit from liquid biopsy due to invasiveness for early detection and monitoring of cancer. Circulating markers include circulating tumor cells (CTC) or circulating tumor DNA (ctDNA), proteins and RNAs in body fluids. Both NGS and single cell analysis can be applied to circulating markers. NGS studies of CTC and ctDNA are more efficient compared to analysis of tumor tissue. On the other hand, single-cell sequencing of CTC can be used for patient specific tumor models including organoids to be served as model for novel or conventional drugs. Up to now there are two FDA approved ctDNA testings in NSCLC and colorectal cancer for EGFR and SEPT9 mutations (Dietz et al. 2016). As seen in Fig. 1, the most common cancer is breast cancer, followed by prostate, lung, colorectal, and cervical cancer, respectively. Breast cancer includes the mutations in epidermal growth factor receptor 2 (HER2) and/or BRCA mutations, and hormone receptors included estrogen receptor (ESR) and progesterone receptor (PR) (Testa et al. 2020). A third of these patients have a poor response to treatment due to either innate or, more frequently, acquired drug resistance. Almost 80% of BCs express hormone receptors, and many patients with such breast cancer are treated with surgery and endocrine therapy. The majority of drug resistance mechanisms involve the acquisition of mutations at the ESR1 gene level and the acquisition of several changes in the genes encoding for the MAPK pathway (such as NF1 deletion, HER2 mutations, EGFR mutations, and KRAS mutations, PIK3CA mutations) (Razavi et al. 2018). Approximately 40% of breast cancers with hormone receptors had PIK3CA mutations. Current findings have shown that PI3K inhibitors, like alpelisib, can effectively treat breast cancer patients (André et al. 2019). Numerous studies have shown that ctDNA analysis can reliably detect PIK3CA mutations with an efficiency of 80% and a significantly high concordance with tissue analysis. Additionally, a stronger response to PIK3 inhibitors was linked to the presence of PIK3CA mutations in ctDNA. According to another study, ctDNA concentrations were lower in patients who had complete pathological responses after neo-adjuvant therapy, and a decline in ctDNA levels was also seen during the course of the treatment (Kodahl et al. 2018). These results support the idea that individualizing ctDNA monitoring throughout neo-adjuvant chemotherapy may help in the real-time evaluation of treatment response and, in conjunction with a pathological response, may help predict the patient's prognosis. Recent studies related to prostate cancer reported that one of the mechanisms underlying resistance to enzalutamide is the ARF877L mutation. Enzalutamide resistance was found to be substantially related with 19 gene alterations found in patients, including prostate cancer-associated AR copy number gains and/or amplifications, multiple AR mutations, RB1 loss, MET gain, and MYC gain (Wyatt et al. 2016; Azad et al. 2015; Choudhury et al. 2018). The treatment for non-small cell lung cancer (NSCLC) has undergone significant change as a result of the identification of various genetic abnormalities in NSCLC patients that are drug-actionable. These alterations include MET amplification, MET exon 14 skipping variants, and NTRK rearrangements as well as point mutations and other alterations in the EGFR, BRAF, and HER2 genes, fusions in the ALK, ROS1, and RET genes (Testa et al. 2018a). Studies characterizing the molecular alterations in colorectal cancers have demonstrated that ~16% of these cancer cases present hypermutation, which is associated with high microsatellite instability and mutations of polymerase epsilon and mismatch repair genes, including ACVR2A, APC, EGFR, TGFBR2, BRAF, MSH3, MSH6, MYO18, TCF7L2, and CASP8. Remaining non-hypermutated cancers show recurrent mutations in various genes, including TP53, SMAD4, PIK3CA, KRAS, SOX9, FAM123B, and ARID1A (Testa et al. 2018b; Cancer Genome Atlas Network 2012). Almost 30% of cervical cancers had genomic alterations in either one or both of the PI3K/MAPK and TGF-beta signaling pathways, which provide multiple targets for therapy. MED1, ERBB3, CASP8, HLA-A, and TGFBR2 are other unique mutant genes that have been linked to cervical cancer, according to current investigations. Furthermore, research that looked at various genetic alterations that cervical cancer patients may be able to cure with already accessible medications revealed that immune system-related genes including CD274 and PDCD1LG2 may be targeted by immunotherapeutic therapies. Another study found that the breast cancer medicine lapatinib may be effective against cancers that have BCAR4 gene fusions (Cancer Genome Atlas Research Network et al. 2017).

By identifying mutations that are druggable, doctors will be able to quickly assess tailored therapy choices by analyzing liquid biopsy samples using next-generation sequencing techniques and even less expensive droplet PCR techniques. In addition, the integration of cancer genomics into clinical trials has the potential for increasing the accuracy of early diagnosis and determine the success of treatment. A focus of precision oncology is the matching of the most effective and accurate treatments to patients based not only on their genetic profiles and their cancer, but also on other unique characteristics that make them unique. A clinical trial can provide this information during patient management. There have been many successful efforts in precision oncology which are summarized in Table 1.

3 Transcriptomic Approaches in Precision Oncology

The transcriptome is known as the whole set of RNA transcripts, including coding and non-coding, in an individual or a population of cells (Milward et al. 2016). Exons and introns constitute a significant portion of eukaryotic genes, as is well known. On the other hand, introns are excised after transcription and only exons are translated into protein products. Splicing is the method used to rejoin exons and introns that have been cut from an mRNA. Transcriptome mainly includes six alternative splice variants including constitutive splicing, mutually exclusive exons, cassette alternative exon, alternative 3′ splice site, alternative 5′ splice site, and intron retention (Wan et al. 2011; Wang et al. 2015). The main goal of performing transcriptomic studies in cancer studies is to determine the genes with altered expression reflecting the phenotype of the disease, as well as to examine the change in the level of these splicing variants. It investigates the whole transcript alterations involved in the function of cells, tissues, or organisms, across a wide range of biological conditions including disease conditions and environmental exposures (such as air pollution, cancer agents, and cancer drugs etc.). The alterations that cannot be detected at the genomics level can be revealed at the transcriptomics level. Then, if these changes have pathogenic importance, they are identified to design more reliable treatments. To develop more effective treatments, these changes are then detected if they are pathogenic. These modifications could then serve as the primary targets of cancer therapy or serve as biomarkers (mRNAs, miRNAs, lncRNAs, and circRNAs) to help with a cancer diagnosis, prognosis, and post-treatment monitoring. In the era of precision medicine, transcriptomics is an approach that fills the gap between proteomics and genomics and fulfills the task of genomics.

Table 1 Summary of some successful therapeutics in five common cancer

Gene/ biomarker	Cancer type	Used technique	Result	References
BRCA	Breast cancer	DNA sequencing	HER2-negative metastatic breast cancer patients with BRCA1 or BRCA2 mutations are eligible to use the PARP inhibitor Olaparib (Lynparza)	Tutt et al. (2021)
BRCA	Prostate cancer	DNA sequencing	Two PARP inhibitors Olaparib (Lynparza) and rucaparib (Rubraca) target the treatment in prostate cancer. Men with BRCA1, BRCA2, RAD54L, PALB2, or RAD54L mutations gain from taking both medications.	Hussain et al. (2020), Abida et al. (2020)
BRAF	Non-small cell lung cancer (NSCLC)	DNA sequencing	Dabrafenib (Tafinlar) and trametinib (Mekinist) are FDA approved for the treatment of patients with advanced solid tumors that have V600E mutation in the BRAF gene.	Planchard et al. (2016)
MET	NSCLC	DNA sequencing	Advanced NSCLC which are ALK- or ROS1-positive can be treated with crizotinib, an approved tyrosine kinase inhibitor (TKI)	Kazandjian et al. (2014)
EGFR	NSCLC	DNA sequencing	In patients who have failed prior TKIs for metastatic EGFR T790M mutation-positive NSCLC osimertinib is a third generation TKI.	Bollinger et al. (2018)
EGFR	Colorectal cancer	DNA sequencing	In patients who have K- and N-RAS wild-type (no mutation), EGFR-expressing tumors, Cetuximab (Erbitux); wild-type RAS (no mutations in either KRAS or NRAS), Panitumumab (Vectibix)	Dean and Kane (2020a), Dean and Kane (2020b)
PD-L1	Cervical cancer	DNA sequencing	In patients who have recurrent or metastatic cervical cancer progressing after chemotherapy and PDL-1 positive patients. Pembrolizumab, a third-line therapy	Chung et al. (2019)

Microarray and RNA-seq are the two major approaches in this area (Wang et al. 2009). Short nucleotide oligomers named probes are the main character of which are coated on a solid substrate like glass. Generally, transcript abundancy can be visually determined by probes which hybridize to fluorescent labelled transcripts and intensity at every probe location implies the transcript abundance (Romanov et al. 2014).

For the creation of a certain probe, prior knowledge about the targeted organism is mandatory (Liu et al. 2010). Microarray for transcriptomes is classified into two groups: low-density spotted arrays or high-density short probe arrays (Tomiuk and Hofmann 2001). In spotted low-density arrays only picolitre drops of cDNA samples need to be on the surface slide (Daiba et al. 2004). For sample controlling various fluorophores use in spotted arrays, and also for measuring abundancy the ratio of fluorescence is calculated (Zou et al. 2008). On the other hand, the high-density approach uses single-channel detection, every single sample is hybridized and identified separately (Sîrbu et al. 2012).

Despite all benefits of the microarray technique, there are some limitations in this area, for instance, there is no powerful standard for data gathering, analyzing, and validating. Also, the quality and quantity of isolated RNA remain a substantial challenge in microarray studies, so this approach is generally performed with abundance and easily obtained samples (Russo et al. 2003). RNA-Seq overcomes the challenges of both high-throughput sequencing approaches with computing techniques to provide copies existing in an RNA sample (Ozsolak and Milos 2011). Based on the methodology the used nucleotide sequences consist of almost 100 bp in length, and variable between 30 bp to nearly 10,000 bp (Salzman et al. 2011). As an alternative to microarray, RNA-seq has a typical dynamic range of five orders of magnitude. In addition, input RNA volumes are significantly lesser for RNA-Seq (ng) in comparison to microarrays (μg) which leads to a greater investigation of cellular structures, and this amount can be at the single-cell level if it is combined with a direct magnification of using cDNA (Stears et al. 2000; Mantione et al. 2014). Generally, transcriptomic approaches can be used in a wide range of studies such as disease-related gene profiling and also diagnosis (Russo et al. 2003) (Table 2). Via the RNA-Seq method it is possible to study epigenetic changes like single nucleotide polymorphisms (SNP), or genetically investigate allele-specific expression or gene fusions which can unmask disease-hidden variants (Mantione et al. 2014; Khurana et al. 2016). Transcriptomes are increasingly being used to introduce novel biomarkers, classify disease based on its severity, and conduct precision medicine investigations, particularly in cancer research. It applies to a wide range of sample analyses, including those of fresh tissue, FFPE tissue, blood and PBMC samples, tumour tissue, and biopsies, all of which fall under the purview of precision medicine in the context of cancer transcriptomics as well as many other diseases and environmental conditions. Web-based tools called UALCAN allow access to TCGA data, which is high-throughput mRNA, miRNA, and lncRNA expression data for diverse cancer types (Chandrashekar et al. 2022, 2017). The study comparing the analysis of patient transcriptome changes to various treatment options, such as chemotherapy, hormone therapy, and anti-HER2 therapy in 3104 breast cancer patients, is an excellent example of a study applying the transcriptomic approach in the context of precision medicine. The results of this study showed that the predictive biomarkers for luminal breast cancer were TP53 for chemotherapy, HER2 for anti-HER2 therapy, and PGR for hormonal therapy (Fekete and Győrffy 2019). Another study found that several specific genes, including KLRG1, BTK, CCR2, and SCML4, were associated with poor immunotherapy results in non-small

Table 2 Transcriptomic approaches in various cancers

Cancer type	Approach	Result	References
Non-small cell lung cancer	Microarray	hsa_circ_0014130 c overexpress in NSCLC tissues and correlated with lymphatic metastasis	Zhang et al. (2018)
Breast cancer	Microarray	microRNA-1246, -1307-3p, and -6861-5p and-4634 and -6875-5p overexpression in breast cancer	Shimomura et al. (2016)
Colorectal cancer	RNA seq	Immunomodulation of CD73 exhibits potential as a novel anticancer immunotherapy for colorectal cancer patients	Kim et al. (2021)
Non-small cell lung cancer	RNA seq	Hyperactivity of naïve B cell, memory CD4+ T cell, T follicular helper cell, T regulatory helper cell and M1 macrophage in lung cancer tissues	Zhong et al. (2021)
Epithelial ovarian cancer	RNA seq	Combined CYR61/RGS5 expression is related to relapse-free-survival of patients and can be early diagnostic marker for epithelial ovarian cancer	Kan et al. (2022)
BRCA1 mutation derived Breast cancers	Whole-exome sequencing and RNA seq	Luminal cells can be assigned as BRCA1 mutation derived breast cancers origin	Hu et al. (2021)

cell lung cancer (NSCLC) (Yang et al. 2021). Additionally, it was shown that the expression of the ART1 gene promotes cytotoxic T-cell death, which drives resistance to immune checkpoint inhibitors (ICPI), and this finding suggests it is a therapeutic target for NSCLC (Wennerberg et al. 2022). Moreover, it was determined that JAK1/JAK2 as potential targets to restore the efficacy of ICPIs (Torrejon et al. 2020). Moreover, Faruki et al. (2017) discovered via a transcriptomic approach that the downregulation of STK11 in patients with lung adenocarcinoma or KEAP1/NFE2L2 alteration in patients with squamous-cell carcinoma is associated with a reduced immune response (Faruki et al. 2017). All these studies demonstrate the potential for the use of transcriptomics in precision medicine. Although sample costs have decreased in recent years, they are still relatively high. How to integrate transcriptomic analysis techniques into the diagnosis and treatment process of cancer patients compared to genomic techniques, and how accreditation and validation processes will be carried out in health institutions are still controversial issues.

4 Proteomics and Phosphoproteomics Approaches in Precision Oncology

The concept of proteomic is investigating about all bioactivity aspect of the protein such as expression, composition, interaction, structure and intracellular activity (Wilkins et al. 1996). In contrast with genomics, proteomics is better choice for studying about functionality of organisms (Meissner et al. 2022), however, changing in protein expression under various condition make this area more complex (Garbis et al. 2005). Proteomic technique can be classified as gel-based approach like one-dimensional (1D) and two-dimensional (2D) gel electrophoresis (2-DE) or gel-free such as multidimensional protein identification technology, stable isotope labeling through amino acids *in vitro*, or in order to achieve numerical results isotope-coded affinity tag and isobaric tagging can be applicable (VerBerkmoes et al. 2009; Dobrowolska Zakaria et al. 2022). Moreover, for high-throughput processing investigations large-scale western blotting and multiple reaction monitoring assays can be suitable choice (Ornberg et al. 2005; Mani et al. 2012). Proteomic applied in several steps and most noticeable step is sample preparation because in this stage all protein content will be solubilize and every interfering and inhibitory factor must be wiped out (Hughes et al. 2019).

Proteomic can provide quantitative and qualitative measurement for cellular response toward concerned protein (Ong and Mann 2005). Having quantitative results can be highly helpful in area of precision medicine because in order to understanding disease mechanism or investigating new biomarker via proteomic analysis all data must be quantitative.

Proteomic is performing in three ways. Expression proteomics approach provides quantitative and qualitative protein expression in order to illustrate difference between two conditions and also determination of disease related or new proteins (Banks et al. 2000). On the other hand, functional proteomic studies provide the determination of molecular mechanism, protein activity included post-translational modifications such as phosphorylation, O-glycation, N-glycation, ubiquitination, and sumoylation, and also protein-protein interaction in order to identify specific protein complex in disease cases (Yanagida 2002). Finally, structural proteomics uses Nuclear magnetic resonance spectroscopy and X-ray crystallography in order to reveal 3D-structure (Renfrey and Featherstone 2002).

Proteomic can be noticeably effective in area of precision medicine studies. Discovering new biomarkers is always being a challenge for scientists, protein profiling in normal and diseased cells can paves a way to introduce updated biomarkers (Hanash et al. 2008; Asleh et al. 2022). Also based on proteomic results the response to specific cancer drug treatment can be realized. For example, expression of human epidermal growth factor receptor 2 can be sing of response to trastuzumab (Hofheinz et al. 2022). Also, proteomic approaches in cancer studies use variety of samples like tissues, serum, blood, and urine. However, diversity and complexity of body fluid proteins can be problematic during proteomic analysis (Lubec et al. 2003).

Oncoproteomics is known as proteomic study of cancers have been performing to anti-cancer drug investigation and also make cancer management more personalize (Meissner et al. 2022; Karcini and Lazar 2022). Many proteomics techniques like aptamer-based molecular probes, cancer immunomics, tissue microarrays and antibody microarrays can be used to introduce biomarkers in cancer (Gold et al. 2010; Caron et al. 2007). One of the most important examples in using oncoproteomic is discovery of leukemia biomarkers for determining type of the disease such as catalase and tropomyosin 3, also proteomic provide post-treatment guideline in chemotherapy or allogeneic stem cell transplantation via screening protein expression changes before and after treatment (Jain 2008).

Phosphorylation one of the most widespread protein regulatory system which generally works via protein kinases and protein phosphatases activity (Manning et al. 2002). The approach for studying phosphorylation changes called as Phosphoproteomics which also known as subtype of proteomics (Reinders and Sickmann 2005; Schmelzle and White 2006). Phosphoproteomics studies kinase activity via high-throughput approaches like peptide arrays, reverse-phase protein arrays and mass spectrometry (Schmelzle and White 2006). Based on great body of documentations, protein phosphorylation is widely involving in some crucial cell process like cell growth, proliferation and differentiation for this reason studying underlying mechanisms like kinase activity via phosphoproteomics can make disease mechanisms more understandable (Hunter 2000; Harsha and Pandey 2010). On the other hand, kinase targeting therapy via newly discovered drugs is one of the other interesting approaches in this area (He et al. 2022). Moreover, kinase activity studies via phosphoproteomics can provide concept of cancer early stage related markers or can be helpful in terms of drug targeting and drug toxicity. All these benefits leads to individualize the patient's treatment (Moran et al. 2006).

Like transcriptomic, proteomic technique has some limitations such as, protein's fluctuated expression in various conditions or among the cell types, shortage of comparable proteomics which make this approach more challenging in area of personalize medicine (Al-Amrani et al. 2021). Additionally, The use of proteomic approaches in precision medicine is needed to reduce costs and to be a more clinically accessible technique in diseases that require urgent diagnosis and treatment targets such as cancer. Although the technique is costly, there is still a great need to identify less costly specific proteins or their targets by further examining the functions of enzymes such as protein kinase involved in cancer-triggering signaling pathways investigated by proteomic approaches and other proteins phosphorylated by these enzymes. More research is needed to gain a deeper understanding of the epigenetic regulatory processes of genes involved in cancer.

5 Metabolomics Approaches in Precision Oncology

Metabolomics is defined as studying intracellular and extracellular small molecule metabolites in tissue or body fluids via analytical chemistry technique. This area provides investigation several classes of metabolites playing critical role during cell bioactivity such as biosynthesis and metabolism of glucose, lipid and nucleotide and also reactions occurring in mitochondria like TCA cycle or fatty acid oxidation (Johnson et al. 2016). Metabolites are also involved in some other crucial cellular functions such as signal transduction and apoptosis (Rudolph et al. 2001). Moreover, according to this fact that metabolites are final product of gene or protein expression, so they can be in close relation with diseases related phenotype (Patti et al. 2012).

Altogether, studying metabolites can make remarkable progress in patient personalized treatment, for example, by performing metabolomics analysis between drug treated and untreated group or disease and control group for better understanding the disease pathology. One of the interesting cohort studies performed by Carraro et al., have shown the importance of metabolomic analysis of amniotic fluid in order to early diagnosis of early onset wheezing as phenotype of pediatric asthma suspected infants (Carraro et al. 2018). On the other hand, understanding the metabolic dependency of cancer cells to specific metabolites such as glucose and lactate or TCA intermediates, such as succinate, fumarate or malate may result in discovery new strategy toward these cells (Liang et al. 2021).

In contrast with genomics or proteomics, metabolomics can measure molecules that have various physical characteristic (Kuehnbaum and Britz-McKibbin 2013). There are two main techniques in metabolomics, nuclear magnetic resonance (NMR) and mass spectrometry (MS). In order to recognize key metabolite modifications and to identify possible biomarkers these two techniques are combining with univariate and multivariate statistical approaches (Griffiths et al. 2010). Both approaches are capable to measure bulk number of metabolites. A wide variety of metabolomic studies have been carried out in common cancer types such as breast cancer, prostate cancer, and lung cancer in samples such as serum, plasma, urine, tumour tissue and stool, using various techniques such as LC-MS, and GC-MS. It shows that the metabolic parameters determined by these studies can be powerful potential targets in the diagnosis, prognosis and monitoring of the post-treatment response of cancer (Table 3). The need to use metabolic abnormalities obtained in samples such as urine and faeces in the diagnosis and treatment of the disease and their integration into precision medicine approaches is an urgent requirement.

In NMR spectroscopy, the samples are located in a magnetic surface. After absorption of the radiation, result will be in various frequency peaks based on the small molecule's location which are specific for each compounds. These NMR results then normalize with internal reference. By NMR, structural information will be available even for unknown compounds, also quantitative results of NMR make this technique useful for diagnostic aims (Ottka et al. 2021). The most important two advantages of the NMR are no sample destruction and no sample preparation. However, low sensitivity is the main drawback of the NMR (Pan and Raftery 2007). Unlike NMR, MS is able to identify any compounds like protons, carbon, nitrogen or oxygen in the samples. For performing MS, the samples are first divided through liquid or gas chromatography and then ionized. After splitting the ionized compounds results as a mass-to-charge (m/z) ratio will be determined by MS (Alseekh et al. 2021). Besides being speedy, MS approach's selectivity and sensitivity are known as its major advantages. However, sample destruction and metabolite lose during the sample preparation are the key drawbacks of this technique (Pan and Raftery 2007). There are various cancer researches based on metabolomics ranging from profiling to drug resistance studies. In the research performed by Nizioł and her co-workers, metabolic profiling of urine samples from 50 kidney cancer cases and 50 healthy controls via high-resolution proton nuclear magnetic resonance spectroscopy (^{1}H-NMR) and silver-109 nanoparticle enhanced steel target

Table 3 Some potential metabolic parameters, which have a diagnostic and/or a prognostic importance, their associated cancer types, analysis methods, sample types, and their use in the most common cancers

Cancer type	Metabolites	Analysis method	Sample type	Usage	References
Breast cancer	Quinone conjugates	HPLC	Tumor tissue (Up)	Diagnosis	Rogan et al. (2003)
Breast cancer	2-OHE1/16-OHE1	ELISA	Urine (down)	Diagnosis	Ruan et al. (2015)
Breast cancer	phosphocholine/choline	LC/ESI-MS	Tumor tissue (Up)	Diagnosis	Mimmi et al. (2011)
Breast cancer	Glutamin, Ornitin, Threonin, Trptophan, Methionin-SO, C2, C3 metabolics panel	UHPLC-MS/MS	Plasma	Diagnosis	Yuan et al. (2019)
Prostate cancer	Spermine and citrate	HR-MAS	Tumor tissue (Up)	Prognosis	Giskeødegård et al. (2013)
Prostate cancer	Pantothenic Acid, Hypoxanthine, CDCA/DCA	SID-MRM-MS	Plasma	Prognosis	Cheema et al. (2019)
Prostate cancer	Methylglyoxal, Dihydroedulan IA, Hexanal, 3-Phenylpropionaldehyde, 4-Methylhexan-3-one, 2,5-Dimethylbenzaldehyde	GC-MS	Urine	Diagnosis	Lima et al. (2019)
Lung cancer	Putrescine	UHPLC-MS	Plasma (Most abundant)	Diagnosis	Xu et al. (2016)
	Spermidine	UHPLC-MS	Plasma		
	S-adenosyl-L-methionine	UHPLC-MS	Urine		
	N-acetylspermidine	UHPLC-MS	Urine (most abundant)		
Lung cancer (NSCLC)	PGE-MUM	Radioimmunoassay	Urine (Up)	Diagnosis	Kawamoto et al. (2019)
Colorectal cancer	Lysophosphatidylcholine (18:2), cephaline, sphingomyelin (42:2), phosphatidylglycerol (34:0), ceramide (44:5)	2D LC-QToF/MS	Plasma (Down)	Diagnosis	Shen et al. (2017)

Colorectal cancer	cerotic acid	GC-MS	Serum (Up)	Diagnosis	Mika et al. (2017)
Colorectal cancer	M2-Pyruvate kinase	ELISA	Stool (Up)	Diagnosis	Tonus et al. (2006)
Cervix cancer	Lactate, proline, glutamate, adenosine monophosphate, aspartate, pyroglutamate and hypoxanthine	UPLC-QTOF-MS	Plasma (Up)	Diagnosis	Khan et al. (2019)

NSCLC: Non-small cell lung cancer, PGE-MUM: Prostaglandin E-Major Urinary Metabolite, UHPLC-MS: Ultra high pleasure liquid chromatography-mass spectrometry, 2D LC-QToF/MS: two-dimensional liquid chromatography coupled with quadrupole time-of-flight mass spectrometry, GC-MS: gas chromatography mass spectrometry, UPLC-QTOF-MS: ultra-performance liquid chromatography coupled with quadrupole time-of-flight mass spectrometry, ELISA: Enzyme linked immunesorbent assay, Up: Upregulation, Down: Downregulation, PGE2: Prostaglandin E2, 2-OHE1/16-OHE1: 2-hydroxyestrone/16-hydroxyestrone laser desorption/ionization mass spectrometry (^{109}AgNPET LDI MS) has done. According to their results some possible urine biomarkers have investigated such as creatine, sucrose, urea and isoleucine (Nizioł et al. 2021). Understanding the mechanisms involving in drug resistance in cancer treatment one of the vital aims of the cancer research. Colorectal cells that are resistant to irinotecan, a crucial medication for treating metastatic colorectal cancer, exhibit higher levels of unsaturated lipids compared to the control group in a single-cell mass spectrometry study (Sun et al. 2022).

Metabolomics is perhaps one of the most widely used techniques after genomic approaches and is also widely used in relatively current clinical approaches. However, there are still a limited number of metabolomic markers and targets that can be put into clinical practice. It is hoped that it will see much development in the future in terms of being a newly developing field and allowing it to work with non-invasive samples. The development of metabolomic approaches at the single-cell level is revolutionary in this field, as in other fields. However, it still seems to have some time before its widespread use in precision medicine.

6 The Other Omics Approaches in Precision Oncology

6.1 Single-Cell Omics Approaches

Despite the advances in cancer research, challenging issues such as drug resistance varying from patient to patient against various cancer drugs and tumour heterogeneity have necessitated the emergence of new techniques and the development of integrative methods. The single-cell transcriptomic method was reported for the first time in 2009 (Tang et al. 2009). It has provided a ray of hope for addressing problems like tumour heterogeneity that have proven challenging to solve with current bulk transcriptomics techniques. Over the past 10 years, we have been able to describe the transcriptional, genomic, proteomic, epigenomic, metabolic, and other aspects of thousands of individual cells using novel techniques in single-cell analysis approaches (Table 4). Additionally, as these procedures advanced, it was discovered that the information about which section of the tissue each cell originated from was lost when cells were isolated from tissue (Kashima et al. 2020). Also, there are a lot of techniques including G&T-seq (for genome and transcriptome), scM&T seq (single cell genome-wide methylome and transcriptome), T-ATAC-seq (open chromatin and TCR), SNARE-seq (open chromatin and transcriptome), and CITE-seq (surface protein and transcriptome) investigating stratified multiomics data of one cell in cancer studies (Kashima et al. 2020) (Table 4). This problem was solved by performing spatial omics analyses of each individual cell that makes up the tissue-on-tissue sections. Since they provide crucial information for resolving complex problems like tumour heterogeneity, spatial single-cell techniques have gained prominence in addition to single-cell approaches over the past few years (Stuart and Satija 2019; Xing et al. 2020). All these technological developments contribute

Table 4 Single-cell omics approaches, their methodology and applications

Omics approaches	Methodology	Application	References
Single-cell transcriptomics	STRT-seq, Smart-seq, CEL-seq2, MARS-seq, Drop-seq, 10× Genomics, SCRB-seq, SPLit-seq, MATQ-seq, ScNaUmi-seq, Indrop, ICELL8, Super-seq	Gene expression analysis in single cells	Adil et al. (2021), Nayak and Hasija (2021)
Single-cell genomics	MDA, DOP-PACR, MALBAC, LIANTI, META-CS	Single-cell genomic structure variation, SNPs, and copy number variations	Kashima et al. (2020)
Single-cell epigenomics	scBS-seq, scRRBS, scAba-seq, scATAC-seq, Drop-ChIP, scChICseq, CUT&Tag, Single-cell Hi-C	The analysis of chromatin accessibility, methylation levels and histone modifications	Kashima et al. (2020), Clark et al. (2016)
Single-cell proteomics	SCoPE-MS, SCoPE2, sc-CyTOF	In-depth proteomic analysis for the analysis of alterations at protein and post-translational level	Kashima et al. (2020), Specht et al. (2021)
Single-cell immunomics	scTCR-seq, scBCR-seq	The interaction analysis between all lymphocytes and tumour cells.	Chen et al. (2019), Bonaguro et al. (2022)

to providing in-depth information about the clonal heterogeneity of the tumour and the complexity of molecules and mechanisms involved in cancer (Lei et al. 2021). Single-cell techniques have a wide range of applications in cancer studies. Among them, important applications are the detection of dynamic alterations in the genome, transcriptome, proteome, and metabolome levels during the development and progression of tumours, the discovery of novel cell subpopulations, the determining of modifications to the state and phenotype of individual cells, the analysis of the regulatory networks of important genes in each cell, and the identification of potential therapeutic targets (Papalexi and Satija 2018).

6.2 Epigenomics Approaches

Since the last decade, epigenetic changes have been increasingly studied in cancer research. Epigenetic changes include DNA hypo- and hyper-methylation, chromatin remodeling by mechanisms including histone modifications and microRNA alterations. These alterations are observed frequently in breast, prostate, colon, and lung cancers. According to TCGA data, it contains hypermethylated and hypomethylated genes in over 100 promoter regions in the four most common cancer types (Gonzalo et al. 2005) (Table 5). In cancer, these changes can influence a variety of cellular processes from regulation of gene transcription to proper chromosome segregation.

Table 5 The gene counts included hyper-and hypomethylated promoters in the most common cancers

Cancer type	Hypermethylated gene counts	Hypomethylated gene counts
Breast cancer	112	76
Prostate cancer	205	62
Lung cancer (LUAD)	36	39
Lung cancer (LUSC)	72	143
Colorectal cancer	250	165

Also, they correlate with tumor suppressor gene silencing and oncogene activation. The retinoblastoma (Rb) tumor suppressor gene family, which plays an important role in the regulation of chromatin structure in cancer studies, maintains histone 4 lysine 20 (H4K20) tri-methylation in heterochromatin regions. However, studies have shown that the loss of Rb reduces H4K20 tri-methylation, resulting in chromosomal segregation defects and abnormal telomere elongation (Langevin et al. 2015).

Since epigenetic changes may be reversible, they are being actively studied as therapeutic targets. In recent years, targeted cancer medicines have increasingly focused on the changes. Epigenetic therapies using DNA methyltransferase inhibitors or histone deacetylase inhibitors may avoid the issue of tumor cell heterogeneity by promoting the expression of silenced tumor suppressor genes (Ansari et al. 2016; Waarts et al. 2022). These therapies exert their effects by regulating epigenome changes that regulate the expression of abnormally altered genes in many cellular signaling pathways involved in cancer development. The therapeutic editing does not result in nucleic acid base changes. For precision medicine, these epigenetic modifiers may provide more effective personalized treatment for many cancers. Most epigenome modifiers with the potential for use in personalized therapy are now being developed, and clinical studies have started.

Currently, seven agents in three epigenetic target classes (DNA methyltransferases (DNMTs), Histon deacetylases (HDACs), and Enhancer of Zeste 2 Polycomb Repressive Complex 2 Subunit (EZH2) inhibitors) have been approved by the US FDA for the treatment of various cancer types and a wide variety of epigenetic-based drugs are undergoing clinical trials. These are 5-azacytidine and Hydralazine as DNMT inhibitors, Ruxolitinib as JAK1/2 inhibitor, and Vorinostat, Romidepsin, Panobinostat, Belinostat, and suberoylanilide hydrocxamic acid (SAHA) (Kronfol et al. 2017; Majchrzak-Celińska et al. 2021). Therefore, in Precision Medicine, epigenomics is invaluable for diseases such as cancer that need immediate diagnosis and identification of targets for treatment. More research is needed to gain a deeper understanding of the epigenetic regulatory processes of genes involved in cancer.

7 Conclusion

Multi-omics techniques provide invaluable high-throughput data for scientists in revealing the formation mechanisms and development of cancer. Processing and stratifying multi-omic signatures of aberrational changes from each biological sample or even a single cell requires complex bioinformatics and systems biology approaches. Performing these approaches by well-trained people with validated standardized approaches is one of the major challenges for precision medicine. One of the other difficulties is that multi-omic techniques are an area that is relatively avoided due to the high cost of implementation and one of the high health expenditures by governments. In recent years, with the development of artificial intelligence systems, there has been a hope that the analysis of multi-omic data and systems biology approaches will be easier and faster. Another challenge is that validation and standardization processes have not yet been developed between ohmic analyzes to be performed in different centres. Despite this, it is an urgent requirement to develop systems that can overcome the above-mentioned difficulties so that multi-omic analyzes can be made very widely within the scope of personalized health approaches.

References

Abida W, Patnaik A, Campbell D, Shapiro J, Bryce AH, McDermott R, Sautois B, Vogelzang NJ, Bambury RM, Voog E, Zhang J, Piulats JM, Ryan CJ, Merseburger AS, Daugaard G, Heidenreich A, Fizazi K, Higano CS, Krieger LE, Sternberg CN, Watkins SP, Despain D, Simmons AD, Loehr A, Dowson M, Golsorkhi T, Chowdhury S, TRITON2 Investigators (2020) Rucaparib in men with metastatic castration-resistant prostate cancer harboring a BRCA1 or BRCA2 gene alteration. J Clin Oncol 38(32):3763–3772

Adil A, Kumar V, Jan AT, Asger M (2021) Single-cell transcriptomics: current methods and challenges in data acquisition and analysis. Front Neurosci 15:591122

Al-Amrani S, Al-Jabri Z, Al-Zaabi A, Alshekaili J, Al-Khabori M (2021) Proteomics: concepts and applications in human medicine. World J Biol Chem 12(5):57–69

Alseekh S, Aharoni A, Brotman Y, Contrepois K, D'Auria J, Ewald J, Ewald JC, Fraser PD, Giavalisco P, Hall RD, Heinemann M, Link H, Luo J, Neumann S, Nielsen J, Perez de Souza L, Saito K, Sauer U, Schroeder FC, Schuster S, Siuzdak G, Skirycz A, Sumner LW, Snyder MP, Tang H, Tohge T, Wang Y, Wen W, Wu S, Xu G, Zamboni N, Fernie AR (2021) Mass spectrometry-based metabolomics: a guide for annotation, quantification and best reporting practices. Nat Methods 18(7):747–756

André F, Ciruelos E, Rubovszky G, Campone M, Loibl S, Rugo HS, Iwata H, Conte P, Mayer IA, Kaufman B, Yamashita T, Lu YS, Inoue K, Takahashi M, Pápai Z, Longin AS, Mills D, Wilke C, Hirawat S, Juric D, SOLAR-1 Study Group (2019) Alpelisib for PIK3CA-mutated, hormone receptor-positive advanced breast cancer. N Engl J Med 380(20):1929–1940

Ansari J, Shackelford RE, El-Osta H (2016) Epigenetics in non-small cell lung cancer: from basics to therapeutics. Transl Lung Cancer Res 5(2):155–171

Asleh K, Negri GL, Spencer Miko SE, Colborne S, Hughes CS, Wang XQ, Gao D, Gilks CB, Chia SKL, Nielsen TO, Morin GB (2022) Proteomic analysis of archival breast cancer clinical specimens identifies biological subtypes with distinct survival outcomes. Nat Commun 13(1): 896

Azad AA, Volik SV, Wyatt AW, Haegert A, Le Bihan S, Bell RH, Anderson SA, McConeghy B, Shukin R, Bazov J, Youngren J, Paris P, Thomas G, Small EJ, Wang Y, Gleave ME, Collins CC, Chi KN (2015) Androgen receptor gene aberrations in circulating cell-free DNA: biomarkers of therapeutic resistance in castration-resistant prostate cancer. Clin Cancer Res 21(10):2315–2324

Banks RE, Dunn MJ, Hochstrasser DF, Sanchez JC, Blackstock W, Pappin DJ, Selby PJ (2000) Proteomics: new perspectives, new biomedical opportunities. Lancet 356(9243):1749–1756

Biswas N, Chakrabarti S (2020) Artificial Intelligence (AI)-based systems biology approaches in multi-omics data analysis of cancer. Front Oncol 10:588221

Bollinger MK, Agnew AS, Mascara GP (2018) Osimertinib: a third-generation tyrosine kinase inhibitor for treatment of epidermal growth factor receptor-mutated non-small cell lung cancer with the acquired Thr790Met mutation. J Oncol Pharm Pract 24(5):379–388

Bonaguro L, Schulte-Schrepping J, Ulas T, Aschenbrenner AC, Beyer M, Schultze JL (2022) A guide to systems-level immunomics. Nat Immunol 23(10):1412–1423

Cancer Genome Atlas Network (2012) Comprehensive molecular characterization of human colon and rectal cancer. Nature 487(7407):330–337

Cancer Genome Atlas Research Network, Albert Einstein College of Medicine, Analytical Biological Services, Barretos Cancer Hospital, Baylor College of Medicine, Beckman Research Institute of City of Hope, Buck Institute for Research on Aging, Canada's Michael Smith Genome Sciences Centre, Harvard Medical School, Helen F. Graham Cancer Center &Research Institute at Christiana Care Health Services, HudsonAlpha Institute for Biotechnology, ILSbio, LLC, Indiana University School of Medicine, Institute of Human Virology, Institute for Systems Biology, International Genomics Consortium, Leidos Biomedical, Massachusetts General Hospital, McDonnell Genome Institute at Washington University, Medical College of Wisconsin et al (2017) Integrated genomic and molecular characterization of cervical cancer. Nature 543(7645):378–384

Caron M, Choquet-Kastylevsky G, Joubert-Caron R (2007) Cancer immunomics using autoantibody signatures for biomarker discovery. Mol Cell Proteomics 6(7):1115–1122

Carraro S, Baraldi E, Giordano G, Pirillo P, Stocchero M, Houben M, Bont L (2018) Metabolomic profile of amniotic fluid and wheezing in the first year of life-a healthy birth cohort study. J Pediatr 196:264–269.e4

CDC (2020) Precision health: improving health for each of us and all of us. https://www.cdc.gov/genomics/about/precision_med.htm. Accessed 11 Nov 2022

Chandrashekar DS, Bashel B, Balasubramanya SAH, Creighton CJ, Ponce-Rodriguez I, Chakravarthi BVSK, Varambally S (2017) UALCAN: a portal for facilitating tumor subgroup gene expression and survival analyses. Neoplasia 19(8):649–658

Chandrashekar DS, Karthikeyan SK, Korla PK, Patel H, Shovon AR, Athar M, Netto GJ, Qin ZS, Kumar S, Manne U, Creighton CJ, Varambally S (2022) UALCAN: an update to the integrated cancer data analysis platform. Neoplasia 25:18–27

Cheema AK, Grindrod S, Zhong X, Jain S, Menon SS, Mehta KY, Suy S, Collins S, Wang Y, Timofeeva O, Bandi G, Pahira J, Kowalczyk K, Lynch J, Dritschilo A (2019) Discovery of metabolic biomarkers predicting radiation therapy late effects in prostate cancer patients. Adv Exp Med Biol 1164:141–150

Chen H, Ye F, Guo G (2019) Revolutionizing immunology with single-cell RNA sequencing. Cell Mol Immunol 16(3):242–249

Choudhury AD, Werner L, Francini E, Wei XX, Ha G, Freeman SS, Rhoades J, Reed SC, Gydush G, Rotem D, Lo C, Taplin ME, Harshman LC, Zhang Z, O'Connor EP, Stover DG, Parsons HA, Getz G, Meyerson M, Love JC, Hahn WC, Adalsteinsson VA (2018) Tumor fraction in cell-free DNA as a biomarker in prostate cancer. JCI Insight 3(21):e122109

Chung HC, Ros W, Delord JP, Perets R, Italiano A, Shapira-Frommer R, Manzuk L, Piha-Paul SA, Xu L, Zeigenfuss S, Pruitt SK, Leary A (2019) Efficacy and safety of pembrolizumab in previously treated advanced cervical cancer: results from the phase II KEYNOTE-158 study. J Clin Oncol 37(17):1470–1478

Clark SJ, Lee HJ, Smallwood SA, Kelsey G, Reik W (2016) Single-cell epigenomics: powerful new methods for understanding gene regulation and cell identity. Genome Biol 17:72

Daiba A, Inaba N, Ando S, Kajiyama N, Yatsuhashi H, Terasaki H, Ito A, Ogasawara M, Abe A, Yoshioka J, Hayashida K, Kaneko S, Kohara M, Ito S (2004) A low-density cDNA microarray with a unique reference RNA: pattern recognition analysis for IFN efficacy prediction to HCV as a model. Biochem Biophys Res Commun 315(4):1088–1096

Dean L, Kane M (2020a) Cetuximab therapy and RAS and BRAF genotype. In: Pratt VM et al (eds) Medical genetics summaries. National Center for Biotechnology Information (US), Bethesda, MD

Dean L, Kane M (2020b) Panitumumab therapy and RAS and BRAF genotype. In: Pratt VM et al (eds) Medical genetics summaries. National Center for Biotechnology Information (US), Bethesda, MD

Dietz S, Schirmer U, Mercé C, von Bubnoff N, Dahl E, Meister M, Muley T, Thomas M, Sültmann H (2016) Low input whole-exome sequencing to determine the representation of the tumor exome in circulating DNA of non-small cell lung cancer patients. PLoS One 11(8):e0161012

Dobrowolska Zakaria JA, Bateman RJ, Lysakowska M, Khatri A, Jean-Gilles D, Kennedy ME, Vassar R (2022) The metabolism of human soluble amyloid precursor protein isoforms is quantifiable by a stable isotope labeling-tandem mass spectrometry method. Sci Rep 12(1): 14985

Dudek AZ, Baxstrom K, Bharadwaj S, Blaes A, Kulkarni A, Lou E, Nehru V, Rabinovich E, Shergill A, Viner M (2020) Genomic strategies for personalized cancer therapy. In: Aydogan B, Radosevich JA (eds) Precision medicine in oncology. Wiley, New York, NY. https://doi.org/10.1002/9781119432487.Ch1

Faruki H, Mayhew GM, Serody JS, Hayes DN, Perou CM, Lai-Goldman M (2017) Lung adenocarcinoma and squamous cell carcinoma gene expression subtypes demonstrate significant differences in tumor immune landscape. J Thorac Oncol 12(6):943–953

Fekete JT, Győrffy B (2019) ROCplot.org: validating predictive biomarkers of chemotherapy/hormonal therapy/anti-HER2 therapy using transcriptomic data of 3,104 breast cancer patients. Int J Cancer 145(11):3140–3151

Ferlay J, Laversanne M, Ervik M et al (2020a) Global cancer observatory: cancer today. International Agency for Research on Cancer, Lyon. https://gco.iarc.fr/today/home. Accessed 11 Nov 2022

Ferlay J, Laversanne M, Ervik M et al (2020b) Global cancer observatory: cancer tomorrow. International Agency for Research on Cancer, Lyon. https://gco.iarc.fr/tomorrow/en. Accessed 11 Nov 2022

Ferlay J, Colombet M, Soerjomataram I, Parkin DM, Piñeros M, Znaor A, Bray F (2021) Cancer statistics for the year 2020: an overview. Int J Cancer 149:778. https://doi.org/10.1002/ijc.33588

Garbis S, Lubec G, Fountoulakis M (2005) Limitations of current proteomics technologies. J Chromatogr A 1077(1):1–18

Giskeødegård GF, Bertilsson H, Selnæs KM, Wright AJ, Bathen TF, Viset T, Halgunset J, Angelsen A, Gribbestad IS, Tessem MB (2013) Spermine and citrate as metabolic biomarkers for assessing prostate cancer aggressiveness. PLoS One 8(4):e62375

Gold L, Ayers D, Bertino J, Bock C, Bock A, Brody EN, Carter J, Dalby AB, Eaton BE, Fitzwater T, Flather D, Forbes A, Foreman T, Fowler C, Gawande B, Goss M, Gunn M, Gupta S, Halladay D, Heil J, Heilig J, Hicke B, Husar G, Janjic N, Jarvis T, Jennings S, Katilius E, Keeney TR, Kim N, Koch TH, Kraemer S, Kroiss L, Le N, Levine D, Lindsey W, Lollo B, Mayfield W, Mehan M, Mehler R, Nelson SK, Nelson M, Nieuwlandt D, Nikrad M, Ochsner U, Ostroff RM, Otis M, Parker T, Pietrasiewicz S, Resnicow DI, Rohloff J, Sanders G, Sattin S, Schneider D, Singer B, Stanton M, Sterkel A, Stewart A, Stratford S, Vaught JD, Vrkljan M, Walker JJ, Watrobka M, Waugh S, Weiss A, Wilcox SK, Wolfson A, Wolk SK, Zhang C, Zichi D (2010) Aptamer-based multiplexed proteomic technology for biomarker discovery. PLoS One 5(12):e15004

Gonzalo S, García-Cao M, Fraga MF, Schotta G, Peters AH, Cotter SE, Eguía R, Dean DC, Esteller M, Jenuwein T, Blasco MA (2005) Role of the RB1 family in stabilizing histone methylation at constitutive heterochromatin. Nat Cell Biol 7(4):420–428

Gray KA, Yates B, Seal RL, Wright MW, Bruford EA (2015) Genenames.org: the HGNC resources in 2015. Nucleic Acids Res 43(Database Issue):D1079–D1085. https://doi.org/10.1093/nar/gku1071

Griffiths WJ, Koal T, Wang Y, Kohl M, Enot DP, Deigner HP (2010) Targeted metabolomics for biomarker discovery. Angew Chem Int Ed Eng 49(32):5426–5445

Hanahan D (2022) Hallmarks of cancer: new dimensions. Cancer Discov 12(1):31–46

Hanash SM, Pitteri SJ, Faca VM (2008) Mining the plasma proteome for cancer biomarkers. Nature 452(7187):571–579

Harsha HC, Pandey A (2010) Phosphoproteomics in cancer. Mol Oncol 4(6):482–495

He S, Li Q, Huang Q, Cheng J (2022) Targeting protein kinase C for cancer therapy. Cancers 14(5): 1104

Hofheinz RD, Merx K, Haag GM, Springfeld C, Ettrich T, Borchert K, Kretzschmar A, Teschendorf C, Siegler G, Ebert MP, Goekkurt E, Mahlberg R, Homann N, Pink D, Bechstein W, Reichardt P, Flach H, Gaiser T, Battmann A, Oduncu FS, Loose M, Sookthai D, Pauligk C, Göetze TO, Al-Batran SE (2022) FLOT versus FLOT/trastuzumab/pertuzumab perioperative therapy of human epidermal growth factor receptor 2-positive resectable esophagogastric adenocarcinoma: a randomized phase II trial of the AIO EGA study group. J Clin Oncol 40(32):3750–3761

Hu L, Su L, Cheng H, Mo C, Ouyang T, Li J, Wang T, Fan Z, Fan T, Lin B, Zhang J, Xie Y (2021) Single-cell RNA sequencing reveals the cellular origin and evolution of breast cancer in BRCA1 mutation carriers. Cancer Res 81(10):2600–2611

Hughes CS, Moggridge S, Müller T, Sorensen PH, Morin GB, Krijgsveld J (2019) Single-pot, solid-phase-enhanced sample preparation for proteomics experiments. Nat Protoc 14(1):68–85

Hunter T (2000) Signaling--2000 and beyond. Cell 100(1):113–127

Hussain M, Mateo J, Fizazi K, Saad F, Shore N, Sandhu S, Chi KN, Sartor O, Agarwal N, Olmos D, Thiery-Vuillemin A, Twardowski P, Roubaud G, Özgüroğlu M, Kang J, Burgents J, Gresty C, Corcoran C, Adelman CA, de Bono J, PROfound Trial Investigators (2020) Survival with olaparib in metastatic castration-resistant prostate cancer. N Engl J Med 383(24):2345–2357

Jain KK (2008) Innovations, challenges and future prospects of oncoproteomics. Mol Oncol 2(2): 153–160

Johnson CH, Ivanisevic J, Siuzdak G (2016) Metabolomics: beyond biomarkers and towards mechanisms. Nat Rev Mol Cell Biol 17(7):451–459

Kan T, Zhang S, Zhou S, Zhang Y, Zhao Y, Gao Y, Zhang T, Gao F, Wang X, Zhao L, Yang M (2022) Single-cell RNA-seq recognized the initiator of epithelial ovarian cancer recurrence. Oncogene 41(6):895–906

Karcini A, Lazar IM (2022) The SKBR3 cell-membrane proteome reveals telltales of aberrant cancer cell proliferation and targets for precision medicine applications. Sci Rep 12(1):10847

Kashima Y, Sakamoto Y, Kaneko K, Seki M, Suzuki Y, Suzuki A (2020) Single-cell sequencing techniques from individual to multiomics analyses. Exp Mol Med 52(9):1419–1427

Kawamoto H, Hara H, Araya J, Ichikawa A, Fujita Y, Utsumi H, Hashimoto M, Wakui H, Minagawa S, Numata T, Arihiro S, Matsuura T, Fujiwara M, Ito S, Kuwano K (2019) Prostaglandin E-major urinary metabolite (PGE-MUM) as a tumor marker for lung adenocarcinoma. Cancers 11(6):768

Kazandjian D, Blumenthal GM, Chen HY, He K, Patel M, Justice R, Keegan P, Pazdur R (2014) FDA approval summary: crizotinib for the treatment of metastatic non-small cell lung cancer with anaplastic lymphoma kinase rearrangements. Oncologist 19(10):e5–e11

Khan I, Nam M, Kwon M, Seo SS, Jung S, Han JS, Hwang GS, Kim MK (2019) LC/MS-based polar metabolite profiling identified unique biomarker signatures for cervical cancer and cervical intraepithelial neoplasia using global and targeted metabolomics. Cancers 11(4):511

Khodadadian A, Darzi S, Haghi-Daredeh S, Sadat Eshaghi F, Babakhanzadeh E, Mirabutalebi SH, Nazari M (2020) Genomics and transcriptomics: the powerful technologies in precision medicine. Int J Gen Med 13:627–640

Khurana E, Fu Y, Chakravarty D, Demichelis F, Rubin MA, Gerstein M (2016) Role of non-coding sequence variants in cancer. Nat Rev Genet 17(2):93–108

Kim M, Min YK, Jang J, Park H, Lee S, Lee CH (2021) Single-cell RNA sequencing reveals distinct cellular factors for response to immunotherapy targeting CD73 and PD-1 in colorectal cancer. J Immunother Cancer 9(7):e002503

Kodahl AR, Ehmsen S, Pallisgaard N, Jylling AMB, Jensen JD, Laenkholm AV, Knoop AS, Ditzel HJ (2018) Correlation between circulating cell-free PIK3CA tumor DNA levels and treatment response in patients with PIK3CA-mutated metastatic breast cancer. Mol Oncol 12(6):925–935

Kosorok MR, Laber EB (2019) Precision medicine. Ann Rev Stat Appl 6:263–286

Kravitz RL, Duan N, Braslow J (2004) Evidence-based medicine, heterogeneity of treatment effects, and the trouble with averages. Milbank Quart 82(4):661–687

Kronfol MM, Dozmorov MG, Huang R, Slattum PW, McClay JL (2017) The role of epigenomics in personalized medicine. Exp Rev Precis Med Drug Dev 2(1):33–45

Kuehnbaum NL, Britz-McKibbin P (2013) New advances in separation science for metabolomics: resolving chemical diversity in a post-genomic era. Chem Rev 113(4):2437–2468

Langevin SM, Kratzke RA, Kelsey KT (2015) Epigenetics of lung cancer. Transl Res 165(1):74–90

Lei Y, Tang R, Xu J, Wang W, Zhang B, Liu J, Yu X, Shi S (2021) Applications of single-cell sequencing in cancer research: progress and perspectives. J Hematol Oncol 14(1):91

Liang L, Sun F, Wang H, Hu Z (2021) Metabolomics, metabolic flux analysis and cancer pharmacology. Pharmacol Ther 224:107827

Lim SM, Kim EY, Kim HR, Ali SM, Greenbowe JR, Shim HS, Chang H, Lim S, Paik S, Cho BC (2016) Genomic profiling of lung adenocarcinoma patients reveals therapeutic targets and confers clinical benefit when standard molecular testing is negative. Oncotarget 7(17): 24172–24178

Lima AR, Pinto J, Azevedo AI, Barros-Silva D, Jerónimo C, Henrique R, de Lourdes Bastos M, Guedes de Pinho P, Carvalho M (2019) Identification of a biomarker panel for improvement of prostate cancer diagnosis by volatile metabolic profiling of urine. Br J Cancer 121(10):857–868

Liu H, Bebu I, Li X (2010) Microarray probes and probe sets. Front Biosci 2(1):325–338

Lonergan M, Senn SJ, McNamee C, Daly AK, Sutton R, Hattersley A, Pearson E, Pirmohamed M (2017) Defining drug response for stratified medicine. Drug Discov Today 22(1):173–179

Longford NT, Nelder JA (1999) Statistics versus statistical science in the regulatory process. Stat Med 18(17–18):2311–2320

Lubec G, Krapfenbauer K, Fountoulakis M (2003) Proteomics in brain research: potentials and limitations. Prog Neurobiol 69(3):193–211

Majchrzak-Celińska A, Warych A, Szoszkiewicz M (2021) Novel approaches to epigenetic therapies: from drug combinations to epigenetic editing. Genes (Basel) 12(2):208. https://doi.org/10.3390/genes12020208

Mani DR, Abbatiello SE, Carr SA (2012) Statistical characterization of multiple-reaction monitoring mass spectrometry (MRM-MS) assays for quantitative proteomics. BMC Bioinform 13 (Suppl 16):S9

Manning G, Whyte DB, Martinez R, Hunter T, Sudarsanam S (2002) The protein kinase complement of the human genome. Science (New York, NY) 298(5600):1912–1934

Mantione KJ, Kream RM, Kuzelova H, Ptacek R, Raboch J, Samuel JM, Stefano GB (2014) Comparing bioinformatic gene expression profiling methods: microarray and RNA-Seq. Med Sci Monit Basic Res 20:138–142

Meissner F, Geddes-McAlister J, Mann M, Bantscheff M (2022) The emerging role of mass spectrometry-based proteomics in drug discovery. Nat Rev Drug Discov 21(9):637–654

Mika A, Kobiela J, Czumaj A, Chmielewski M, Stepnowski P, Sledzinski T (2017) Hyper-elongation in colorectal cancer tissue - cerotic acid is a potential novel serum metabolic marker of colorectal malignancies. Cell Physiol Biochem 41(2):722–730

Milward EA, Shahandeh A, Heidari M, Johnstone DM, Daneshi N, Hondermarck H (2016) Transcriptomics. In: Bradshaw RA, Stahl PD (eds) Encyclopedia of cell biology. Academic Press, Waltham, MA, pp 160–165. https://doi.org/10.1016/B978-0-12-394447-4.40029-5

Mimmi MC, Picotti P, Corazza A, Betto E, Pucillo CE, Cesaratto L, Cedolini C, Londero V, Zuiani C, Bazzocchi M, Esposito G (2011) High-performance metabolic marker assessment in breast cancer tissue by mass spectrometry. Clin Chem Lab Med 49(2):317–324

Moorcraft SY, Gonzalez D, Walker BA (2015) Understanding next generation sequencing in oncology: a guide for oncologists. Crit Rev Oncol Hematol 96(3):463–474

Moran MF, Tong J, Taylor P, Ewing RM (2006) Emerging applications for phospho-proteomics in cancer molecular therapeutics. Biochim Biophys Acta 1766(2):230–241

Nayak R, Hasija Y (2021) A hitchhiker's guide to single-cell transcriptomics and data analysis pipelines. Genomics 113(2):606–619

Nizioł J, Ossoliński K, Tripet BP, Copié V, Arendowski A, Ruman T (2021) Nuclear magnetic resonance and surface-assisted laser desorption/ionization mass spectrometry-based metabolome profiling of urine samples from kidney cancer patients. J Pharm Biomed Anal 193:113752

Ong SE, Mann M (2005) Mass spectrometry-based proteomics turns quantitative. Nat Chem Biol 1(5):252–262

Ornberg R, Harper T, Liu H (2005) Western blot analysis with quantum dot fluorescence technology: a sensitive and quantitative method for multiplexed proteomics. Nat Methods 2:79–81

Ottka C, Vapalahti K, Puurunen J, Vahtera L, Lohi H (2021) A novel canine nuclear magnetic resonance spectroscopy-based metabolomics platform: validation and sample handling. Vet Clin Pathol 50(3):410–426

Ozsolak F, Milos PM (2011) RNA sequencing: advances, challenges and opportunities. Nat Rev Genet 12(2):87–98

Pan Z, Raftery D (2007) Comparing and combining NMR spectroscopy and mass spectrometry in metabolomics. Anal Bioanal Chem 387(2):525–527

Papalexi E, Satija R (2018) Single-cell RNA sequencing to explore immune cell heterogeneity. Nat Rev Immunol 18(1):35–45

Patti GJ, Yanes O, Shriver LP, Courade JP, Tautenhahn R, Manchester M, Siuzdak G (2012) Metabolomics implicates altered sphingolipids in chronic pain of neuropathic origin. Nat Chem Biol 8(3):232–234

Planchard D, Besse B, Groen HJM, Souquet PJ, Quoix E, Baik CS, Barlesi F, Kim TM, Mazieres J, Novello S, Rigas JR, Upalawanna A, D'Amelio AM Jr, Zhang P, Mookerjee B, Johnson BE (2016) Dabrafenib plus trametinib in patients with previously treated BRAF(V600E)-mutant metastatic non-small cell lung cancer: an open-label, multicentre phase 2 trial. Lancet Oncol 17(7):984–993

Pritchard CC, Salipante SJ, Koehler K, Smith C, Scroggins S, Wood B, Wu D, Lee MK, Dintzis S, Adey A, Liu Y, Eaton KD, Martins R, Stricker K, Margolin KA, Hoffman N, Churpek JE, Tait JF, King MC, Walsh T (2014) Validation and implementation of targeted capture and sequencing for the detection of actionable mutation, copy number variation, and gene rearrangement in clinical cancer specimens. J Mol Diagn 16(1):56–67

Razavi P, Chang MT, Xu G, Bandlamudi C, Ross DS, Vasan N, Cai Y, Bielski CM, Donoghue MTA, Jonsson P, Penson A, Shen R, Pareja F, Kundra R, Middha S, Cheng ML, Zehir A, Kandoth C, Patel R, Huberman K, Smyth LM, Jhaveri K, Modi S, Traina TA, Dang C, Zhang W, Weigelt B, Li BT, Ladanyi M, Hyman DM, Schultz N, Robson ME, Hudis C, Brogi E, Viale A, Norton L, Dickler MN, Berger MF, Iacobuzio-Donahue CA, Chandarlapaty S, Scaltriti M, Reis-Filho JS, Solit DB, Baselga J (2018) The genomic landscape of endocrine-resistant advanced breast cancers. Cancer Cell 34(3):427–438.e6

Reinders J, Sickmann A (2005) State-of-the-art in phosphoproteomics. Proteomics 5(16): 4052–4061

Renfrey S, Featherstone J (2002) Structural proteomics. Nat Rev Drug Discov 1(3):175–176

Rogan EG, Badawi AF, Devanesan PD, Meza JL, Edney JA, West WW, Higginbotham SM, Cavalieri EL (2003) Relative imbalances in estrogen metabolism and conjugation in breast tissue of women with carcinoma: potential biomarkers of susceptibility to cancer. Carcinogenesis 24(4):697–702

Romanov V, Davidoff SN, Miles AR, Grainger DW, Gale BK, Brooks BD (2014) A critical comparison of protein microarray fabrication technologies. Analyst 139(6):1303–1326

Ruan X, Seeger H, Wallwiener D, Huober J, Mueck AO (2015) The ratio of the estradiol metabolites 2-hydroxyestrone (2-OHE1) and 16α-hydroxyestrone (16-OHE1) may predict breast cancer risk in postmenopausal but not in premenopausal women: two case-control studies. Arch Gynecol Obstet 291(5):1141–1146

Rudolph IL, Kelley DS, Klasing KC, Erickson KL (2001) Regulation of cellular differentiation and apoptosis by fatty acids and their metabolites. Nutr Res 21(1–2):381–393

Russo G, Zegar C, Giordano A (2003) Advantages and limitations of microarray technology in human cancer. Oncogene 22(42):6497–6507

Salzman J, Jiang H, Wong WH (2011) Statistical modeling of RNA-Seq data. Stat Sci 26(1):62. https://doi.org/10.1214/10-STS343

Schmelzle K, White FM (2006) Phosphoproteomic approaches to elucidate cellular signaling networks. Curr Opin Biotechnol 17(4):406–414

Shen S, Yang L, Li L, Bai Y, Cai C, Liu H (2017) A plasma lipidomics strategy reveals perturbed lipid metabolic pathways and potential lipid biomarkers of human colorectal cancer. J Chromatogr B Anal Technol Biomed Life Sci 1068–1069:41–48

Shimomura A, Shiino S, Kawauchi J, Takizawa S, Sakamoto H, Matsuzaki J, Ono M, Takeshita F, Niida S, Shimizu C, Fujiwara Y, Kinoshita T, Tamura K, Ochiya T (2016) Novel combination of serum microRNA for detecting breast cancer in the early stage. Cancer Sci 107(3):326–334

Sîrbu A, Kerr G, Crane M, Ruskin HJ (2012) RNA-Seq vs dual- and single-channel microarray data: sensitivity analysis for differential expression and clustering. PLoS One 7(12):e50986

Sørensen TI (1996) Which patients may be harmed by good treatments? Lancet 348(9024):351–352

Specht H, Emmott E, Petelski AA, Huffman RG, Perlman DH, Serra M, Kharchenko P, Koller A, Slavov N (2021) Single-cell proteomic and transcriptomic analysis of macrophage heterogeneity using SCoPE2. Genome Biol 22(1):50

Srivastava R (2022) Applications of artificial intelligence multiomics in precision oncology. J Cancer Res Clin Oncol 149:503. https://doi.org/10.1007/s00432-022-04161-4

Stears RL, Getts RC, Gullans SR (2000) A novel, sensitive detection system for high-density microarrays using dendrimer technology. Physiol Genomics 3(2):93–99

Stuart T, Satija R (2019) Integrative single-cell analysis. Nat Rev Genet 20(5):257–272

Sun M, Chen X, Yang Z (2022) Single cell mass spectrometry studies reveal metabolomic features and potential mechanisms of drug-resistant cancer cell lines. Anal Chim Acta 1206:339761

Tang F, Barbacioru C, Wang Y, Nordman E, Lee C, Xu N, Wang X, Bodeau J, Tuch BB, Siddiqui A, Lao K, Surani MA (2009) mRNA-Seq whole-transcriptome analysis of a single cell. Nat Methods 6(5):377–382

Testa U, Castelli G, Pelosi E (2018a) Lung cancers: molecular characterization, clonal heterogeneity and evolution, and cancer stem cells. Cancers 10(8):248

Testa U, Pelosi E, Castelli G (2018b) Colorectal cancer: genetic abnormalities, tumor progression, tumor heterogeneity, clonal evolution and tumor-initiating cells. Med Sci 6(2):31

Testa U, Castelli G, Pelosi E (2020) Breast cancer: a molecularly heterogenous disease needing subtype-specific treatments. Med Sci 8(1):18

Tomiuk S, Hofmann K (2001) Microarray probe selection strategies. Brief Bioinform 2(4):329–340

Tonus C, Neupert G, Sellinger M (2006) Colorectal cancer screening by non-invasive metabolic biomarker fecal tumor M2-PK. World J Gastroenterol 12(43):7007–7011

Torrejon DY, Abril-Rodriguez G, Champhekar AS, Tsoi J, Campbell KM, Kalbasi A, Parisi G, Zaretsky JM, Garcia-Diaz A, Puig-Saus C, Cheung-Lau G, Wohlwender T, Krystofinski P, Vega-Crespo A, Lee CM, Mascaro P, Grasso CS, Berent-Maoz B, Comin-Anduix B, Hu-Lieskovan S, Ribas A (2020) Overcoming genetically based resistance mechanisms to PD-1 blockade. Cancer Discov 10(8):1140–1157

Tutt ANJ, Garber JE, Kaufman B, Viale G, Fumagalli D, Rastogi P, Gelber RD, de Azambuja E, Fielding A, Balmaña J, Domchek SM, Gelmon KA, Hollingsworth SJ, Korde LA, Linderholm B, Bandos H, Senkus E, Suga JM, Shao Z, Pippas AW, Nowecki Z, Huzarski T, Ganz PA, Lucas PC, Baker N, Loibl S, McConnell R, Piccart M, Schmutzler R, Steger GG, Costantino JP, Arahmani A, Wolmark N, McFadden E, Karantza V, Lakhani SR, Yothers G, Campbell C, Geyer CE Jr, OlympiA Clinical Trial Steering Committee and Investigators (2021) Adjuvant olaparib for patients with BRCA1- or BRCA2-mutated breast cancer. N Engl J Med 384(25):2394–2405

VerBerkmoes NC, Denef VJ, Hettich RL, Banfield JF (2009) Systems biology: functional analysis of natural microbial consortia using community proteomics. Nat Rev Microbiol 7(3):196–205

Vora NL, Norton ME (2022) Prenatal exome and genome sequencing for fetal structural abnormalities. Am J Obstet Gynecol 228(2):140–149

Waarts MR, Stonestrom AJ, Park YC, Levine RL (2022) Targeting mutations in cancer. J Clin Invest 132(8):e154943

Wan Y, Kertesz M, Spitale RC, Segal E, Chang HY (2011) Understanding the transcriptome through RNA structure. Nat Rev Genet 12(9):641–655

Wang Z, Gerstein M, Snyder M (2009) RNA-Seq: a revolutionary tool for transcriptomics. Nat Rev Genet 10(1):57–63

Wang Y, Liu J, Huang BO, Xu YM, Li J, Huang LF, Lin J, Zhang J, Min QH, Yang WM, Wang XZ (2015) Mechanism of alternative splicing and its regulation. Biomed Rep 3(2):152–158

Wennerberg E, Mukherjee S, Spada S, Hung C, Agrusa CJ, Chen C, Valeta-Magara A, Rudqvist N-P, Van Nest SJ, Kamel MK et al (2022) Expression of the mono-ADP-ribosyltransferase ART1 by tumor cells mediates immune resistance in non-small cell lung cancer. Sci Transl Med 14:eabe8195

White House Office of The Press Secretary (2015) Fact sheet: President Obama's precision medicine initiative. Press release

Wilkins MR, Sanchez JC, Gooley AA, Appel RD, Humphery-Smith I, Hochstrasser DF, Williams KL (1996) Progress with proteome projects: why all proteins expressed by a genome should be identified and how to do it. Biotechnol Genet Eng Rev 13:19–50

Wyatt AW, Azad AA, Volik SV, Annala M, Beja K, McConeghy B, Haegert A, Warner EW, Mo F, Brahmbhatt S, Shukin R, Le Bihan S, Gleave ME, Nykter M, Collins CC, Chi KN (2016) Genomic alterations in cell-free DNA and enzalutamide resistance in castration-resistant prostate cancer. JAMA Oncol 2(12):1598–1606

Xing QR, Cipta NO, Hamashima K, Liou YC, Koh CG, Loh YH (2020) Unraveling heterogeneity in transcriptome and its regulation through single-cell multi-omics technologies. Front Genet 11:662

Xu H, Liu R, He B, Bi CW, Bi K, Li Q (2016) Polyamine metabolites profiling for characterization of lung and liver cancer using an LC-Tandem MS method with multiple statistical data mining strategies: discovering potential cancer biomarkers in human plasma and urine. Molecules 21(8):1040

Yanagida M (2002) Functional proteomics; current achievements. J Chromatogr B Anal Technol Biomed Life Sci 771(1–2):89–106

Yang X, Zheng Y, Han Z, Zhang X (2021) Functions and clinical significance of KLRG1 in the development of lung adenocarcinoma and immunotherapy. BMC Cancer 21:752

Yates LR, Seoane J, Le Tourneau C, Siu LL, Marais R, Michiels S, Soria JC, Campbell P, Normanno N, Scarpa A, Reis-Filho JS, Rodon J, Swanton C, Andre F (2018) The European Society for Medical Oncology (ESMO) Precision Medicine Glossary. Ann Oncol 29(1):30–35

Yip S, Christofides A, Banerji S, Downes MR, Izevbaye I, Lo B, MacMillan A, McCuaig J, Stockley T, Yousef GM, Spatz A (2019) A Canadian guideline on the use of next-generation sequencing in oncology. Curr Oncol 26(2):e241–e254

Yuan B, Schafferer S, Tang Q, Scheffler M, Nees J, Heil J, Schott S, Golatta M, Wallwiener M, Sohn C, Koal T, Wolf B, Schneeweiß A, Burwinkel B (2019) A plasma metabolite panel as biomarkers for early primary breast cancer detection. Int J Cancer 144(11):2833–2842

Zhang F, Gu W, Hurles ME, Lupski JR (2009) Copy number variation in human health, disease, and evolution. Annu Rev Genomics Hum Genet 10:451–481

Zhang S, Zeng X, Ding T, Guo L, Li Y, Ou S, Yuan H (2018) Microarray profile of circular RNAs identifies hsa_circ_0014130 as a new circular RNA biomarker in non-small cell lung cancer. Sci Rep 8(1):2878

Zhong R, Chen D, Cao S, Li J, Han B, Zhong H (2021) Immune cell infiltration features and related marker genes in lung cancer based on single-cell RNA-seq. Clin Transl Oncol 23(2):405–417

Zou S, He HJ, Zong Y, Shi L, Wang L (2008) DNA microarrays: applications, future trends, and the need for standardization. In: Resch-Genger U (ed) Standardization and quality assurance in fluorescence measurements II. Springer series on fluorescence, vol 6. Springer, Berlin. https://doi.org/10.1007/4243_2008_036

Translational Therapeutic Targets: From Bench to Clinical Trials

Kanupriya Vashishth, Samriddhi Ranjan, Hemant Joshi, Tejveer Singh, Mükerrem Betül Yerer Aycan, and Hardeep Singh Tuli

1 Role of Translational Research in Fighting Cancer

1.1 Gaps in Current Therapy for Cancer

Cancer is one of the growing causes of concern for the medical community and researchers. In 2020 alone, 19 million people across the globe were affected, and ten million individuals died. This figure is likely to increase to 28 million by 2040, and death predicated as many as 16 million (Mao et al. 2022). The development of cancer itself is a very complex phenomenon. Several attributable factors lead to cancer, like exposure to carcinogens in the form of chemicals, UV rays, radiation, and biological agents (viruses, bacteria, etc.). Hanahan and Weinberg majorly described six altered cancer cell physiology determining the malignant phenotype of a cell. This included

K. Vashishth
Advance Cardiac Centre Department of Cardiology, Post Graduate Institute of Medical Education and Research (PGIMER) Chandigarh, Chandigarh, India

S. Ranjan
College of Public Health, George Mason University, Fairfax, VA, USA
e-mail: sranjan@masonlive.gmu.edu

H. Joshi
School of Biotechnology, Jawaharlal Nehru University, New Delhi, India

T. Singh
Translanatal Oncology Laboratory, Department of Zoology, Hansraj College, Delhi University, New Delhi, India

M. B. Yerer Aycan
Department of Pharmacology, Erciyes University, Kayseri, Kayseri, Turkey

H. S. Tuli (✉)
Department of Biotechnology, Maharishi Markandeshwar Engineering College, Maharishi Markandeshwar (Deemed to be University), Ambala, Haryana, India

H. S. Tuli, M. B. Yerer Aycan (eds.), *Oncology: Genomics, Precision Medicine and Therapeutic Targets*, https://doi.org/10.1007/978-981-99-1529-3_10

self-sufficiency of growth signals, insensitivity towards antigrowth signals, apoptosis, unlimited replication, sustained angiogenesis, tissue invasion, and metastasis. Thus, the progression of cancer results in the development of a self-sustainable proliferation of tumor cells in a limitless fashion. The proliferation is attributed to the re-activation of cellular programs spotted during cell development and by modifying the existing cellular programs (Mao et al. 2022; Chakraborty and Rahman 2012; Luo et al. 2009).

The conventional treatment approach utilized for the treatment of cancer is one-size-fits-all. This is based upon the assumption; all malignancies originate from somatic cells with similar malignant outcomes. With this methodology, the treatment approach is either based on the origin of the tumor or histomorphogenic features. It leads to a standardized therapy of care for cancer patients, which might provide an intermittent effect in reducing the spread of the tumor. However, when observed from a broader perspective, it fails to eliminate cancer stem cells, resulting in more severe tumours reoccurrence. Thus, lack of specificity in the treatment approach leads to poor prognosis and reappearance of the disease severity. Moreover, this treatment approach is too aggressive and does not deliver the expected longevity of human life; in addition to this, it has a high treatment cost. Imatinib's use in treating chronic myeloid leukemia is one of the most typical examples that could be narrated for such conventional therapy. Imatinib is a kinase inhibitor and prevents the production of BCR-ABL fused protein. Although an effective drug, it lacks curative properties, and its long-term use results in the reoccurrence of the disease (Mao et al. 2022; Chakraborty and Rahman 2012; Zugazagoitia et al. 2016; Goldblatt and Lee 2010; US Food and Drug Administration 2018; National Library of Medicine 2023).

Furthermore, the development of a cancer drug is a cumbersome process. Pharmaceutical companies undertake multiple randomized clinical trials, with the majority delivering negative results. The one with promising outcomes is often based on smaller sample sizes. Eventually, these drugs can be more beneficial for an average patient, but when it comes to an individual, seldom are the results assuring. The one-size-fits-all approach to cancer drugs has led to a bottle-neck approach with limited patient benefits. Consequently, it is essential to identify targets within tumor cells that have increased antitumor properties with minimal side effects to the patients and lesser resistance to treatments. These targets could include oncogene addiction, non-oncogene addiction, synthetic lethality (damage to cancer hallmark), genetic screens, and data integration. This will provide a more precise and targeted approach to cancer treatment. Therefore, future treatment modalities could be discovering genotype-directed precision oncology and adapting it as per the case need, in addition to this aiming to the micro and macro environment of tumor concerning the immune system and antitumor immunity (Fig. 1) (Zugazagoitia et al. 2016; Goldblatt and Lee 2010; de Bono and Ashworth 2010).

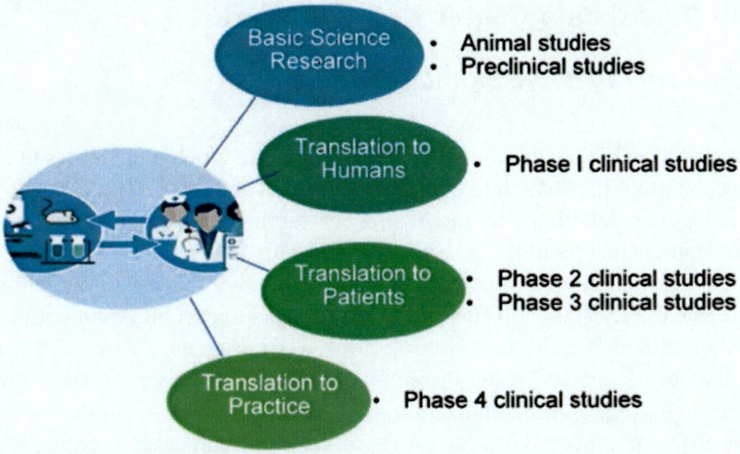

Fig. 1 Translation research from bench to clinical trial

1.2 Benefits of Translation Cancer Research

In the present scenario, approaches toward cancer treatment have shifted from nonspecific and cytotoxic therapies towards more targeted rational approaches with minimal side effects and high potency and efficacy (de Bono and Ashworth 2010). However, despite the ongoing tremendous amount of cancer research throughout the world, the target-specific drug development process is still painfully low due to numerous reasons, such as the complex nature of different types of cancers, recruiting a large number of patients for several years, lack of suitable targets, lack of proper validation of targets selected, commercial pressures, etc. (Goldblatt and Lee 2010; Jönsson and Sullivan 2019). With the push to develop novel therapeutic and diagnostic targets, the trend of translational cancer research is flourishing (Lee et al. 2021). The ongoing genomic and proteomic research and cancer technologies have generated enormous amounts of critical information in the field of cancer biology (Mao et al. 2022; Chakraborty and Rahman 2012; Goldblatt and Lee 2010). The current scenario demands a continuous, improved, and accelerated effort translate pre-clinical findings to improve therapeutic approaches and strategies for cancer patients. Collaboration and dedicated efforts among researchers, clinicians, and pharmaceutical companies are crucial in translating the pre-clinical findings into clinical trials (Eggermont et al. 2019; Drapkin and Rudin 2021; Ringborg 2019).

2 How Translation Cancer Research Work

2.1 *Spectrum of Translational Research*

Translational research means connecting the progress in molecular studies to clinical trials by examining the "bench to bedside" (Goldblatt and Lee 2010; Toi et al. 2005; Woolf 2008). Translational research aims to fabricate potential candidates and treatment approaches to improve effectiveness with fewer adverse effects. It can also be used to predict the biological response to the drug using various biomarkers and segregate the resistant and susceptible tumor cells to current treatment intervention (Goldblatt and Lee 2010; Toi et al. 2005; Hermiston and Kirn 2005). It also enables therapeutic agents' faster development by expediting the process from lead identification to drug commercialization.

Since different cancer types have the similar gene and protein expression, the biological response to treatment regimen will differ, demanding the development of individualized treatment. In addition, the significant barrier hindering targeted therapy's efficacy is that multiple genetic mutations are responsible for tumor formation (Goldblatt and Lee 2010). So, there is a need for a mixture of therapeutic drugs against numerous targets or a single drug that possesses chemoprevention activity against different targets. Unfortunately, innumerable drugs have shown adverse effects due to their mode of action, or they may emanate from non-specific effects of the drug (Fojo and Menefee 2007; Joshi et al. 2023). Advancements in multi-omics technology, artificial intelligence, and preclinical studies facilitate us to characterize the tumor's patient-specific molecular and cellular characteristics that information aid in developing precision medicine (For a specific patient, at an exact time, of a particular dose).

2.2 *Translation of Cancer Research into Targeted Therapy*

2.2.1 Pharmacodynamics (PD) Biomarkers

PD biomarkers measure the drug effect post-treatment on the targeted tumor in a patient. It correlates with exposure to therapeutic intervention and its pharmacological response at the target (Sarker and Workman 2006; Pezo and Bedard 2015). The pharmacological drug response is usually measured using various parameters such as imaging, molecular, cellular, and histopathological (Sarker and Workman 2006). It is helpful in the early phases of drug development by providing the safety and feasibility of treatment intervention and biological response of drug on tumor and patient. It plays an essential role throughout the drug development process, from initial lead identification in preclinical studies to clinical trials (Workman 2003; Hidalgo 2004). These biomarkers can be implicated in various applications such as anticipating biological outcomes, gaining a proposed mechanism of action of medication, fabricating potential drug combinations, providing information on optimal dose selection and drug administration schedule, unraveling the pharmacokinetics and unwanted side effects of the drug, and helps in the identification of drug resistance and response mechanisms (Sarker and Workman 2006). Various PD biomarkers exist that aid in measuring the drug response in cancer therapy, such as apoptosis biomarkers, spindle checkpoint biomarkers, plasma biochemical biomarkers, toxicity biomarkers, protein phosphorylation biomarkers, and cytokinetic biomarkers (Jackson 2012). Identification of these biomarkers improves the designing of effective drug combinations resulting in better therapeutic outcomes and rapid formulation of lead compounds for clinical purposes in the future.

2.2.2 Predictive/Putative Predictive or Enrichment Biomarkers

Predictive biomarkers are useful and entail possible responses to a particular drug regimen before treatment. The survival estimate relies on therapy in individuals with a biomarker linked with a specific targeted therapy outcome. At the same time, those individuals who don't possess biomarkers indicate no change in survival between treated and untreated individuals (Pezo and Bedard 2015). For instance, ALK protein expression is a predictive biomarker in non-small cell lung cancer (NSCLC) associated with outcomes following treatment with kinase inhibitors such as xalkori, zykadia, alecensa, and lorbrena (Bedard et al. 2013; Batis et al. 2021). Another example of a predictive biomarker is kras gene mutations in colorectal cancer linked with susceptibility to EGFR (epidermal growth factor receptor) antagonists like erbitux and vectibix (Bedard et al. 2013; Batis et al. 2021). Identifying and developing potential biomarkers helped assess gene and protein expression changes. They enabled screening only those individuals that would gain from targeted therapy (Pezo and Bedard 2015). Tumor and liquid biopsies are commonly used to get tumor-specific multi-omics information from which these biomarkers would derive (Batis et al. 2021). However, several barriers limit the clinical translation of these predictive biomarkers, including inadequate sample collection and processing protocols, difficulty collecting post-treatment samples, lack of harmonized assays, tumor microenvironment, intra-tumoral heterogeneity, poorly designed clinical trials, and difficulty in conducting robust clinical studies.

3 Currently Available Promising Targets from Bench to Clinical Trials

3.1 NCCN

The National Comprehensive Cancer Network (NCCN) is a philanthropic alliance of 32 leading cancer centers dedicated to patient care, research, and education. NCCN programs aim to improve cancer care's effectiveness and efficiency globally continuously (National Comprehensive Cancer Network 2023). Experts from NCCN member institutions treat patients with a wide range of cancers and are renowned for dealing with complex, aggressive, or rare cancers. Further, NCCN develops resources that present valuable information to various stakeholders in the health care provision. NCCN promotes the importance of a multidisciplinary team approach and innovative research about understanding, diagnosing, and treating cancer. Defining and advancing high-quality cancer care encourages the importance of developing clinical practice guidelines for patients or health care professionals globally (National Comprehensive Cancer Network 2023; Wood 2004).

3.2 TAPUR Study

The American Society of Clinical Oncology originally carried out the non-randomized Targeted Agent and Profiling Utilization Registry (TAPUR) Study (ASCO). It aims to describe the efficacy and safety of commercially accessible, FDA-approved anticancer medications used to treat patients with advanced cancer who have potentially therapeutic genetic variants. The TAPUR Study is a clinical trial for individuals with advanced cancer that examines if particular targeted therapies can help more patients and pave the way for more individualised treatment. The TAPUR Study is a clinical study specifically designed for adults with advanced cancer (or children ages 12 and older for some treatments) for whom normal treatment has failed, stopped working, or for whom there is no standard treatment for that type and stage of cancer (Mangat et al. 2018).

3.3 The NCI Community Oncology Research Program (NCORP)

The National Cancer Institute (NCI) proposed the NCI Community Oncology Research Program (NCORP), a community-based initiative, to fund clinical trials for disease surveillance, prophylaxis, therapy, and monitoring as well as to broaden its research focus to include studies of cancer care delivery. Forty-six community sites and seven research bases make up NCORP. The designation of 14 of those 46 community sites as minority/underserved community sites (NCI Community Oncology Research Program (NCORP) 2022). The cornerstone of NCORP is the execution of multi-site cancer clinical trials and other human subjects research in the main focal areas of cancer control, prevention, and care provision. The National Clinical Trials Network (NCTN), which the NCI sponsors for cancer clinical trials, includes NCORP as a crucial part. About 30–35% of the participants engaged in NCTN clinical trials, including precision medicine studies, are at NCORP facilities.

The NCORP Community Sites contribute to the comprehensive national network by enhancing the involvement of community oncologists, other medical specialists, and their patients (NCI Community Oncology Research Program (NCORP) 2022; McCaskill-Stevens et al. 2013).

3.4 NCI Match and Paediatric Match Trial

Cancer treatment clinical trial NCI-MATCH (Molecular Analysis for Therapy Choice) uses precision medicine. In this trial, cancer patients are assigned to receive treatment based on the genetic alterations discovered in their tumours by genomic sequencing and other procedures (solid tumors, including non-Hodgkin lymphomas, brain tumors, and histiocytoses) (National Cancer Institute 2022). Pediatric MATCH is a phase II worldwide precision medicine cancer treatment study that examines the efficacy of targeted medicines in treating solid tumours in children, adolescents, and young adults with particular gene mutations, ranging in age from 1 to 21 years (Children's Oncology Group 2022). In order to match patients with medicines targeted at the genetic abnormalities in their tumours that no longer respond to traditional treatment or have returned after treatment, Pediatric MATCH studies various research medications, each targeting a specific collection of gene mutations (Children's Oncology Group 2022).

4 Importance of Precision Cancer Medicine and Therapeutics

In order to make a diagnosis, plan a course of treatment, evaluate the effectiveness of that treatment, or determine a patient's prognosis, precision medicine in the field of cancer employs particular genomic, transcriptomic, or proteomic data about a person's tumour. The assumption that precision oncology may help more patients than conventional chemotherapies is firmly supported by the evidence. In Phase I clinical trials, patients were chosen using a certain biomarker as opposed to those who were not, and the individualised treatment strategy significantly increased patient response rate and progression-free survival. To continue advancing the idea of precision medicine and make it a viable sector that benefits everyone, a large collaborative effort by scientists, pharmaceutical companies, physicians, regulators, and patients will be required.

5 Newer Avenues and Role of Big Data in Translational Cancer Research

Cancer is a multistage process; its multiplication process is complex and interwound with many proteins and pathways involved. Annually, the cancer research community worldwide generates massive data and significant findings, both phenotypic and molecular, to study cancer's hallmarks comprehensively (Amirmahani et al. 2021). The era of omics and omics data has given rise to the concept of "big data" in cancer, which is defined by two basic properties such as containing abundant information giving insights into prominent and essential questions and the analysis of information demanding a large computer and software infrastructure (Amirmahani et al. 2021). A well-known example is the cancer genome atlas (TCGA), containing more than 10,000 articles and 11,000 NIH grants with petabytes of data requiring specialized computers and storage infrastructure (Weinstein et al. 2013). Various data sources such as DNA mutations, transcriptomics, chromatin accessibility, and histone modifications are available in the big datasets further large consortium with complex analysis tools such as TCGA, Clinical Proteomic Tumour Analysis Consortium (CPTAC), Therapeutically Applicable Research to Generate Effective Treatments (TARGET), Cancer Cell Line Encyclopaedia (CCLE), copy number alteration (CNA), etc. have yielded significant results and outcomes. It is now proven that with the support of existing data, new discoveries on different dimensions of cancer can be leveraged (Jiang et al. 2022; Joos et al. 2019). Many important clinical decisions can be made with the help of data-enriched translational studies. One of the critical breakthroughs and focus area of translational big data studies in cancer research is the development of genomic testing for predicting disease risks, some of which has already been approved by the US FDA for commercial purposes. Many clinical trials are underway with the guidance of available molecular data (Murphy et al. 2017). Many researchers are putting forward the concept of data-driven cancer diagnostics with the help of artificial intelligence using genomics datasets. The multifunctional approach of big data is already revolutionizing the world; with the available tumour data, many biomedical breakthroughs in cancer are expected, with better cancer diagnostics and the development of new therapies (Capobianco 2019; Schachtschneider et al. 2017).

6 Conclusion/Way Forward

Cancer treatment is a long battle, with treatment options ranging from radiotherapy, chemotherapy, surgical options to the use of specific monoclonal antibodies. Cancer research is an ongoing process with different hits and miss. Cancer research paved way in our understanding of cancer and its therapeutics. Since the last few decades many newer therapeutic and diagnostic options have surfaced. Today for some cancer if detected early the prognosis is good with the available treatment options, however cancer treatment is cost with limited success in many cancers. Even today the scientific community has not been able to translate the preclinical finding into clinical settings. Persistent and dedicated efforts are needed among the researchers, clinicians and the pharma companies to translate findings into affordable and efficacious therapeutic options.

References

Amirmahani F et al (2021) Approaches for the integration of big data in translational medicine: single-cell and computational methods. Ann N Y Acad Sci 1493(1):3–28

Batis N et al (2021) Lack of predictive tools for conventional and targeted cancer therapy: barriers to biomarker development and clinical translation. Adv Drug Deliv Rev 176:113854

Bedard PL et al (2013) Tumour heterogeneity in the clinic. Nature 501(7467):355–364

de Bono JS, Ashworth A (2010) Translating cancer research into targeted therapeutics. Nature 467(7315):543–549

Capobianco E (2019) Next generation networks: featuring the potential role of emerging applications in translational oncology. J Clin Med 8(5):664

Chakraborty S, Rahman T (2012) The difficulties in cancer treatment. Ecancermedicalscience 6: ed16

Children's Oncology Group (2022) National Cancer Institute–children's oncology group pediatric MATCH trial. https://childrensoncologygroup.org/pediatricmatch

Drapkin BJ, Rudin CM (2021) Advances in small-cell lung cancer (SCLC) translational research. Cold Spring Harb Perspect Med 11(4):a038240

Eggermont AM et al (2019) Cancer Core Europe: a translational research infrastructure for a European mission on cancer. Mol Oncol 13(3):521–527

Fojo T, Menefee M (2007) Mechanisms of multidrug resistance: the potential role of microtubule-stabilizing agents. Ann Oncol 18:v3–v8

Goldblatt EM, Lee W-H (2010) From bench to bedside: the growing use of translational research in cancer medicine. Am J Transl Res 2(1):1

Hermiston TW, Kirn D (2005) Genetically based therapeutics for cancer: similarities and contrasts with traditional drug discovery and development. Mol Ther 11(4):496–507

Hidalgo M (2004) New target, new drug, old paradigm. J Clin Oncol 22:2270–2272

Jackson RC (2012) Pharmacodynamic modelling of biomarker data in oncology. Int Schol Res Not 2012:590626

Jiang P et al (2022) Big data in basic and translational cancer research. Nat Rev Cancer 22:1–15

Jönsson B, Sullivan R (2019) Mission-oriented translational cancer research–health economics. Mol Oncol 13(3):636–647

Joos S et al (2019) German Cancer Consortium (DKTK)–a national consortium for translational cancer research. Mol Oncol 13(3):535–542

Joshi H et al (2023) Inhibition of cancer cell metastasis by nanotherapeutics: current achievements and future trends. In: Tuli HS (ed) Nanotherapeutics in cancer. Jenny Stanford Publishing, Singapore, pp 161–209

Lee MW et al (2021) Current methods in translational cancer research. Cancer Metastasis Rev 40(1):7–30

Luo J, Solimini NL, Elledge SJ (2009) Principles of cancer therapy: oncogene and non-oncogene addiction. Cell 136(5):823–837

Mangat P et al (2018) Rationale and design of the targeted agent and profiling utilization registry (TAPUR) study. JCO Precis Oncol 2018:122

Mao JJ et al (2022) Integrative oncology: addressing the global challenges of cancer prevention and treatment. CA Cancer J Clin 72(2):144–164

McCaskill-Stevens W et al (2013) The NCI Community Oncology Research Program: what every clinician needs to know. Am Soc Clin Oncol Educ Book 33(1):e84–e89

Murphy S, Castro V, Mandl K (2017) Grappling with the future use of big data for translational medicine and clinical care. Y B Med inform 26(1):96–102

National Cancer Institute (2022) NCI-MATCH trial (molecular analysis for therapy choice). https://www.cancer.gov/about-cancer/treatment/clinical-trials/nci-supported/nci-match

National Comprehensive Cancer Network (2023) About. https://www.nccn.org/home/about

National Library of Medicine (2023) PubChem Imatinib (Compound). https://pubchem.ncbi.nlm.nih.gov/compound/Imatinib#section=3D-Conformer

NCI Community Oncology Research Program (NCORP) (2022) About NCORP. https://ncorp.cancer.gov/about/

Pezo R, Bedard P (2015) Definition: translational and personalised medicine, biomarkers, pharmacodynamics. ESMO handbook of translational research. ESMO, Lugano

Ringborg U (2019) Translational cancer research–a coherent cancer research continuum. Mol Oncol 13(3):517–520

Sarker D, Workman P (2006) Pharmacodynamic biomarkers for molecular cancer therapeutics. Adv Cancer Res 96:213–268

Schachtschneider KM et al (2017) The oncopig cancer model: an innovative large animal translational oncology platform. Front Oncol 7:190

Toi M, Takebayashi Y, Chow LW (2005) Translational research in breast cancer. Breast Cancer 12(2):86–90

US Food & Drug Administration (2018) Precision medicine. https://www.fda.gov/medical-devices/in-vitro-diagnostics/precision-medicine

Weinstein JN et al (2013) The cancer genome atlas pan-cancer analysis project. Nat Genet 45(10):1113–1120

Wood EH (2004) The National Comprehensive Cancer Network (NCCN). J Med Lib Assoc 92(3):382

Woolf SH (2008) The meaning of translational research and why it matters. JAMA 299(2):211–213

Workman P (2003) How much gets there and what does it do?: the need for better pharmacokinetic and pharmacodynamic endpoints in contemporary drug discovery and development. Curr Pharm Des 9(11):891–902

Zugazagoitia J et al (2016) Current challenges in cancer treatment. Clin Ther 38(7):1551–1566